Geology in Engineering

Geology in Engineering

JOHN R. SCHULTZ, Ph.D.
Chief, Geology Branch
Waterways Experiment Station
Vicksburg, Mississippi

ARTHUR B. CLEAVES, Ph.D.
Professor of Geology
Washington University
St. Louis, Missouri

With a Chapter on Soil Mechanics
by E. J. YODER, Research Engineer
Joint Highway Research Project
Purdue University, Lafayette, Indiana

New York · JOHN WILEY & SONS, *Inc.*
London · CHAPMAN & HALL, *Ltd.*

Library of Congress Catalog Card Number: 55–7317

PRINTED IN THE UNITED STATES OF AMERICA

Preface

Practicing engineers come into almost daily contact with geological matters, and there has long been need for a book which introduces them to the phases of geology that experience shows to be the most useful in their work. There is, perhaps, an even greater need for a book acquainting geologists employed in engineering work with engineering practice. To satisfy completely the needs of both would require a volume considerably longer than the present one, and no attempt is made to give a complete account of civil engineering practice. However, it is impossible to separate engineering applications of geology from engineering practice, and engineering matters have been introduced where necessary. The discussions of various types of engineering projects by no means exhaust the applications of geology to civil engineering, but it is believed that they serve to illustrate most of the basic principles which, when thoroughly understood, are readily adapted to omitted topics such as building foundations, bridge piers, and levees.

This book may be used as the basis of a one- or two-term course. Selection and order of topics presented in class depend largely on the purpose of the course and its place in the curriculum. Therefore, it is expected that the instructor will use his own judgment in the selection and order of material presented. Class work should be supplemented with practical problems and suitable laboratory work. It is hoped that if the book proves successful in the class room some one may be motivated to supplement it with a suitable laboratory manual. Visual aids with emphasis on motion pictures and field excursions are also desirable adjutants to a course of this type.

By permitting adaptations of basic studies, the following authors have helped to make this book much more authoritative than it otherwise would have been: Beno Gutenberg and C. F. Richter on earthquakes; Warren J. Mead on dam sites; Siemon

W. Muller on permafrost; R. V. Proctor and T. L. White on tunnels; Parry Reiche on rock weathering; Roger Rhoades on concrete aggregates; C. F. Stewart Sharpe on landslides and related phenomena; F. P. Shepard on shore lines and beaches; and Karl Terzaghi on tunnels and landslides. We are also indebted to the following organizations and publishers for generously permitting use of materials appearing under their imprints: Columbia University Press; Commercial Shearing and Stamping Company; Dover Publications, Inc.; Geological Society of America; Harper and Brothers; Harvard University Graduate School of Civil Engineering; McGraw-Hill Book Company; William Morrow and Company; National Geographic Society; National Research Council; Prentice-Hall, Inc.; U. S. Geological Survey; and the Bureau of Reclamation. Geological studies sponsored by the Mississippi River Commission, Corps of Engineers, U. S. Army, have been woven into Chapter 7, and experience gained from investigations by other organizations of the Corps of Engineers, U. S. Army, and the Pennsylvania Turnpike Commission has been rather extensively used. Other acknowledgments of source materials and assistance are made in the body of the text.

Chapters 2, 3, and 4 were much improved as the result of a reading by Dr. Alonzo Quinn of the Geology Department, Brown University. Chapters 5, 6, and 7 were read by Mr. W. G. Shockley, Chief, Embankment and Foundation Branch, Waterways Experiment Station. Chapters 7 and 10 owe much to comments and suggestions by Dr. J. P. Schafer of Brown University. Chapters 7, 8, 10, 11, and 16 have been read by Mr. Charles R. Kolb of the Geology Branch, Waterways Experiment Station. Chapter 18 has benefited from a reading by Dr. C. E. Adams. Chapter 21 has been read by Mr. Charles R. Foster, Chief, Flexible Pavement Branch, Waterways Experiment Station. Dr. Philip C. Rutledge read the preliminary draft of the manuscript, and the completed product is immeasurably indebted to his precise and penetrating criticism. Dr. M. J. Hvorslev furnished many valuable suggestions pertaining to all parts of the book, and Chapter 1 owes much to his advice and criticism. Professor K. B. Woods of Purdue University furnished numerous illustrations and valuable advice. Finally, we should like to express our appreciation to friends and colleagues who extended moral and intellectual support while the book was in preparation.

JOHN R. SCHULTZ
ARTHUR B. CLEAVES

Vicksburg, Mississippi
St. Louis, Missouri
January, 1955

Contents

C H A P T E R   1

Geology and Its Relationship to Civil Engineering

Geology is the scientific study of the earth, especially its more or less accessible outer shell, or crust. It seeks to trace the history of the planet from the time that it first became a separate entity down to the present day. In so doing it enlists the aid of other sciences, particularly astronomy, physics, chemistry, and biology, and without them geology would be little more than a description of rocks, mountains, river valleys, etc. No natural feature is too small or too large to escape the attention of geology. It is interested in the origin of continents and ocean basins on the one hand; on the other it takes cognizance of the arrangement of atoms in the crystalline substances that form much of the earth's crust. Geology includes not only the study of minerals and rocks, but is also concerned with ocean waves, flowing water, glaciers, and other agencies modifying the surface of the earth, as well as the study of the past development and distribution of living things. A subject so broad must necessarily be divided into numerous specialties and no less than twenty-five are currently recognized. Ten of these specialties are touched upon in this book.

HISTORICAL DEVELOPMENT

The scope and methods of modern geology can, perhaps, be made clearer by sketching the historical development of the subject. We know that even primitive man had a certain interest in rocks, for they furnished the source materials of his tools and weapons. It was not until a certain degree of civilization had been reached, however, that men began to inquire into the nature and origin of the earth. Most early speculations concluded that the earth was created once and for all at no very remote date, and no appreciable changes were thought to have since taken place. Geologic processes generally operate very slowly, and their importance was only very gradually realized. It is now known that nature is not static, but subject to slow inexorable transformations that have changed, and probably will continue to change, the landscapes, climates, and life of the earth. The change in outlook brought about by recognition and investigation of the geologic processes influences many branches of thought and deserves to be ranked as one of the greatest intellectual revolutions in the history of mankind.

The beginnings of modern geology are to be found in mining. Geologic features such as veins, faults, and joints are closely connected with mining practice, and a certain amount of geologic knowledge thus became a part of the practical mining man's working equipment. This essentially tradesman's knowledge was collected by a German, best known under his Latinized name of Agricola, whose treatise entitled *De Re Metallica* was published in 1556. This book parts company with the then prevailing views that minerals and veins grow in the earth under the influence of the stars and even more occult agencies and is essentially modern in its rationality and appeal to experience. Agricola did not found a school, however, and for the next 200 years progress of geologic science was limited largely to flashes of insight on the part of a very few.

During the eighteenth century it began to be appreciated that the earth's crust bears witness to vast and recurrent changes. Mountain ranges were seen to have been formed only to be eroded and covered by thick layers of sediment, and old types of organisms were found to have perished and given place to new. The old views were accordingly modified to make room for these changes, which were conceived as taking place in a violent and speedy manner. The once fairly generally accepted view that nature moves from one catastrophe to another has long since been discarded. It is of interest only because it appears to represent an essential transitional stage between the ancient static and modern evolutionary points of view.

Modern geology was founded by a native of Scotland named James Hutton, who observed the natural processes in operation: rivers transporting silt and cutting valleys, waves breaking along coasts and building beaches. He reached the then revolutionary conclusion that processes in operation today are sufficient to account for all the vast changes the earth is known to have undergone during its immensely long history. This concept was published in 1785 and soon became known as the theory of uniformitarianism. It is simply a statement that natural laws operated in the past exactly as they do today, and given enough time the very slow natural forces at present in operation are capable of forming a mountain range, eroding the newly formed mountains to sea level, and covering their eroded remnants with thick layers of sedimentary rock. It does not exclude the catastrophes favored by earlier workers; it merely makes them unnecessary.

During much of the nineteenth century geology was still handicapped by old conceptions and prejudices. Few were prepared to accept the vast antiquity of the earth indicated by the new science, and to many the philosophical implications of the subject seemed downright dangerous. Consequently, during this period geologists were forced to defend their views and to spend a great deal of time in the study of what today would be regarded as natural history and philosophy. Sweeping generalizations were highly valued, and relatively little really detailed work was done. The efforts of this period were not in vain, for they furnished a background of observation on which present-day work is ultimately based.

Modern geology was profoundly influenced by the publication of Darwin's *Origin of Species*, which occurred in 1859. The concept of orderly and gradual change in the biologic world set forth in this book not only found ample support from the fossil record compiled by geologists, but ultimately served to revolutionize concepts pertaining to the inorganic world as well. Mountains, rivers, lakes, all natural features in fact, including the earth itself and the atoms composing it, are now thought to originate, grow, decline, and vanish according to natural laws. A considerable part of geologic research is devoted to the understanding and application of these natural laws.

Once freed from the weight of tradition and provided with a proper working basis, geology has made fairly rapid progress. Much of the earth's surface has been mapped, more or less accurately, and at present the comparatively inaccessible depths of the ocean are being actively explored. The rapid progress of physics and chemistry has had a corresponding effect on geology. For example, rocks are now dated in terms of years by application of the rates of disintegration of radioactive elements such as radium and carbon 14. The rapid expansion of aviation has also made many formerly remote regions readily accessible, and aerial photography has provided a new and important tool for geologic work. Geologists are no longer content merely to describe and map a certain area of the world; they are striving with increasing success to introduce quantitative methods. The rapid development of geophysics, which makes possible a fairly precise determination of the size, shape, and depth of concealed geologic bodies, is a case in point. Bagnold's study [1] of the aerodyamic principles governing the origin, growth, and decay of sand dunes is perhaps an even more striking example of the current progress toward this goal. The modern student of geology must, therefore, be well grounded in the basic sciences. The civil engineer is thus well prepared to undertake the study of geology, and in these pages a certain familiarity with the basic sciences is assumed. However, in order to avoid giving the impression that geology will soon become a branch of applied physics, it is necessary to look rather closely into its problems and methods.

METHODS OF GEOLOGICAL INVESTIGATION

Geology is guided by experiment and theory, but it is not basically an experimental science. Many geologic processes are as yet inaccessible to experiment, and the vast lapses of time required for the working out of many of them largely preclude the experimental method. As a result, geology is based largely on observation. The geologist is faced with somewhat the same situation as the detective. Both have to deal with events that have transpired in their absence and which can be reconstructed only by carefully piecing together isolated fragments of evidence. The accuracy of the reconstruction depends in both instances on the amount of evidence observed and the validity of the inferences made therefrom. Errors arise more frequently from insufficient evidence than from fallacies in the logical process by which the past events are reconstructed. In other words, geology and crime detection are based on induction, or proceeding from the individual facts to a generalization, and thus stand at opposite poles to mathematics which is essentially deductive. Both geology and crime detection share the inherent difficulty of the inductive procedure, for no matter how numerous the observed facts on which a generalization is based an unobserved fact may invalidate it. There is thus no certainty in a generalization derived from induction. It must be added, however, that there is often a very high probability. It should be noted in this connection that mathematical deduction alone does not constitute an adequate scientific method, for a certain amount of observation and experiment is necessary to establish the mathematical expressions applying to natural phenomena. Thus all sciences progress by a more or less haphazard combination of induction and deduction. The chief differences between geology and the more exact sciences are that in geology observation predominates over experiment and geologic processes are often so complex as to defy mathematical treatment.

It has been suggested that inference plays a very important part in geology. This is necessarily so, for the geologist is never in possession of more than a small fraction of the information he would like to have. No man, for example, has ever looked inside an active volcano and survived to tell his experience. The volcanologist, like geologists in general, is forced to study his subject largely from the outside. He notes that very hot and often molten material is ejected periodically, and from this he infers that in the interior of the volcano the temperature must be very high. He notes further that the ejected material hardens and accumulates around the crater, where it gradually builds up a characteristic conical-shaped mass that may eventually reach mountainous proportions. From these simple observations he infers that mountain masses having similar forms and built of materials resembling cooled volcanic products were also formed by volcanic activity, although at present there may not be the slightest direct evidence of volcanic action. Similarly, through a long and careful study of the grooves and markings left by re-treating glaciers glaciologists have inferred that glacial ice moves mainly by plastic flow closely resembling that of a highly viscous fluid. This inference has since been substantiated by observations in tunnels under glaciers in Switzerland. Geology thus consists mainly of a body of facts derived from observation and the in-ferences made from these facts. The facts are permanent and unalterable; the inferences are subject to error and are often dis-carded or modified as more facts, or different ways of evaluating the facts, become available.

The ways of geology and the ways of civil engineering may thus seem to be wide apart. The civil engineer is generally able to make use of experiment, at least in the modified form of model studies, and is often able to analyze his problems mathematically. As a result, he tends to look rather coldly on inference. How-ever, an inference rightly followed through may save the engi-neer much embarrassment and many dollars. A fault bringing crushed and weak materials into a tunnel line may exist as an inference in the mind of a geologist, but when encountered in the tunnel it becomes an unpleasant fact. The engineer, if he is wise, will employ a geologist more for the inferences he may be able to make than for his factual knowledge. The facts are more or less open to everyone; worthwhile inferences can be made only by the experienced specialist.

APPLICATION OF GEOLOGY TO CIVIL ENGINEERING

Despite their differences in viewpoint, geology and civil engi-neering have been rather closely connected from the very be-ginnings of geology, and it is perhaps not too great an exaggera-tion to say that engineering has contributed more to geology than geology has contributed to engineering. The first geologic map of England was made by William Smith, a civil engineer. Engi-neering works often provide rigorous tests of geologic inferences and thus contribute much to pure geology. The tunnels under the Alps are cases in point. Geologists were asked to furnish the engineers in charge with their conception of conditions to be found along the tunnel routes, and they presented their clients with sections showing extremely complicated and twisted masses of rock. It is doubtful that others, including many geologists, would really have believed these sections to be true had not the actual conditions encountered corresponded fairly closely with the more or less hypothetical preliminary concepts. Engineering and geology thus go more or less hand in hand; application of geology to exploration for engineering works benefits engineering, and the excavations and borings made by engineers benefit geology. It has been well said that civil engineers are to a greater or less extent engaged in the practice of applied geology whether they know it or not.

The earlier applications of geology to engineering dealt mainly with hard rocks and had to do chiefly with mines, tunnels, deep excavations, etc. Exploration for construction materials has always been an important part of engineering geology. Since the early 1940's, however, there has been a rapid development of the application of geology to the comparatively new engineering field of soil mechanics, which deals with the engineering behavior of unconsolidated materials such as gravels, sands, silts, and clays that the engineer calls soils. Soil mechanics began with an inadequate conception of the complexity of soils, and only gradually did it become apparent that an adequate interpretation of the data furnished by soil exploration requires an understanding of how these materials were formed in nature. For example, clays laid down under marine and lacustrine conditions differ considerably in engineering properties from those formed by weathering of volcanic ash. Furthermore, fresh- and salt-water clays also differ considerably in engineering properties. In view of the growing importance of this branch of engineering geology, the geology of soils is treated rather thoroughly in this book. It is also becoming apparent that soil mechanics has much to contribute to pure geology. Rominger and Rutledge's demonstration [6] that Lake Agassiz was drained and subsequently refilled is a good illustration. This conclusion was reached through an investigation of the engineering properties of the clays laid down in this former lake basin.

In the practice of engineering geology accurate and detailed geologic work goes hand in hand with application of the results to the engineering problem at hand. The engineer is seldom qualified to carry out geologic work, and not all geologists are able to apply geologic knowledge to engineering problems. This gap is generally filled by the so-called engineer-geologist, who has considerable competence in both fields. Experience shows, however, that despite the existence of the engineer-geologist the civil engineer cannot afford to be entirely ignorant of geology, for in order to understand engineering-geologic reports a geologic background is necessary. The primary purpose of this book is to furnish this background and not to make geologists out of civil engineers. Although the geologic material has been selected with an eye to its engineering applications, space permits mention of only the more important applications of geology to engineering projects. Geologic subjects having little or no application to civil engineering have been either omitted or treated only in passing. Nevertheless, the subject is so large that it is almost impossible to cover it thoroughly in a discourse of reasonable length. Therefore, selected lists of references have been added for use by those who may wish a more detailed knowledge.

REFERENCES

1. R. A. Bagnold, *The Physics of Blown Sand and Desert Dunes,* William Morrow & Co., New York. Copyrighted by Methuen & Co., Ltd., London, 1941.

2. C. P. Berkey, The Recent Development of Geology as an Applied Science, *Proc. Am. Phil. Soc.,* Vol. 72, 1933.

3. Edward B. Burwell, Jr., and George D. Roberts, The Geologist in the Engineering Organization, in *Application of Geology to Engineering Practice* (Berkey Volume), Geol. Soc. Amer., 1950.

4. Robert F. Legget, *Geology and Engineering,* McGraw-Hill Book Co., New York, 1939.

5. Warren J. Mead, Engineering Geology, *Geol. Soc. Amer., 50th Anniv. Vol.,* 1941.

6. J. F. Rominger and P. C. Rutledge, Use of Soil Mechanics Data in Correlation and Interpretation of Lake Agassiz Sediments, *J. Geol.,* Vol. 60, No. 2, March, 1952.

7. Karl Terzaghi, Foreword, *Géotechnique,* Vol. 1, No. 1, June, 1948.

8. W. J. Turnbull and H. N. Fisk, Relation of Soil Mechanics and Geology in Foundation Exploration, Lower Mississippi Valley, *Proc. 2nd Intern. Conf. Soil Mech. and Foundation Eng.,* Vol. 3, Rotterdam, 1948.

C H A P T E R   2

Minerals

A mineral is an inorganic solid occurring free in nature the chemical composition of which is either definite or varies within definite limits. Products of direct organic origin such as coal and petroleum are not considered minerals, but are often spoken of as mineral fuels. Mineral aggregates known as rocks form the basic units of the earth's crust, and it is impossible to acquire an adequate understanding of geology without at least an elementary knowledge of mineralogy. In addition to their geologic importance, minerals affect civil engineering in several ways. The physical and chemical properties of concrete aggregates are influenced by their mineral composition, and identification of the mineral constituents of aggregates constitutes an important branch of modern concrete technology. Furthermore, the mineral composition of fine-grained soils may have an important bearing on their physical properties, and research along these lines constitutes a new and promising phase of soil mechanics. Fortunately, of the 1400 odd mineral species known less than a hundred are commonly found in rocks, and of these the engineer needs to be familiar with hardly more than a score. Owing to the complexities of the subject, even this small number cannot be treated adequately in a short discourse. For a more comprehensive grasp the references listed at the end of this chapter should be consulted.

PHYSICAL PROPERTIES OF MINERALS

Naturalists of the eighteenth century discovered that corresponding interfacial angles of the same mineral species are always the same, and reasoned correctly that constancy of the interfacial angles reflected a definite and regular arrangement of the atoms. During the nineteenth century mathematicians deduced that based on symmetry there are 32 possible crystal classes, and it is

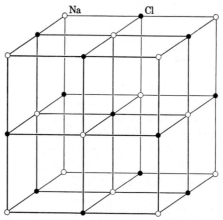

Fig. 1.   Crystal structure of sodium chloride.

interesting to note that X-ray techniques of the twentieth century have demonstrated that all the 32 classes are found in nature. Figure 1 illustrates the crystal structure of sodium chloride, or ordinary table salt. This is one of the simpler crystal structures. The 32 crystal classes can be grouped into 6 systems known as isometric, tetragonal, hexagonal, orthorhombic, monoclinic, and triclinic, distinguished by their respective elements of symmetry.

Crystal structure not only aids in identifying minerals, but is of great physical importance. All the physical properties of a crystalline material reflect its internal structure. For example, the hardness, crushing strength, and response to heat and light may vary with the crystal faces provided that the grade of symmetry is sufficiently low. Isometric crystals of the holohedral, or highest, class of symmetry possess the same physical properties on all faces and in all directions and are said to be isotropic, but this is not true for the remaining classes. Substances that possess different physical properties in different directions and/or on different faces are anisotropic.

Cleavage. Many minerals possess the property of breaking more readily along certain directions than others. This property

Fig. 2. Cleavage fragment of calcite showing cleavage planes paralleling the cleavage faces. Compare this illustration with the crystal form, illustrated in Fig. 7. (Washington University Collection.)

is known as cleavage, and gives rise to smooth faces bounding the cleavage fragments. Although cleavage is related to crystal structure and symmetry, cleavage planes should not be confused with crystal faces (see Fig. 2). Some minerals lack this property, whereas in others it is exceedingly well developed. The direction of cleavage is described in crystallographic terms such as prismatic and rhombohedral, and the manner and ease with which cleavage is accomplished is defined as perfect, imperfect, good, distinct, indistinct, or easy, as the case may be. Cleavage is not only important in identifying minerals but also has considerable bearing on the physical properties of mineral aggregates occurring

Fig. 3. Conchoidal fracture in chert. (Washington University Collection.)

in mass. Various terms used to describe the fracture of minerals which lack cleavage are conchoidal, denoting an irregular break consisting of the intersections of numerous concave surfaces (Fig. 3), hackly, and uneven.

Luster. The appearance of a mineral in reflected light is known as luster, and is important in sight identification. Two main types are distinguished: metallic, and non-metallic. As the names imply, the former is seen most frequently on minerals such as iron sulphide that contain one or more metals in their composition. Other more or less self-explanatory terms are vitreous, resinous, greasy, silky, pearly, adamantine, and dull.

Color. The color of minerals with a metallic luster is usually fairly constant, but it may be extremely variable in minerals of non-metallic luster. When used with care, it may be an important aid in sight identification. Usually the color of the fine powder, or streak, is more reliable for this purpose. The streak of a mineral is best obtained by rubbing a sharp point of the substance over a flat piece of white unglazed porcelain.

Specific Gravity. The specific gravity of a substance is its weight compared with that of an equal volume of water. It is determined by weighing the specimen first in air and then when immersed in water. The difference in the two weights represents the weight of an equal volume of water. The weight of the mineral measured in air divided by the weight of an equal volume of water is its specific gravity. The specific gravity of minerals ranges from 1 (water) to 23 (iridium), but it is usually about 2.65 for silicates and between 4.5 and 10 for those of metallic luster.

Hardness. This term is used to designate a complex combination of physical properties which is usually measured by the ability of one substance to scratch another. Hardness is very useful in sight identification, and has been standardized in what is known as Mohs' scale of hardness, which is given in Table I. The intervals are not of equal value, that between corundum and diamond being greater than any other. Several substances, most of which are artificial, fall in the interval between 9 and 10— boron carbide and silicon carbide, for example.

TABLE I. MOHS' SCALE OF HARDNESS

1. Talc	6. Feldspar
2. Gypsum	7. Quartz
3. Calcite	8. Topaz
4. Fluorite	9. Corundum
5. Apatite	10. Diamond

For field identification it is convenient to keep in mind the hardness of some common objects. The finger nail has a hardness of about 2.5, an iron nail about 5, glass varies considerably but is usually about 5.5 to 6, steel also varies from about 6 for a knife blade to nearly 7 for a file. In testing the hardness of minerals care should be exercised not to confuse the powder left from the softer mineral for a scratch made by the harder substance. It should be noted, furthermore, that the Mohs' scale is not the one ordinarily used for metals. As has been previously mentioned, the hardness of minerals may not be the same on all crystal faces.

Optical Properties. Minerals may be either transparent, translucent, or opaque. All species that transmit light possess the power of changing the direction of its propagation, a phenomenon known as refraction. The degree to which light is bent when passing through a substance is chiefly a function of the wave length of the light and the composition and thickness of the material. The index of refraction is defined as the sine of the angle of incidence over the sine of the angle of the refracted ray. The index of refraction of a transparent crystal grain may be determined by comparing it to a liquid of known index.

Minerals of certain grades of symmetry possess the property of double refraction, or birefringence; i.e., they break up the plane of polarized light into two mutually perpendicular rays each of which has a different index of refraction. (Polarized light is defined as light that vibrates in a single plane at right angles to the line of propagation.) It is this and similar physical properties that make the polarizing microscope of value in the study of thin sections of minerals and rocks. Application of the microscope to a study of thin sections (0.03 millimeter in thickness) of rocks and minerals is based primarily on the relationship between the optical properties and chemical compositions of crystalline materials. Because no two chemical compounds ever possess exactly the same optical properties, the latter are in effect characteristic of any given compound. Consequently, if determined with sufficient accuracy, they may be used as a short cut for a chemical analysis. Furthermore, in all but those rocks that consist of a single mineral, the bulk composition does not disclose the mineral constitution, which is best determined by a study of thin sections. By measuring and comparing areas occupied by the various mineral grains in a number of thin sections of the same rock, a reasonably accurate quantitative analysis can be made (see Fig. 4).

X-ray Analysis. The wave length of X-rays is so short compared with the dimensions of atoms that the spatial arrangement of the atoms in a mineral, or the crystal structure to use a more precise term, can be photographed by this means. Crystal structure is characteristic of each mineral substance; hence it follows that to identify the crystal structure is to determine the mineral

Fig. 4. Thin section of granite magnified approximately 44 times. Characteristic grid structure of microcline feldspar shown in upper left; triangular-shaped quartz crystal in center; rectangular feldspar crystal in lower right; and dark fragment of biotite in extreme lower right. (Photo by Joseph E. Cox, Jr.)

itself. X-ray techniques are extremely useful in determining the crystal structure and mineralogical composition of very fine-grained aggregates such as clays.

Amorphous State. Not all minerals are crystalline; some like opal and volcanic glass, or obsidian, do not possess a regular arrangement of the atoms and are said to be amorphous or colloidal. Amorphous substances are isotropic, and they seldom occur in as pure a form as crystalline materials. Technically, amorphous substances such as glass should be classified as super-cooled liquids rather than solids.

CHEMICAL PROPERTIES OF MINERALS

Some minerals like halite ($NaCl$) are readily soluble in water, and are seldom found near the ground surface except in arid regions. Others like calcite ($CaCO_3$) are soluble in acids, whereas some like silica (SiO_2) are soluble in basic solutions. Most silicates are only very slightly soluble in water. In addition to solubility, all minerals react chemically in a manner determined by their compositions. In view of the fact that minerals are chemical compounds, they may be classified accordingly. Some such as gold and sulphur consist of a single element, others like hematite (Fe_2O_3) are oxides, some are carbonates, and others sulphides. Grouping of minerals according to their chemical composition greatly simplifies the study of mineralogy. The composition of minerals may be determined by chemical analysis, but, as previously explained, this is seldom done when the optical properties are known.

Polymorphism. It may happen that substances of the same chemical composition are capable of crystallizing in more than one class of symmetry. This phenomenon is known as poly-morphism. Aragonite (orthorhombic) and calcite (hexagonal) are examples of different crystal modifications of calcium car-bonate. Polymorphism is analogous to isomerism in organic chemistry and possesses about the same significance for inorganic materials as does the latter for organic substances. It should be noted that, although the chemical composition of polymorphous minerals is the same, their physical properties are not.

Isomorphism. If two elements have about the same atomic radius they may displace each other in the crystal lattice. Thus the calcium and sodium atoms, which have nearly the same dimen-sions, may occupy interchangeable positions in the plagioclase feldspar lattice and give rise to all gradations in composition be-tween pure sodium aluminum silicate ($NaAlSi_3O_8$), or albite, to pure calcium aluminum silicate ($CaAl_2Si_2O_8$), or anorthite.

Several intermediate varieties such as oligoclase and andesine are distinguished by rather arbitrary proportions of the albite and anorthite molecules. Oligoclase, for example, has 90–70 per cent of the albite and 10–30 per cent of the anorthite molecule. Needless to say, its chemical formula is rather complex. A similar phenomenon, called replacement, is known to organic chemistry.

Stability. Like all other substances minerals possess limited ranges of stability, and if subjected to changes of temperature and pressure they will either break down or undergo a change in phase. The latter is well illustrated in polymorphic substances, of which silicon dioxide is a good example. Quartz, the hexagonal crystalline form of this compound, changes from one crystal modification (*alpha* quartz) to another (*beta* quartz) at a temperature of 573° centigrade. At a temperature of about 870° *beta* quartz changes to tridymite (orthorhombic), and at about 1470° tridymite changes to cristobalite (tetragonal?). The latter melts at 1725°. Owing to their response to temperature, substances such as silicon dioxide form what are known as geologic thermometers. It should be noted that, although tridymite should revert to *beta* quartz at temperatures below 870°, the rate of this reaction is very slow. *Beta* quartz does, however, immediately revert to the *alpha* form.

A similar condition is seen in the case of diamond and graphite. It has been shown that the former is stable only at very high pressure, and under atmospheric conditions diamonds should revert to graphite. However, this reversion is so slow that diamonds are known to have existed at, or near, the surface of the earth for the greater portion of geologic time without a noticeable production of graphite.

Mineral Formation. Minerals may be formed in a number of ways the more important of which are crystallization from molten masses and precipitation from solutions, either by chemical reactions or by organisms. They may also be formed directly from gases (sublimation), and they may be produced by the action of heat and pressure on previously existing minerals. They may also originate from reactions of liquid and/or gaseous solutions on pre-existing substances. Reactions taking place near the earth's surface in the presence of water, oxygen, and other substances are known as chemical weathering.

COMMON ROCK-FORMING MINERALS

Minerals may be placed in groups on the basis of their chemical composition: elements, oxides, sulphides, etc. Each group can be divided into subgroups by using a more restricted range of characteristics, and the subgroups may be still further divided into species, which may be still further differentiated into varieties. Because it is seldom necessary for an engineer to consider species and varieties, they will be mentioned only when justified by their engineering importance.

Elements

As might be expected, only the more stable elements tend to occur free in nature. Among them native gold, platinum, silver, copper, and sulphur are the most common. Even iron is known in the elemental form, but native iron is one of the rarest of minerals.

Oxides

Oxides are very common in nature and are exceeded in order of abundance only by the silicates. Among the more important are silica, iron oxide, aluminum oxide, and oxides of copper and zinc. Only the first two are of much engineering importance. Most naturally occurring oxides contain a greater or less amount of water, a portion or all of which is held in the crystal lattice and is therefore a part of the mineral composition.

Silica. Silica (SiO_2) is one of the most abundant substances in nature and occurs in a variety of forms, some of which are crystalline and others amorphous. Among these are the various types of quartz (anhydrous) and opal (hydrated). Quartz is the only one deserving detailed attention. The various crystal modifications of quartz have already been discussed.

Quartz. Quartz, one of the most common minerals, occurs in a great variety of rocks where it is often associated with feldspar. The composition is SiO_2; the crystallization is hexagonal; there is no cleavage; the fracture is conchoidal; hardness, 7; and specific gravity, 2.65. The color varies from white, colorless, gray, brown to black, but may be yellow, pink, red, green, or blue.

Individual crystals generally have a vitreous luster, but when found in rocks quartz usually presents a rather dull to greasy appearance. It is fusible at a temperature of 1725° centigrade, and is insoluble in all acids except hydrofluoric. It is very resist-

Fig. 5. Quartz crystals. (Washington University Collection.)

ant to chemical weathering, but it may break up under the action of physical weathering.

Hematite. The principal ore of iron, hematite, is widely distributed in nature where it occurs in many types of rocks formed under a variety of conditions. It is commonly associated with limonite and other iron oxides. The composition is Fe_2O_3; crystallization, hexagonal; there is no cleavage but a well-developed rhombohedral parting is sometimes present. The fracture is conchoidal or uneven; hardness, 5.5 to 6.5; specific gravity, 5.3; and the streak, cherry-red to reddish brown. The color is dull red or bright gray with metallic luster (specular hematite). It fuses with difficulty, and is slowly soluble in hydrochloric acid. It becomes magnetic when heated in a reducing flame. The principal alteration product is limonite.

Limonite. Limonite is a non-crystalline hydrated form of iron oxide ($Fe_2O_3 \cdot nH_2O$), but it is often impure and mixed with other oxides of iron. It usually forms earthy masses, although botryoidal and stalactitic forms may also occur (Fig. 6). There is no cleavage; the luster is dull; hardness, 5 to 5.5; specific gravity, 3.6 to 4.0. It may be distinguished from hematite by its brown streak. The color is yellow, brown, or black.

Magnetite. Magnetite is a common mineral, especially in the igneous and metamorphic rocks, but it is not so widely used as an iron ore because of the frequent association of titanium in solid solution. The composition is Fe_3O_4; the crystallization is isometric; it has an indistinct cleavage and octahedral parting. The fracture is subconchoidal to uneven; hardness, 5.5 to 6.5; specific gravity, 4.97 to 5.18. It is opaque; black in color; streak also black; and is strongly magnetic. Magnetite alters to hematite, limonite, and occasionally to siderite ($FeCO_3$).

Carbonates

Carbonates are very abundant and are usually formed by weathering of other minerals, chiefly silicates. They may be deposited in the place of origin, or they may be carried in solution and deposited in seas or lakes by chemical or organic agencies. Only the three carbonate minerals discussed below are of much engineering importance.

Calcite. Calcite, the chief mineral of limestones, may occur as a cementing material in other sedimentary rocks. The composition is $CaCO_3$; the crystallization, hexagonal; cleavage, rhombohedral and perfect; hardness, 3; specific gravity, 2.72. The color is usually white, but it may show a great variety of tints caused by impurities. The luster is vitreous to earthy. It is usually transparent, but when impure may be opaque. Calcite dissolves readily in acids, to which it reacts with effervescence. The cleavage and crystal form are illustrated in Figs. 2 and 7. In rocks calcite often occurs in a finely disseminated form, but it can be readily detected by its reaction to acid.

Fig. 6. Botryoidal (top) and stalactitic forms of limonite. (Washington University Collection.)

21

Fig. 7. Calcite crystals. (Washington University Collection.)

Aragonite. Aragonite is similar to calcite in nearly all respects, but its crystal structure is orthorhombic. It is less stable, and therefore rarer than calcite to which it tends to alter.

Dolomite. Dolomite is a mixed calcium magnesium carbonate $[CaMg(CO_3)_2]$, occurring chiefly in thick sedimentary beds also known as dolomite, or dolomitic limestones. Under certain conditions calcite may alter to dolomite, causing a decrease of about 8 per cent in volume of the mass. The crystallization is hexagonal but the faces are often curved; the cleavage is like that of calcite; the hardness is 3.5 to 4; specific gravity, 2.80 to 2.99. The color varies from white to pink; the luster is vitreous or pearly; and the mineral is translucent to opaque. Unlike calcite, dolomite is only very slightly soluble in cold dilute acids, but it dissolves readily with effervescence in hot dilute acidic solutions.

Sulphates

A fairly large number of sulphates are found in nature, but only two, gypsum and anhydrite, are common enough to be important to engineers. Like the carbonates, the sulphates are usually formed by weathering of other minerals, and they are deposited chemically in seas or lakes.

Gypsum. Gypsum is well known in the calcined form called plaster of Paris. The composition is $CaSO_4 \cdot 2H_2O$. The crystallization is monoclinic, and the cleavage is perfect in one direction and less perfect in two others that intersect at angles of 66° and 114°. The hardness is 2, the specific gravity, 2.31 to 2.33. Gypsum is colorless or white, but it may show red, yellow, brown, or

Fig. 8. Twinned gypsum crystal. (Washington University Collection.)

black tints caused by various impurities. The luster is pearly on some faces, vitreous on others. It may also be satin-like (satin spar). The streak is white, and the mineral is transparent, translucent, or opaque. When heated it loses water and becomes opaque. It is soluble in water and in hydrochloric acid.

Anhydrite. As the name suggests, anhydrite differs from gypsum in lacking water. The composition is therefore $CaSO_4$. The crystallization is orthorhombic; the cleavage is in three directions, forming rectangular to cube-like forms. The hardness is 3 to 3.5; the specific gravity, 2.95. The color is variable as in gypsum, but it is usually white. The luster is pearly, greasy, or vitreous, depending on the crystal face; in massive varieties it may be dull. In the presence of water anhydrite alters to gypsum with a considerable increase in volume.

Sulphides

Sulphides are very common in nature and form the greater portion of the metallic ore minerals, but they are usually rather rare in rocks. They are, therefore, of only minor engineering importance, and only two will be discussed.

Pyrite. Pyrite, or "fool's gold," is a common constituent of many rocks. The composition is FeS_2, and the crystallization is

Fig. 9. Pyrite crystals. (Washington University Collection.)

isometric. There is no cleavage; the fracture is conchoidal; the hardness, 6 to 6.5; and the specific gravity, 5.01. The color is brass-yellow, but pyrite may become darker because of tarnish. The luster is metallic; the streak, greenish to brownish black. It is opaque even in thin section. Pyrite gives off sulphur fumes when heated, and it is soluble in nitric acid with the separation of sulphur. It is unstable and oxidizes rapidly to hematite and limonite with the liberation of heat, hydrogen sulphide, and SO_3. In the presence of water the latter forms sulphuric acid.

Marcasite. Marcasite differs from pyrite chiefly in its ortho-rhombic crystallization. The hardness is 6 to 6.5; the specific gravity, 4.6 to 4.8. Marcasite has much the same associations as pyrite, but it tends to alter more readily to limonite and melanter-ite ($FeSO_4 \cdot 7H_2O$).

Silicates

The silicates find their true home in the igneous and meta-morphic rocks, and they are a very abundant group of minerals. They are very stable and also occur abundantly in the clastic sediments.

Feldspars. The name feldspar is applied to a group of minerals that constitute the most abundant substances in the earth's crust. The feldspars may be divided into orthoclase (cleavage at right angles) and plagioclase members (cleavage at an angle of 86°). As previously mentioned, the plagioclases form an isomorphous series that is of great importance in the detailed classification of the igneous rocks.

Orthoclase. The composition of orthoclase is $KAlSi_3O_8$; the crystallization is monoclinic; cleavage, good in two directions at right angles to each other; fracture, uneven; hardness, 6; specific gravity, 2.50 to 2.62. The luster is vitreous to pearly; the streak, white but not characteristic; color, usually red to pink. Thin fragments are generally transparent to translucent. Orthoclase is insoluble in ordinary acids. In the presence of water and CO_2 orthoclase and other feldspars may alter to kaolin, muscovite, and other complex hydrated silicates.

Plagioclase. Plagioclase is the name given to a completely isomorphous series that may vary from pure sodium aluminum silicate ($NaAlSi_3O_8$), or albite, to pure calcium aluminum silicate ($CaAl_2Si_2O_8$), or anorthite. Intermediate varieties such as oligo-clase, andesine, labradorite, and bytownite are distinguished by arbitrarily designated proportions of the albite and anorthite mole-cules. Plagioclases crystallize in the triclinic system and are fre-quently twinned so as to give rise to parallel grooves, or striations, on the cleavage faces. The cleavage is good in two directions, intersecting at about 86° (distinguishes plagioclase from ortho-

Fig. 10. Orthoclase feldspar crystal, showing right-angle cleavage at right-
and left-hand edges. (Washington University Collection.)

clase). The hardness is 6. The specific gravity varies with composition, being 2.62 for pure albite and 2.76 for pure anorthite.
The color is commonly gray to white, but colorless, green, and
iridescent varieties are known. Albite is insoluble in ordinary
acids, but the members high in calcium (labradorite and anorthite)
are slowly soluble. The alteration products of plagioclase are
similar to those of orthoclase.

Mica Group. The micas comprise a large group of minerals
which crystallize in the monoclinic system. The crystals have
hexagonal outlines, and they are marked by a very well-defined
basal cleavage along which the mineral splits into very thin plates
which are tough and elastic. It is this property that gives the
group its chief commercial value. The two most important

species are: muscovite [$H_2KAl_3(SiO_4)_3$], and biotite [$(H,K)_2$ $(Mg,Fe)2Al_2(SiO_4)_3$]. Crystals are often observed in igneous and metamorphic rocks, but the mineral usually occurs as flakes, scales, and shreds. The hardness is 2 to 3; the luster is vitreous or pearly; the streak is colorless; the specific gravity, 2.7 to 3.2. Muscovite is transparent; biotite is translucent or opaque. Most micas are insoluble in hydrochloric acid, but when boiled in

Fig. 11. Mica crystal. (Washington University Collection.)

sulphuric acid biotite decomposes into a milky solution. Most species are subject to alteration, and this is especially true of biotite. Silvery white scales of mica, called sericite, are an alteration product of feldspar. On further alteration sericite changes to clay.

Pyroxene Group. The pyroxene group comprises several minerals that crystallize in the orthorhombic, monoclinic, and triclinic systems. The more important members are: orthorhombic section—enstatite ($MgSiO_3$) and hypersthene [$(Mg,Fe)SiO_3$]; monoclinic section—diopside, a complex calcium magnesium iron silicate; augite, composition similar to diopside; and aegirite [$NaFe(SiO_3)_2$]. Pyroxenes are distinguished by cleavage in two directions parallel to the prism faces, forming angles that intersect at 87° and 93°. The color varies with the amount of iron present and may be white, gray, green, brown, or black. The luster is vitreous, resinous, or pearly. The streak varies from white to brown to grayish green. The hardness is 5 to 6, and the specific

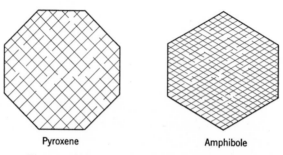

Pyroxene   Amphibole

Fig. 12. Cleavage of pyroxene and amphibole.

gravity, 2.8 to 3.7. Pyroxenes occur mainly in igneous rocks and are important in their classification. They alter to amphiboles, serpentine, chlorite, various carbonates, and iron oxides, chiefly limonite.

Amphibole Group. The amphiboles resemble the pyroxenes in many respects, but they can be readily distinguished by prismatic cleavages that intersect at angles of 124° and 56°. Like the pyroxenes, the crystallization is orthorhombic, monoclinic, and triclinic, but only the monoclinic varieties are of much geologic importance. The more common members of the group are: tremolite $[CaMg_3(SiO_3)_4]$; actinolite, composition similar to tremolite; hornblende, a complex calcium, magnesium, iron, sodium, aluminum silicate; and arfedsonite, composition similar to hornblende. The color varies according to the amount of iron present from white or gray (tremolite) to bright green in actinolite, to dark green or black in hornblende. The luster is vitreous to pearly on cleavage faces, but it is often silky in fibrous varieties. The streak is white, gray-green, or brownish. The hardness is 5 to 6, the specific gravity varies from 2.9 to 3.5, increasing with the amount of iron present. Iron-rich varieties are slightly soluble in ordinary acids. Amphiboles are abundant in the igneous and metamorphic rocks and alter in much the same manner as the pyroxenes.

Olivine Group. The members of the olivine group form a series that may be regarded as mixtures of Mg_2SiO_4, or forsterite, and Fe_2SiO_4, or fayalite. The crystallization is orthorhombic; cleavage, indistinct; fracture, conchoidal; hardness, 6.5 to 7; specific gravity, 3.19 to 4.14. The color is green, olive-green, or bottle-green. The luster is vitreous, streak uncolored or rarely yellowish. Olivines are soluble in acids, yielding a gelatinous precipitate when evaporated. Minerals of this group are common constituents of the basic igneous rocks, but they may also be found in schists and metamorphosed magnesium limestones. They alter to serpentine and iron oxide.

Talc. The substance talc is usually the product of alteration of other minerals such as enstatite, olivine, or tremolite, and occurs mainly in metamorphosed rocks, where it is frequently found as a white soapy material known as talc schist. The composition is $H_2Mg_3(SiO_3)_4$, and the crystallization probably monoclinic. The

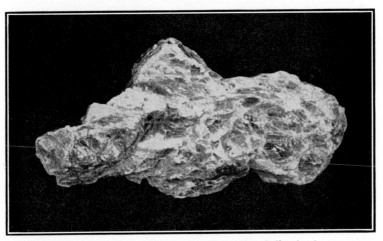

Fig. 13. Talc. (Washington University Collection.)

cleavage is perfect in one direction; the hardness, 1; and the specific gravity, 2.7 to 2.8. It has a characteristic greasy feel. The luster is pearly on cleavage faces, and the streak light. The color varies from silvery white to green. It is not soluble in acids. Steatite, or soapstone, is an impure massive form of talc that is usually green in color and fine to coarsely granular in texture.

Serpentine. Like talc, serpentine is a secondary mineral, formed by alteration of olivine, pyroxenes, and amphiboles. Its chief occurrence is in the metamorphic rocks. The composition is $H_4Mg_3Si_2O_9$, and the crystallization is probably monoclinic. The cleavage is of little importance in sight identification. The

Fig. 14. Asbestos veins (gray) in serpentine. (Washington University Collection.)

hardness is 2.5 to 5.0; the specific gravity varies from 2.50 to 2.65. Serpentine is usually some shade of green or yellow, but brown, red, and black tints are also known to occur. It is translucent to opaque; it has a smooth or greasy feel, greasy or wax-like luster, white streak, and conchoidal fracture. It is soluble in hydrochloric acid and yields water on ignition.

Chlorite. Although actually a group rather than a single mineral, the species of chlorite are very similar in megascopic characteristics and will be discussed together under a single name.

The chlorites are hydrous silicates of aluminum, magnesium, and ferrous iron. The crystallization is monoclinic, and, as in mica, the basal cleavage is perfect. Thin plates of chlorite are flexible, but unlike mica the cleavage plates are not elastic. The hardness is 2 to 2.5, and the specific gravity varies from 2.60 to 3.00. The color is usually dark green, and the luster is sometimes pearly. Chlorite is insoluble in hydrochloric acid, but it is soluble in boiling sulphuric acid, forming a milky solution. Like talc and serpentine, it is a secondary mineral and may be formed by alteration of pyroxenes, amphiboles, and micas. It is a common constituent of the metamorphic rocks and often occurs in well-defined masses known as chlorite schist.

Zeolite Group. The minerals of the zeolite group are hydrous silicates of aluminum, calcium, and sodium. The crystallization of the various species is orthorhombic (natrolite), isometric (analcine), and monoclinic (stilbite and heulandite). Zeolites are usually colorless or white, but they may exhibit shades of yellow and red. The luster is vitreous; hardness, 3.5 to 5.5; the specific gravity, 2 to 2.4. They dissolve in hydrochloric acid, and some species are soluble in water. Many species possess the property known as base exchange; i.e., they will take calcium or other bases into their composition in the place of sodium and are often used as water softeners. Because of this property, materials containing significant amounts of zeolites are unsuitable for use as concrete aggregates. Zeolites are formed by alteration of feldspars by steam or by circulating waters and are often found in cavities in igneous rocks. They frequently fill gas cavities in lavas giving rise to the so-called amygdaloidal texture (Fig. 15).

Clay Minerals. The large group of clay minerals is of great importance both to geology and civil engineering. All members consist of more or less hydrated and complex aluminum silicates, and they are derived chiefly from the weathering of feldspathic rocks. Under certain conditions, volcanic ash may alter to a material known as bentonite, which possesses the property of increasing up to ten times the original volume when immersed in water. Individual clay particles are usually too small to be studied by ordinary means, but the electron microscope and X-rays have greatly increased knowledge of the group.[1] Naturally occurring clays are usually more or less intimate mixtures of several clay minerals and various impurities, some of which are colloidal. Although the clay minerals are crystalline, crystallization is of little importance in megascopic identification. Clay is a common constituent of soils, and it may also occur in well-defined and fairly thick beds derived from the weathering of under-

Fig. 15. Radiating zeolite crystals in cavities in basalt, forming what is known as the amygdaloidal texture. (Washington University Collection.)

lying rocks. Various clay minerals may also be produced by the action of hot ascending waters.

Formerly, the term "kaolin" was used to cover all residual clays, but it is now known that kaolin is not composed entirely of the mineral kaolinite, but consists of many other clay minerals as well. The better-known clay minerals are: kaolinite ($Al_2O_3 \cdot 2SiO_2 \cdot 2H_2O$); nacrite and dickite ($Al_2O_3 \cdot 2SiO_2 \cdot 2H_2O$), usually of hydrothermal origin; halloysite ($Al_2O_3 \cdot 2SiO_2 \cdot nH_2O$), resulting from surface weathering; montmorillonite [$Al_2O_3(Ca,Mg)$-$0.5SiO_2 \cdot nH_2O$], resulting from the weathering of volcanic ash and forming the chief constituent of bentonite; and beidellite

($Al_2O_3 \cdot 3SiO_2 \cdot 2H_2O$), also belonging to the montmorillonite group. Only pure clay minerals are white; they are usually colored red, buff, or brown by various foreign substances. The luster is dull or earthy; hardness, 2 to 2.5; specific gravity, 2.6 to

Fig. 16. Kaolinite. (Washington University Collection.)

2.63. Nearly all clays have an unctuous feel and are more or less plastic when wet. Most clay minerals are insoluble in acids and when heated give off various quantities of water.

Attapulgite is a comparatively little-known clay mineral found mainly in former lake basins. It is often associated with montmorillonite and gypsum. The electron microscope indicates that attapulgite crystallizes in minute tubes that tend to form hair-like masses. Its physical and engineering properties are not well known, but it probably has a high base-exchange capacity, high absorption, and high compressibility. Attapulgite is probably much more abundant and widespread than is presently believed.

REFERENCES

1. Ralph E. Grim, *Clay Mineralogy*, McGraw-Hill Book Co., New York, 1953.
2. E. H. Kraus, W. F. Hunt, and L. S. Ramsdell, *Mineralogy*, McGraw-Hill Book Co., New York, 1936.
3. L. V. Pirsson and Adolph Knopf, *Rocks and Rock-Minerals*, John Wiley & Sons, New York, 1947.
4. C. S. Ross and S. B. Hendricks, Minerals of the Montmorillonite Group: Their Origin and Relation to Soils and Clays, *U. S. Geol. Survey, Prof. Paper* 205-B, 1945.

CHAPTER 3

Rocks

Both geologists and civil engineers are concerned with rocks, but from different points of view. To the geologist, rocks are essential units of the earth's crust the origin, classification, history, and spatial relationships of which possess fundamental significance. Their physical properties form no part of his definition, and from a geological point of view ice, sand, granite, and coal are rocks. To the civil engineer, the physical properties are of primary importance, and in engineering the term "rock" is restricted to materials that cannot be excavated without blasting. All other naturally occurring substances are known as "soil" or "earth." In this book, the engineering usage is the one most generally followed.

In geology rocks are classified according to origin, composition, and texture. Civil engineers are constrained to follow the geological usage, but owing to differences in viewpoints of the two professions a certain amount of confusion results. For example, to the engineer the term "sandstone" indicates a hard, strong material that is satisfactory for foundations of heavy structures and many other engineering purposes. As used by geologists, however, the term includes many friable and very weak materials. Similarly, the wide range in physical behavior of materials classified as shale is a perpetual source of difficulty to engineers. On the other hand, nearly all intrusive igneous rocks possess nearly the same engineering properties, and the civil engineer cannot but be annoyed by a system that classifies these materials under 25 or more names. In fact, from an engineering point of view shales probably exhibit much wider variations in significant properties than all the intrusive igneous rocks combined. It is probable that when engineering geology reaches its full maturity an engineering classification of rocks will be worked out that satisfies the needs of both the civil engineer and the engineer-geologist. Such an attempt would be premature at the present time, and in the following pages an effort is made to present data of value to civil engineers within the framework of the established geological classification. For a more detailed treatment of the engineering properties of rocks the treatise of von Moos and de Quervain [19] should be consulted.

IGNEOUS ROCKS

As the name indicates, this class of rocks constitutes materials that have crystallized from a previously molten condition. With regard to mode of occurrence two major types are known: (1) intrusive rocks that have crystallized within the earth, and (2) extrusive, or volcanic, rocks formed by cooling of molten material at the surface. The consolidated fragmental products of volcanic eruptions form still another group known as pyroclastic rocks. In addition to their modes of occurrence, intrusive and extrusive types can usually be distinguished by differences in texture. Slow cooling promotes the growth of large crystals, whereas rapid cooling tends to give rise to smaller grains, or even to a glassy material. Consequently, intrusive rocks are usually coarsely crystalline (constituent particles large enough to be seen with the unaided eye), whereas extrusive bodies are generally fine grained (constituent particles usually too small to be distinguished by the unaided eye). Molten material occurring in the depths of the earth is known as magma, but liquid rock reaching the surface is usually referred to as lava. Since extrusive bodies must be connected with deep-seated molten masses, lavas must reach the surface along some form of conduit. Crystalline material filling such conduits constitutes a fairly well-defined intermediate category known as dike rocks, but they are usually considered as belonging in the intrusive group.

Classification. Igneous rocks are classified according to texture, mode of occurrence, and mineral composition. As shown in Table II, rocks of the same composition consequently have

TABLE II. CLASSIFICATION OF THE COMMON IGNEOUS ROCKS

Name	Texture and Usual Mode of Occurrence	Essential Minerals	Common Accessory Minerals
Granite	Coarse; intrusive	Quartz, orthoclase	Mica, plagioclase, pyroxenes, amphiboles
Rhyolite	Fine; extrusive	Quartz, orthoclase	Mica, plagioclase, pyroxenes, amphiboles
Syenite	Coarse; intrusive	Orthoclase, plagioclase	Mica, pyroxenes, amphiboles
Trachyte	Fine; extrusive	Orthoclase, plagioclase	Mica, pyroxenes, amphiboles
Diorite	Coarse; intrusive	Plagioclase, amphiboles	Mica, pyroxenes
Andesite	Fine; extrusive	Plagioclase, amphiboles	Mica, pyroxenes
Gabbro	Coarse; intrusive	Plagioclase, pyroxenes	Biotite, magnetite
Basalt and trap *	Fine; extrusive	Plagioclase, pyroxenes	Biotite, magnetite
Peridotite	Coarse; intrusive	Pyroxenes, magnetite	Chromite

* The terms "basalt" and "trap" are often used to designate almost any fine-grained basic igneous rock but, as has sometimes been done, should not be extended to include other types such as gneiss and schist.

different names, depending on their texture and manner of occurrence. For example, both granite and rhyolite contain orthoclase and quartz as essential minerals, but granite is a coarse-grained intrusive whereas rhyolite is a fine-grained extrusive rock. In geological parlance rhyolite is said to be an extrusive, or volcanic, equivalent of granite. With respect to composition, a distinction is made between essential and accessory minerals. To return to the example of granite, although orthoclase and quartz are the only essential minerals of this rock, they are usually associated with mica, plagioclase, pyroxenes, and amphiboles. The latter occur in subordinate and variable amounts and are regarded as accessory minerals. Because silica (SiO_2) forms an acid and iron a base, rocks high in silica are often referred to as acidic, and those high in iron are designated as basic. This usage is not in accord with ordinary chemical nomenclature, and for this reason is to be discouraged. It will be noted that in Table II there is a complete gradation from acidic to ultrabasic rocks. When the constituent grains are too small to be distinguished with a hand lens, igneous rocks are frequently classified on the basis of color. Light-colored, extremely fine-grained rocks are called felsites, but no generally accepted term has been proposed for the dark-colored rocks. Attempts have been made to systematize the classification of igneous rocks by application of chemical principles, but at present none of them seems to have gained wide acceptance.

Textures. The term texture applies to the relative sizes and shapes of the mineral grains forming a rock. Textures of igneous rocks may be coarsely crystalline, finely crystalline, cryptocrys-

Fig. 17. Porphyritic granite. Large crystals (phenocrysts) are orthoclase feldspar. (Washington University Collection. Photo by Joseph E. Cox, Jr.)

talline (grains visible only under the microscope), or glassy (entirely non-crystalline). Cryptocrystalline and glassy textures are limited largely to extrusive types. The texture may be equigranular or porphyritic, depending on the relative size of particles. Porphyritic rock consists of a mass of large grains (phenocrysts) set in a finer-grained matrix. Porphyries are thought to be caused by a change in the rate of cooling, the large crystals being formed during a period of slow cooling and the finer grains during a period of more rapid heat radiation. In addition to those already mentioned, a great number of textures are visible under the microscope, and many rock varieties are distinguished on this basis. The chief significance of texture to civil engineering is its influence on the strength and permeability of a rock. The greatest strength is imparted by a mass of interlocking crystal grains, whereas a non-interlocking texture is the weakest arrangement.

INTRUSIVE MASSES

Although intrusive rocks often occur in fairly well-defined bodies of characteristic shape that have received descriptive names, it should not be assumed that all intrusive bodies can be fitted into one of the types described below. The shape is often so irregular as to defy classification; it must simply be worked out by detailed mapping and core drilling, if necessary.

Batholiths. As originally described and defined, batholiths were considered to be large intrusive masses extending indefinitely downward, but it has been suggested that they are intruded along a definite lower surface or "floor" and that they make room for themselves by arching and lifting the overlying rocks.[4] For all practical purposes, however, they may be considered as bottomless. Batholiths are usually associated with folded regions, and in a sense they may be regarded as forming the cores or "roots" of mountains. Two types may be distinguished: one which cuts across the regional structure, known as discordant; the other which parallels the structural trend of the enclosing rocks, referred to as concordant. Depth at which batholiths are intruded is not known with any degree of precision, but it is estimated at thousands of feet to many miles. Consequently, a batholithic mass is exposed at the surface only after prolonged erosion, and the area of surface outcrop is chiefly a function of the extent to which it has been uncovered. Inclusions of the surrounding rocks and roof pendants are common features of batholithic bodies (see Fig. 18).

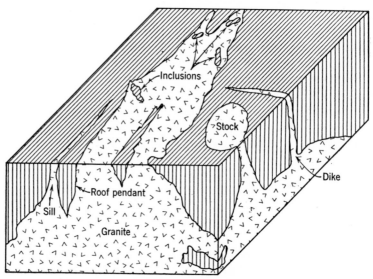

Fig. 18. Block diagram showing idealized conception of a concordant batholith and associated features.

From a geological point of view batholiths are of extraordinary interest, the source of heat, origin of the magma, and mechanics of intrusion being largely unsolved problems. Batholithic masses are usually granitic in composition, and many modern workers maintain that many, if not all, so-called batholiths originated from alteration of sediments by liquids and/or gases, a phenomenon known as granitization or ultrametamorphism. The largest batholith in the United States is in central Idaho; it covers an area of approximately 16,000 square miles. Other large masses are the Sierra Nevada and Coast Range batholiths.

Stocks. Stocks are smaller intrusive bodies without known floors. They are rather arbitrarily distinguished from batholiths on the basis of size, an intrusive mass less than 40 square miles in area being designated a stock. They are distinguished from volcanic necks by lack of evidence indicating the former existence of a volcano. Since this evidence may be destroyed by erosion, it is possible that many "stocks" are actually volcanic necks.

Laccoliths. Laccoliths are plano-convex or doubly convex intrusive masses which are thought to rest on a definite floor and to have domed up the overlying strata. Excellent examples occur in the Henry Mountains of Utah where they were first described.[6] Laccoliths range from about 2 to 4 miles in diameter, and attain a thickness of thousands of feet. It is usually assumed that the magma forming these bodies gained access to the overlying strata along some sort of conduit and that, upon reaching a favorable level, spread out along the bedding planes causing concomitant arching and uplift of the overlying rocks. However, some geologists question this view, and, until more is known concerning the "floors" and "feeders" of these masses, it is unwise to generalize about their origin. An idealized example is illustrated in Fig. 19.

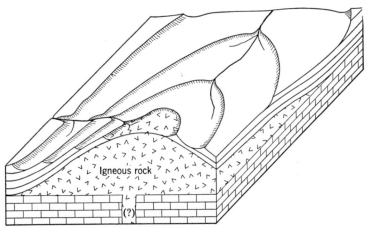

Fig. 19. Block diagram showing idealized conception of a laccolith.

Lopoliths. A lopolith is a concordant intrusion associated with a structural basin.[7] In the simplest case the sediments above and below the intrusive body dip toward a central point. The diameters of lopoliths range from a few tens to hundreds of miles. The thickness is normally measured in thousands of feet. As in

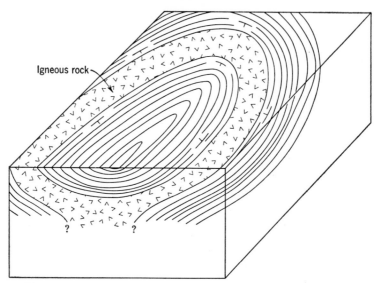

Igneous rock

Fig. 20. Block diagram showing idealized conception of a lopolith.

the case of laccoliths, it seems reasonable to suppose that they are connected with a "feeder" at depth. It is thought that the basin-like structure is formed contemporaneously with the intrusion and is the result of subsidence of the floor, following withdrawal of a part of the magma from the underlying reservoir. Examples of lopoliths occur in many parts of the world. At Sudbury, Ontario, nickel ores are associated with what is generally regarded as a lopolith.

Phacoliths. After their deposition, sedimentary rocks may be deformed by earth movements, which form flexures known as folds. During folding igneous material may force its way, or be

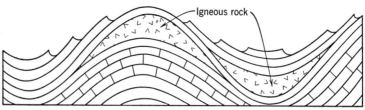

Igneous rock

Fig. 21. Section through idealized phacoliths.

forced, into the crests of upwarped folds (anticlines) or into the troughs of downwarps (synclines). Such concordant intrutive bodies have been termed phacoliths.[9] These bodies are commonly associated with plunging folds, and are consequently usually crescentic both in plan and in cross section. They are seldom more than a mile long, and are usually hundreds to a few thousands of feet thick.

Volcanic Necks. Volcanic necks, or vents or plugs, are channels through which the magma that fed volcanoes gained access to the surface, and are now filled with a rather heterogeneous assemblage of rock materials. They are circular to irregular in plan; they vary from about 100 feet to a mile or so in diameter. In compound necks the diameter may be considerably larger. A feature often associated with volcanic necks or other intrusive bodies is the presence of ring-dikes. These are roughly concentric sheets of igneous material, which either are vertical or dip steeply toward or away from a central point. Ring-dikes have been described from New Hampshire,[11] and other regions. Concentric dikes that dip inward at moderately high angles are known as cone sheets.

Dikes. Tabular masses of igneous rocks that cut the bedding of sedimentary strata, or the structures of intruded formations, are known as dikes. They vary from a few feet to hundreds of

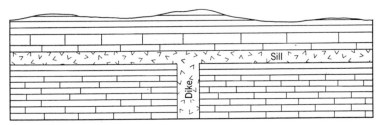

Fig. 22. Section through idealized dike and sill.

miles in length. The thickness varies from a fraction of an inch to several miles, the great dike of Rhodesia being 2 to 7 miles in width. Dikes may present a variety of surface patterns, a common one being the radial arrangement observed on the peripheries of volcanic necks.

Sills. Sills, or sheets, are tabular-shaped igneous masses that have been intruded parallel to the bedding (more or less parallel zones of different composition, grain size, color, etc., formed during deposition of the material and approximately parallel to the depositional surface) or schistosity (nearly parallel planes of relatively easy breakage caused by heat and pressure subsequent to the original formation of the rock) of the enclosing rocks. They may be horizontal, inclined, or vertical, the distinguishing characteristic being the relationship of their upper and lower contacts with respect to the bedding planes of the enclosing rocks. Hence sills will always partake of the structure of the rocks into which they have been intruded, whereas dikes will not. They are distinguished from lava flows by the presence of heating and baking effects on both their lower and upper surfaces. Lava flows exhibit these effects only on their lower boundaries. Sills may extend over considerable areas, and they may attain thicknesses of hundreds of feet or more. An idealized example is shown in Fig. 22.

Apophyses. An apophysis is an irregularly shaped dike which has been derived from some nearby intrusive body. The term "stringer" is more or less synonymous with apophysis.

EXTRUSIVE MASSES

Molten rock that reaches the surface generally spreads out to form relatively thin sheets called lava flows. Two types of flows can be distinguished: (1) more or less local and irregular tongues found in the vicinity of volcanoes, and (2) roughly tabular sheets often of great thickness and lateral extent which are associated with no known volcano, and are assumed to have reached the surface along fissures. Flows of the latter type tend to be basaltic in composition and are often referred to as "plateau basalts." Extrusive rocks often harden around included gas bubbles, giving rise to vesicular or scoriaceous texture (see Fig. 23). Should these cavities become filled with mineral matter, the texture is said to be amygdaloidal. Lava flows may be the products of several periods of volcanic activity, and it may happen that an older flow was deeply weathered and eroded before an overlying sheet was laid down. In tunneling and excavation work the presence of softer weathered zones may be of considerable importance. Furthermore, recently extruded lavas may contain appreciable amounts of carbon dioxide and other gases. Extrusive rocks may also contain long hollow tubes, or tunnels, formed

Fig. 23. Scoriaceous basalt. (Washington University Collection. Photo by Joseph E. Cox, Jr.)

by withdrawal of molten material from under a surficially hardened crust. Lavas may occur at considerable distances from any recently active volcano; they are, of course, invariably present in volcanically active regions. Extensive bodies of plateau basalts, sometimes exceeding 200,000 square miles in area, are found in northwestern United States, Siberia, India, and other parts of the world.

ENGINEERING PROPERTIES OF IGNEOUS ROCKS

When fresh and unweathered, all intrusive rocks are characterized by high crushing and shearing strengths; and, unless too minutely fractured, they are usually satisfactory for all types of engineering construction and operations. They are often good sources of concrete aggregates and other types of construction materials. In operations involving intrusive rocks attention should be given to jointing and possible faulting, and especially to possible mineral alteration along such fractures. If alteration occurs, strength of the rock may be greatly impaired. Despite the generally favorable character of intrusive rocks, it is unwise to assume after only a cursory examination that they will be satisfactory for the purpose intended. Sufficient tests to prove this should always be made.

Extrusive rocks exhibit considerable variation in physical properties and usually require extended examination before their engineering characteristics can be evaluated. Many lavas are quite as satisfactory as intrusive rocks, but volcanic breccias and

Fig. 24. Columnar jointing in basalt, Orange, New Jersey. (J. P. Iddings, U. S. Geological Survey.)

scoriaceous or vesicular materials may be very unreliable. The physical properties of the rock underlying the lava should also be taken into consideration, especially where the flows are thin. Lava flows frequently exhibit columnar jointing caused by contraction during cooling (see Fig. 24). Presence of joints tends to lower the strength of the mass as a whole.

PYROCLASTIC ROCKS

These materials are formed by explosive volcanic activity, and consist of fragments torn loose by such explosions. Although the constituent particles are predominantly igneous in character, usually they have been hurled into the air and are in a sense of sedimentary origin. Some investigators consider the pyroclastics as a subdivision of the sedimentary rocks, but in this volume it has been thought best to regard them as a subclass of the igneous rocks. Owing to their great range in characteristics and generally heterogeneous constitution, pyroclastic rocks are difficult to classify. The most satisfactory classification offered so far is that of Wentworth and Williams [20] given in Table III. This scheme distinguishes between the unconsolidated fragmental products and their consolidated equivalents. Usually the coarser materials are limited to a rather small area near the volcanic vent, but the finer materials, such as ash and dust, may be carried for great distances by the wind to form widely distributed tuff beds. Pyroclastic rocks are common occurrences in all volcanic areas; they are especially abundant in western United States, Central America, and other regions of more or less recent volcanic activity. They may also occur at considerable distances from any recent centers of volcanism. Where pyroclastic materials have settled through water, an almost complete hybrid between a sedimentary and igneous rock results. Owing to the selective effects of transportation through the air, nearly all pyroclastic rocks show stratification, which is usually better developed in finer materials than in agglomerates.

Engineering Properties. Because of their great range in physical characteristics, it is impossible to generalize concerning the engineering properties of pyroclastic rocks. Most tuff beds are soft and are characterized by low compressive and shearing strengths. Furthermore, under some conditions they may become altered to a very soft and plastic clay material known as bentonite, which when saturated with water may increase up to ten times its original volume. On the other hand, agglomerates

TABLE III. CLASSIFICATION OF THE PYROCLASTIC ROCKS
(After Wentworth and Williams) [20]

Name	Unconsolidated Material	
	Size of Fragments	Character of Fragments
Volcanic dust	Mostly less than 0.25 mm in diameter	Ash and dust
Volcanic ash	Mostly 0.25 to 4.0 mm in diameter	May contain fragments of molten material hardened in flight
Lapilli	Mostly 4.0 to 32.0 mm in diameter	May contain bombs (molten material hardened in flight)

Name	Consolidated Material
	Constituents
Tuff	Volcanic dust and volcanic ash
Volcanic breccia	Lapilli and blocks
Tuff-breccia	Same as above except that much of the matrix is less than 4.0 mm in diameter
Agglomerate	Rounded or subangular fragments larger than 4.0 mm in diameter set in finer matrix
Volcanic conglomerate	Similar to volcanic breccia but the fragments are rounded by action of running water

may be nearly as strong and reliable as the average intrusive rock, their strength being largely determined by character of the matrix. Agglomerates are generally not satisfactory for concrete aggregates, because of the large and variable proportion of fines that results when the rock is crushed. As a rule, large-scale engineering operations in pyroclastic rocks will tax the abilities of engineers and geologists to the utmost.

SEDIMENTARY ROCKS

Rocks exposed on the surface are subjected to natural agencies that tend to disintegrate and destroy them: the action of running water, ice, wind, and waves, to mention only a few outstanding examples. Materials broken from the parent rock by these and other agencies are known as sediments, and the rocks formed by disintegration, transportation, and deposition of materials derived from the parent mass are known as sedimentary rocks. Two major types of sedimentary rocks can be distinguished: (1) chemical sediments in which the material has been transported in solution and later precipitated, and (2) mechanical, or clastic, sediments in which transportation and deposition have been mainly by mechanical means.

CHEMICAL SEDIMENTS

The chemical sediments form a large and varied class of rocks; they cannot be treated adequately in a work of this size. The more important members of this class of materials are listed in Table IV. It should be noted that chemical deposits may be precipitated by organic and/or inorganic means. The inorganic sediments of the class are known as evaporites, i.e., materials deposited by evaporation from chemically charged waters; they involve rather complex considerations that cannot be discussed in an elementary treatment. It should be noted that, owing to the processes involved in their origin, chemical sediments are nearly always consolidated rocks, i.e., the constituent particles are tightly bound together to form a single coherent mass.

Limestone. Limestones are rocks consisting mainly of calcium carbonate in the form of calcite or aragonite. Aragonite is un-

Table IV. Classification of the More Important Chemical Sediments

Name	Chemical Composition	Mode of Precipitation
Limestone	Calcium carbonate	Organic and inorganic
Dolomite	Calcium-magnesium carbonate	Organic and inorganic
Diatomite	Silica	Organic
Radiolarite	Silica	Organic
Chert, flint	Silica	Organic and inorganic
Iron deposits	Iron oxide, silicates, sulphides, and carbonates	Mainly inorganic
Manganese deposits	Mainly oxides and carbonates	Inorganic
Peat, lignite, and coal	Carbon, oxygen, nitrogen, hydrogen, sulphur	Organic
Rock salt	Sodium chloride	Inorganic
Gypsum	Calcium sulphate, hydrated	Inorganic
Anhydrite	Calcium sulphate	Inorganic

Fig. 25. Jointed limestone, Drummond Island, Michigan. (I. C. Russell, U. S. Geological Survey.)

stable and is rare compared with calcite. Most limestones are of marine origin, the calcium carbonate being precipitated either by organisms or by inorganic processes. Some limestones, known as coquina, consist of little more than a loosely cemented mass of shells. Limestones may also be precipitated inorganically by a great variety of chemical reactions. Calcareous deposits formed by the evaporation of lime-charged waters are known as tufa or travertine and are a source of ornamental building stone. In arid regions calcareous waters may be brought to the surface by capillary action, where evaporation gives rise to a deposit known as caliche. When unaltered, limestones are soft (3 in Mohs' scale of hardness) and are soluble in acidic waters to which they react with effervescence. Under certain conditions they may become impregnated with silica to give rise to a very dense and hard rock known as silicified limestone. Because of their soluble character limestones may be very porous, and they frequently possess large cavities. Common impurities found in limestone are clay, iron oxide, aluminum oxide, and quartz. Limestones are known to have been deposited in all geological periods, are often hundreds to thousands of feet thick, and may cover vast areas. The Niagara limestone, for example, underlies such widely separated places as Chicago and Niagara Falls.

Dolomite. Dolomite consists of the mixed calcium magnesium carbonates $[(CaMg)(CO_3)_2]$. It may originate through deposition of the mixed carbonates, or through the alteration of limestone by magnesium-charged waters. It closely resembles limestone in occurrence and physical properties.

Diatomite. The term diatomite is applied to deposits formed from accumulation of diatom shells. The latter are minute plants of marine or fresh-water habitat that secrete siliceous shells. The rock formed by deposition of diatom shells is light in weight, fairly soft, and usually porous. Diatoms are known only in relatively late geological time, and deposits of diatomite are therefore limited to relatively young strata. In the Coast Ranges of California diatomite deposits are fairly abundant but are rather uncommon in other parts of the world.

Radiolarite. Rocks formed by the accumulation of siliceous shells of microscopic marine animals called Radiolaria are known as radiolarite. Unlike diatoms, Radiolaria have been in existence since very early geological time and deposits formed by these organisms occur in many parts of the world.

Chert and Flint. The terms chert and flint are applied to siliceous sediments of organic or inorganic origin, which usually form bands and nodules in limestone and shales. They are very hard substances (7 in Mohs' scale) and are thus difficult to drill. From a geological viewpoint chert and flint are of considerable interest, as are the silicified shales with which they are often associated.[2] They are of widespread occurrence, and it is difficult to find a limestone or a shale that does not contain at least minor quantities of chert or flint.

Iron and Manganese Deposits. Although of great commercial importance, iron and manganese deposits are too limited in occurrence to be of much engineering significance. The sedimentary varieties of iron ore consist mainly of hematite (Fe_2O_3) and limonite ($Fe_2O_3 \cdot nH_2O$). Precipitation is probably in the main by inorganic agencies, but certain bacteria are known to be instrumental in deposition of iron oxide from solution. In some instances, the iron ores of the Birmingham, Alabama, region for example, sedimentary deposits of iron oxide consisting of minute spherical grains occur. Such deposits are known as oölites, because of their resemblance to a mass of fish eggs. Many limestones are also oölitic. Deposits of manganese are even rarer than iron ores, and consequently of practically no engineering importance. Most commercial deposits of manganese are of residual origin; i.e., they have resulted from the accumulation of relatively insoluble manganese compounds through solution and removal of soluble constituents of the parent rock.

Peat, Lignite, and Coal. As defined by Stutzer and Noé [16] "coal is a combustible rock which had its origin in the accumulation and partial decomposition of vegetation." The process of coal formation is complex and involves both bacteriological and physical agencies. The initial stages of coal formation are represented by peat, which consists of little more than a mass of vegetable fibers having a high moisture and volatile content, and a relatively low proportion of fixed carbon (either free or combined carbon which is non-volatile and not driven off by heating at a temperature of 950° centigrade). Lignite, or brown coal, differs from peat in having a somewhat lower percentage of volatile matter. The relative abundance of fixed carbon and volatile matter determines the rank of a coal. The percentage of fixed carbon increases in passing from peat to coal, and a similar transition is found in passing from the lower-rank bituminous to anthracite coals. The grade of a coal is determined by the proportion of ash and other objectionable impurities, the chief of which is sulphur. The higher the percentage of impurities, the lower is the grade of the coal. When the ash content exceeds 50 per cent, it has been suggested that the material be called "coaly rock" rather than coal. In some parts of the world coals containing as much as 30 per cent ash are mined. Coal beds vary from a fraction of an inch to over 200 feet in thickness; they usually extend over fairly large areas. The Pittsburgh coal bed, for example, is estimated to underlie an area of about 21,000 square miles. Usually, more than one coal bed is found in any given region, and in some German districts as many as 400 seams are known, but the majority are not minable. Although coal beds were formed chiefly in fresh water, they are commonly overlain by marine sediments. The rocks most commonly associated with coal are clay, shale, slate, sandstone, and limestone. Coal beds always partake of the structure of the enclosing rocks, and the field relationships tend to support the view that the change from peat to anthracite is in a large part the result of folding and/or pressure of the overlying strata.

Rock Salt. The most abundant saline occurring in nature is sodium chloride, or rock salt. It is probable that salt deposits originated from evaporation of sea water, but there are many features of natural salt deposits that are difficult to explain by this hypothesis, the chief of which are the great thickness and purity of many occurrences. Rock salt occurs in flat-lying beds intercalated with marine strata, or in plug-like masses known as salt domes (see Fig. 45). All sodium and potassium salts are soft (2 to 2.5 in Mohs' scale of hardness) and are soluble in water.

Gypsum and Anhydrite. Pure calcium sulphate is known as anhydrite, whereas the hydrated compound ($CaSO_4 \cdot 2H_2O$) is known as gypsum. They are believed to result from precipitation from ocean water, and are usually associated with beds of other marine sediments. Anhydrite and gypsum are soft (2 to 3.5 in Mohs' scale) and are fairly soluble in water. Anhydrite may be transformed into gypsum by addition of water causing about 30 per cent increase in volume of the mass. Consequently, anhydrite is an unreliable rock for most engineering purposes.

Engineering Properties. Most chemical sediments are soluble, relatively soft, have fairly low crushing and shearing strengths, and may be subject to volume changes when hydrated. As a result, they are often unsatisfactory for most engineering purposes; but some, especially limestones and dolomites, can often be made acceptable by proper treatment. Rock salt is especially objectionable, but it is of rather rare occurrence. Amorphous, or opaline, varieties of chert react with cement, and when used as aggregates they are deleterious to the concrete. However, crystalline and cryptocrystalline varieties are satisfactory as aggregates.

CLASTIC, OR MECHANICAL, SEDIMENTS

As the name implies, clastic sediments have been broken loose from the parent rock and transported mechanically a greater or less distance and deposited. The amount of wear shown by the constituent particles is roughly proportional to the distance of travel. When the distance of transportation has been great, only the harder and more resistant constituents tend to survive. Ow-

TABLE V. CLASSIFICATION OF CLASTIC SEDIMENTS

(By permission from *Principles of Sedimentation*, by W. H. Twenhofel. Copyright, 1939, McGraw-Hill Book Company, Inc.)

Limiting Dimensions, in Millimeters	Particles	Aggregate	Lithified Product
256 and larger	Boulder	Boulder gravel	Boulder conglomerate
64 to 256	Cobble	Cobble gravel	Cobble conglomerate
4 to 64	Pebble	Pebble gravel	Pebble conglomerate
2 to 4	Granule	Granule gravel	Granule conglomerate
1 to 2	Very coarse sand grain	Very coarse sand	Very coarse sandstone
$\frac{1}{2}$ to 1	Coarse sand grain	Coarse sand	Coarse sandstone
$\frac{1}{4}$ to $\frac{1}{2}$	Medium sand grain	Medium sand	Medium sandstone
$\frac{1}{8}$ to $\frac{1}{4}$	Fine sand grain	Fine sand	Fine sandstone
$\frac{1}{16}$ to $\frac{1}{8}$	Very fine sand grain	Very fine sand	Very fine sandstone
$\frac{1}{256}$ to $\frac{1}{16}$	Silt particle	Silt	Siltstone
$\frac{1}{256}$	Clay particle	Clay	Claystone

ing to the almost universal presence of inorganic and organic precipitates in nature, the clastic sediments usually contain more or less of chemically deposited material. In classifying the clastic sediments a dual system is used which distinguishes between the unconsolidated and the consolidated phases of materials possessing the same range of grain size (see Table V). Grouped with the claystones are laminated sediments, known as shales, which are composed mainly of clay-sized particles. Sediments may be cemented by a variety of materials, the most important of which are various oxides, carbonates, and silica.

Gravel and Conglomerate. As indicated in Table V, gravel and conglomerate consist of particles ranging from 2 to 256 millimeters in size. Commonly the coarser grains tend to be more

Fig. 26. Conglomerate near Kittatinny Gap, Pennsylvania. (Pennsylvania Geological Survey.)

angular and not so well sorted as the finer materials; they are usually less widespread in occurrence. Rocks composed of cemented angular fragments are known as breccias; they may be formed in a variety of ways.[12] It is obvious that breccias have not been transported any considerable distance, for, if they had, the angularity would be lost.

Till and Tillite. Till consists of a heterogeneous assemblage of materials ranging in size from very large boulders to clay that have been deposited by glaciers. A lithified till is known as a tillite. Till is very widespread in the northern hemisphere and is discussed in detail in Chapter 10.

Sand and Sandstone. Sand and sandstone consist of particles ranging from $\frac{1}{16}$ to 2 millimeters in size. Most sands have been transported considerable distances; as a result they consist mainly of hard and resistant minerals, the chief of which is quartz. However, magnetite, monazite, and even rarer minerals may occasionally constitute the predominate material. Feldspathic sands are known as arkoses. Rocks high in amphiboles and pyroxenes are often referred to as graywackes. Sands and sandstones may consist of angular to well-rounded particles. The constituent grains may be of uniform size or may exhibit a wide range of sizes. Sands and sandstones usually occur in rather thick beds extending over wide areas.

Silt and Siltstone. Silts are composed of clay minerals, iron oxides, silica, and particles of common and many rather uncommon minerals produced chiefly by rock abrasion. They usually contain some water in capillary interstices and varying amounts of colloidal material. Their physical properties tend to resemble those of clays and shales in most important respects. Like sands and sandstones, silts may occur over fairly wide areas, but they are usually found in relatively thin beds associated with clays and shales.

Loess. Loess consists of clay and silt particles, which are believed to have been transported by wind. The most common mineral found in loess is quartz, but calcite, feldspar, mica, and other minerals may also occur. As a rule, the constituent particles are subangular, and the deposits are unstratified. Loess deposits are characterized by the presence of numerous vertically elongated interstices, which are interpreted as having been formed by roots and stems of plants. These interstices are often filled with secondary calcium carbonate; they are thought to be largely responsible for the ability of the material to stand for considerable periods of time on nearly vertical slopes (Fig. 27). It should also be added that loess is one of the few natural materials whose vertical permeability may exceed the horizontal permeability. Loess deposits tend to occur in periglacial areas, or zones that were immediately adjacent to and in front of maximum glacial advances, to which they seem to have been genetically related. Several hundred thousand square miles of the earth's surface are covered with these deposits, which may reach hundreds of feet in thickness. The engineering characteristics of loess have been discussed by Scheidig [13] in an exhaustive treatise, which should be consulted by those working in areas where this material is found.

Fig. 27. Vertical cut in loess, Warren County, Mississippi.

Clay and Shale. Clay and shale consist of clay minerals, various oxides, silica, fine particles of ordinary minerals, and a greater or less amount of colloidal and organic matter. They contain a large amount of water, some of which is held in subcapillary openings and has the properties of a solid. Two types of shale are known: (1) cementation shale, and (2) compaction shale. In the former the constituent particles are held together chiefly by various types of cementing materials, of which silica and calcium carbonate are the most common. The compaction shales are bound together chiefly by molecular attraction of the particles. Cementation shales are usually rather stable, and they do not disintegrate readily when subjected to alternate wetting and drying and other agencies of weathering. However, both cementation and com-

Fig. 28. Shale (dark) overlain by dolomite (light). Downward percolating water deflected by the shale has frozen to form the conspicuous white streaks. (Missouri Geological Survey.)

paction varieties may slake and disintegrate with prolonged exposure to atmospheric agencies. Clay and shales are formed in a number of ways; they usually occur in beds of considerable thickness extending over fairly large areas.

SEDIMENTARY STRUCTURES

It is a matter of common observation that nearly all sediments are stratified, i.e., separated into numerous more or less parallel bands caused by variations in composition, grain size, texture, or change in rate of deposition. The smallest unit that can be distinguished from the overlying and underlying units is known as a stratum, the sum total forming strata. Stratification planes 1 centimeter or less apart are known as laminae; they may or may not parallel the bedding, or stratification. It may happen that the bedding planes and/or laminae may not maintain even approximate parallelism for any considerable portion of the rock but truncate each other in a complicated manner. In this event the rock is said to be cross-bedded, or cross-laminated. Cross-lamination originates under conditions which are subject to rapid variation, causing local deposition at angles different from the regional angle of deposition. If confused with true bedding, a totally erroneous interpretation of the inclination of the strata may result. Stratification furnishes a datum from which the time and spatial relationships of adjacent beds can be determined and is also useful in investigating conditions under which the sediments were deposited. In civil engineering stratification is important because of its influence on the physical properties of the material. For example, a well-laminated shale will generally possess lower shearing and crushing strength parallel to the laminae than at some angle thereto. In other words, stratified materials are anisotropic. Not only is this true with respect to shearing and crushing strengths but also it holds for porosity and permeability as well.

When the initial surface of deposition is horizontal, or nearly so, the stratification tends to form parallel to this surface, and the planes of stratification are parallel to each other. When the initial surface is steeply inclined or undulatory, this is not true. Considerable attention has been given to the maximum initial slopes on which sediments can be deposited without slumping, and it has been found that for subaqueous deposition of coarse-grained angular to subangular sands the maximum is about 43°. Rounding of the grains and reduction of grain size tend to reduce the limiting angle of slope. It is believed that many of the contorted structures observed in fine-grained sediments were caused by slumping of the materials contemporaneously with deposition on very steep slopes rather than by later crustal movements. However, it is probable that most dips of over a few degrees have been caused by folding rather than by original deposition.

Unconformities. An unconformity is any break in the continuity of deposition and gives rise to a time break, or hiatus, between adjacent strata. Two types of unconformity have been distinguished: (1) non-conformity in which the underlying strata were deformed and eroded prior to deposition of the overlying beds, and (2) disconformity in which the overlying and underlying strata are parallel. Disconformities may arise either from erosion without deformation of the strata, or by non-deposition. Unconformities are of great geologic importance, but estimates

Nonconformity

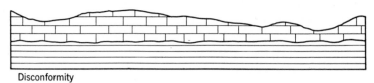

Disconformity

Fig. 29. Sections illustrating two types of unconformity.

of the time interval they represent are often difficult to make. Those interested in this phase of the subject should consult the work of Blackwelder listed at the end of the chapter.[1]

Miscellaneous Structures. A great variety of phenomena of some geological but little engineering importance are included in miscellaneous structures. The most important are: ripple marks, mud cracks, clay galls, mud pebbles and boulders, and wave and rill marks. Most of these terms are self-explanatory. Clay galls are cylinders of clayey materials, which are thought to have been formed by the rolling of clay sheets along the surface. Contortions are frequently observed in gypsum beds and are generally attributed to volume changes resulting from hydration of anhydrite to gypsum. Many sedimentary structures furnish evidence relating to the top and bottom of the beds at the time of deposition, and they are thus of great importance in the detection of subsequent overturning by earth movements. The criteria useful for this purpose have been brought together in a single volume [15] which is very useful to the field geologist.

SEDIMENTARY TEXTURES

The texture of clastic rocks is fragmental; i.e., it consists of fragments showing a wide variety of sizes and shapes, which in the case of tuff and volcanic ash (pyroclastics) resemble shards of broken pottery. Crystalline textures are commonly developed in many chemical sediments. Crystalline textures may be crypto-crystalline (very fine to microscopic in size), fine-grained (not apparent to the unaided eye), and coarse-grained (individual crystals easily visible). Occasionally, sediments may contain spheroidal and ellipsoidal grains that possess an amorphous or radiating internal structure. Such particles are known as spherulites, and the texture of the rock is said to be spherulitic. Spherulites resemble oölites, but the latter differ in having a concentrically laminated internal structure. Oölitic particles larger than 2 millimeters in diameter are called pisolites. It has already been mentioned that many limestones and iron ore deposits are oölitic, but silica, phosphates, bauxite, and barite deposits may also exhibit this texture. Origin of oölites and spherulites has not been satisfactorily explained. It is probable that they result from a variety of conditions.

ENGINEERING PROPERTIES

The hardness, crushing strength, and shearing resistance of clastic sediments depend mainly on the degree of consolidation. Consequently sands have a very low shearing strength, but quartzites (very thoroughly cemented sandstones) may exceed granite in this respect. Stratified materials are always anisotropic in their physical properties. All the coarser clastics (sands, gravels, etc.) may be regarded as practically incompressible when confined, but shales and clays are often highly compressible. Compaction shales are unreliable for most engineering purposes, and should be carefully investigated in advance of construction operations. Because of their tendency to slake when exposed to air, most shales require special protective measures prior to the pouring of concrete. Many clastic sediments lose their coherence when exposed to the action of moisture, freezing and thawing, and other weathering agencies, and should always be tested for their resistance to these agencies before being put to use. Failure of the St. Francis dam in southern California is thought to have been caused by the breaking down and removal of the cementing material of the conglomerate underlying a part of the foundation.

METAMORPHIC ROCKS

Strictly speaking, any change which takes place in a rock subsequent to its formation is metamorphism. However, the present tendency is to consider as metamorphism only those changes that take place under the action of heat, pressure, or both. All other changes are considered as alteration or weathering. Consequently, metamorphic rocks may be defined as naturally occurring substances that have been formed from the parent rock (either igneous or sedimentary) by the action of heat, pressure, or both. Two types of metamorphism are known: (1) contact metamorphism resulting primarily from the action of heat and solutions, and (2) dynamic metamorphism resulting primarily from the action of differential pressure. Contact metamorphism takes place on the boundaries of intrusive masses, and may also be observed on the lower contacts of lava flows. The temperatures are not known precisely, but probably they vary from about 1000° to 600° centigrade at the contact. In addition to purely thermal effects, the action of hot solutions and gases is important. Dynamic metamorphism takes place at considerable depth below the earth's surface; it is most marked in areas of extreme crustal deformation. In view of their origin, it is evident that metamorphic rocks are exposed on the surface only as the result of uplift and prolonged erosion.

CONTACT METAMORPHISM AND CONTACT ROCKS

Contact metamorphism is the term applied to the changes produced in the intruded rock by an intrusive body, and in the intrusive body by interaction with the intruded rock. The first class of changes is known as exomorphic and the second as endomorphic. The first is of greater importance, and in these pages the endomorphic effects will not be discussed.

Exomorphic Effects. Contact metamorphism may be caused by the action of heat, gases, and solutions, or a combination thereof. In cases where the gases and solutions were either not present or negligible in amount, contact action expresses itself chiefly in baking and recrystallization of the enclosing rocks. Sandstones may be altered to quartzites, and limestones recrystallized to form marbles. Shales may be baked and otherwise altered. A frequent change in argillaceous rocks is the segregation of organic materials into nuclei giving rise to "spotted shales." In the presence of high temperatures the mineral grains of shales may recrystallize to form a rock known as hornfels. In purely thermal metamorphism the bulk chemical composition of the rocks remains unchanged.

Contact metamorphism may be accompanied by addition of fluorine, chlorine, iron, silicon, copper, sulphur, and other elements to the enclosing rocks. A change brought about by the action of hot gases and solutions is known as pneumatolysis; it is especially marked where the enclosing rocks consist of limestone or dolomite. Valuable metallic ores are often found in pneumatolytic deposits.

In addition to pneumatolytic effects, magma may be injected bodily along the bedding planes and fissures, giving rise to what is known as injection contact metamorphism. An example of this process is found in the so-called injection gneisses—rocks of granitic composition in which the component minerals are segregated so as to impart a banded structure to the mass. Contact metamorphism is a complex phenomenon, and investigators are not entirely agreed as to the relative importance of the processes outlined above. Fortunately, it is not of great engineering significance.

DYNAMIC METAMORPHISM AND ROCK PRODUCTS

Dynamic metamorphism has been used with varying shades of meaning, but it usually indicates the changes in rocks produced by the action of shearing forces under a confining pressure. It has been shown by subjecting materials to pressures as high as 700,000 pounds per square inch accompanied by a simultaneous shearing stress that not only do chemical reactions occur but they take place at relatively low temperatures.[3]   The mechanisms of such reactions seem to be a rearrangement of the molecular structure to form denser atomic groupings.   Many of the new and denser minerals formed by this process possess well-marked tabular cleavages and tend to impart this property to the rock as a whole.   There is little or no addition of material during dynamic metamorphism, and rocks of the most diverse chemical and physical characteristics tend to assume similar physical properties when subjected to the process.   Attempts have been made to distinguish between depth zones and to group dynamically metamorphosed rocks into classes, depending on whether alteration took place in the upper, middle, or lower zones of the earth's crust.[8]   However, there are reasonable grounds for doubting that intensity of metamorphism is entirely a function of depth.

Gneiss.   This name is applied to rocks of granitic composition which possess a well-marked banded structure (Fig. 30).   Gneisses

Fig. 30.   Contorted gneiss in road cut, Montgomery County, Pennsylvania.
(Pennsylvania Geological Survey.)

may be formed by contact as well as by dynamic metamorphism. Primary, or flow, gneisses are thought to be formed by differential movement of a still molten magma, causing segregation and orientation of the crystallizing minerals.   It is a common error to suppose that, simply because gneiss is mineralogically similar to granite, the parent rock must also have been a granite. Conglomerates, sandstones, shales, and a great variety of materials may give rise to this rock, the only requirement being that the bulk composition of the parent rock include all the elements found in gneiss. Gneisses are fairly abundant in folded regions, and they form the cores of many mountain ranges. They are especially common in the so-called "shield areas"—masses of ancient metamorphic, igneous, and sedimentary rocks upon which younger and less metamorphosed rocks have been deposited.

Schist. A metamorphic rock possessing a well-marked cleavage is known as a schist (Fig. 31). This property is a reflection

Fig. 31. Mica schist. Cleavage parallels the horizontal banding. (Washington University Collection. Photo by Joseph E. Cox, Jr.)

of the uniform orientation of the cleavage planes of the constituent minerals. The orientation is caused by recrystallization of the material under differential pressure, resulting in alignment of the new minerals at right angles to the direction of greatest pressure, rather than by crushing and rotation of the grains. Many varieties of schist are known of which mica, amphibole, and talc are the most common.

Slate. Slate is a fine-grained rock formed from the metamorphism of shale (Fig. 32). It has a well-marked tabular cleavage, a noteworthy feature of which is the relationship to the bedding planes. Folds are usually formed by lateral pressure, and their axes are normal to the direction in which the pressure was ap-

Fig. 32. Slate with cleavage paralleling axis of a synclinal fold. (Arthur Keith, U. S. Geological Survey. From *Outlines of Physical Geology*, by Longwell, Knopf, and Flint, John Wiley & Sons, New York, 1941.)

plied. The cleavage of slate is the result of pressure causing recrystallization with the development of tabular minerals the long axes of which are oriented at right angles to the direction of the applied pressure. The cleavage planes of these tabular minerals (usually mica and its relatives) parallel the elongation of the crystals, and the cleavage planes thus tend to parallel the axes of folds. Thus on the flanks of folds slaty cleavage and bedding are nearly parallel, but on the crests they are mutually perpendicular. The relationships between slaty cleavage and bedding planes are of considerable importance in working out the geologic structure of complex regions.

Marble. Marbles are formed by the recrystallization of limestone and dolomite under the influence of heat, pressure, or both. As a result, the original sedimentary structures are largely lost, but fossils and other remnants of the original rock are occasionally preserved, indicating, in some instances at least, that differential movement was not very extensive.

Quartzites. A sandstone that is so thoroughly cemented as to break through, rather than around, the constituent grains is known as a quartzite. Quartzites may be formed by both contact and dynamic metamorphism, or by deposition of silica from circulating ground waters. Rocks formed by deposition of siliceous material from the ground water are sometimes referred to as orthoquartzites and are not of metamorphic origin in the usual sense of the word. Although conglomerates may also become so firmly cemented as to break like quartzites, no generally accepted term has been proposed for such rocks.

Textures

The texture of metamorphic rocks is usually coarsely to finely crystalline, and in many instances is platy; i.e., it consists of tabular-shaped grains more or less parallel in orientation. The chief importance of platy texture is that it imparts a highly anisotropic character to the rock, of which cleavage is an outstanding example. Marbles constitute an important exception to the above rule in that they are usually roughly equigranular in texture and do not, of course, exhibit cleavage.

Engineering Properties

Owing to their wide range in physical and chemical properties, it is impossible to generalize concerning the engineering properties of metamorphic rocks. Most gneisses are hard and tough and are characterized by high crushing and shearing strengths. Owing to their well-marked cleavage, schists are highly anisotropic, and attention should always be given to cleavage orientation. Many schists, talc and chlorite for example, are soft (1 to 2.5 in Mohs' scale of hardness) and are, therefore, unsuitable for high unit

Fig. 33. Interbedded massive quartzite and shale, excavation for Narrows Dam, Pike County, Arkansas. (W. B. Steinriede, Jr.)

loading. They may also be rather permeable in a direction parallel to the schistosity. The engineering properties of marble are similar to those of the parent rock (limestone or dolomite), and solution cavities should be anticipated.

REFERENCES

1. Eliot Blackwelder, The Valuation of Unconformities, *J. Geol.*, Vol. 17, 1909.

2. M. N. Bramlette, The Monterey Formation of California and the Origin of Its Siliceous Rocks, *U. S. Geol. Survey, Prof. Paper* 212, 1946.

3. P. W. Bridgman, Shearing Phenomena at High Pressure of Possible Importance for Geology, *J. Geol.*, Vol. 44, 1936.

4. Hans Cloos, Das Batholithenproblem, *Fortschr. Geol. Pal.*, Heft 1, 1923.

5. P. Eskola, The Mineral Facies of Rocks, *Norsk Geol. Tidsskr.*, Vol. 6, 1920.

6. G. K. Gilbert, Report on the Geology of the Henry Mountains, *U. S. Geog. & Geol. Survey of the Rocky Mountain Region*, 1880.

7. F. F. Grout, The Lopolith; an Igneous Form Exemplified by the Duluth Gabbro, *Am. J. Sci., 4th Ser.*, Vol. 46, 1918.

8. U. Grubenmann, *Die Kristallinen Schiefer II*, Specieller Teil, Gebrüder Borntraeger, Berlin, 1907.

9. A. Harker, *The Natural History of Igneous Rocks*, MacMillan Co., New York, 1909.

10. A. Michel Lévy, Contribution a l'étude du granit de Flamanville et des granites français en général, *Bull. services carte géol. France* 36, 1893.

11. D. Modell, Ring-Dike Complex of the Belknap Mountains, New Hampshire, *Bull. Geol. Soc. Amer.*, Vol. 47, 1936.

12. W. H. Norton, A Classification of Breccias, *J. Geol.*, Vol. 25, 1917.

13. Alfred Scheidig, *Der Löss und seine geotechnischen Eigenschaften*, Theodor Steinkopft, Dresden und Leipzig, 1934.

14. J. J. Sederholm, On Migmatites and Associated Pre-Cambrian Rocks of Southwest Finland, *Bull. comm. géol. Finlande* 77, 1926.

15. Robert R. Shrock, *Sequence in Layered Rocks*, McGraw-Hill Book Co., New York, 1948.

16. Otto Stutzer and A. C. Noé, *Geology of Coal*, Univ. of Chicago Press, 1940.

17. W. H. Twenhofel, *Principles of Sedimentation*, McGraw-Hill Book Co., New York, 1939.

18. C. R. Van Hise, A Treatise on Metamorphism, *U. S. Geol. Survey, Mono.* 47, 1904.

19. A. von Moos and F. de Quervain, *Technische Gesteinskunde*, Birkhäuser, Basel, 1948.

20. C. K. Wentworth and H. Williams, The Classification and Terminology of the Pyroclastic Rocks, *Bull. Natl. Research Council* 89, 1932.

C H A P T E R   4

Geologic Structure

In geology, the term "structure" generally refers to spatial relationships of rocks. For example, sedimentary strata may lie nearly horizontal, as originally deposited, or they may be bent into more or less regular folds; they may be offset along fractures known as faults; or they may be traversed by planes of parting known as joints. Geologic structure includes fractures ranging from almost microscopic to continental in size, and it comprises all degrees of complexity ranging from the simplicity of a series of flat-lying sedimentary rocks to almost incredibly intricate combinations of folding and faulting.

Geologic structure influences engineering projects in many ways. Folds and faults obviously have much to do with the selection of dam sites, and even such seemingly unimportant matters as the spacing of joints may have a vital bearing on uplift pressures and safety of dams.[12] Crushed and chemically altered rocks contiguous to faults may cause great difficulties in tunneling operations, and earthquakes originating along faults may damage or destroy engineering structures. The design of deep cuts in rock is greatly influenced by geologic structure, and the suitability of quarry sites is largely a matter of joint spacing. Perhaps the most important of all is the influence of geologic structure on circulation of the ground water.

Forces involved in the formation of geologic structures are of tremendous magnitude and obscure origin. The mechanisms involved are also somewhat obscure, and much remains to be done in the field of theoretical structural geology. Fortunately, these matters are of only minor civil engineering importance and need not be discussed. However, the civil engineer is well qualified to study theoretical phases of the subject, and future progress in this field depends to some extent on application of engineering principles such as the scale effect used in the application of model studies to engineering structures.

FOLDS

Sediments probably can be deposited without slumping on initial slopes as steep as 43°, but in most instances the original inclination is probably much flatter and practically negligible. In consequence, unless acted on by some outside force, sediments should generally lie nearly horizontal. In large areas of the world this condition exists, but in others it is very apparent that the strata have been deformed in various ways subsequent to their deposition. Perhaps the most common type of deformation is folding. As the name implies, folds are undulations, or flexures, in sedimentary rocks analogous in many ways to ocean waves.

The co-ordinate system used in describing the spatial relationships of sedimentary beds and other tabular geologic bodies is furnished by latitude, longitude, and a plane parallel to sea level. The strike of a bed is the direction of the line formed by intersection of the stratification and a horizontal plane (Fig. 34). The

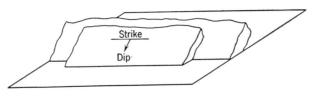

Fig. 34. Diagram illustrating dip and strike.

angle between the bedding planes and the horizontal measured at right angles to the strike is known as the dip. On geologic maps dip and strike are shown by the following symbol ⊤, in which the longer line represents the strike and the shorter the dip.

GEOMETRY OF FOLDS

On the basis of dip relationships two major types of folds can be distinguished: (1) anticlines, in which the strata dip in opposite directions from a more or less centrally located plane, or axis; and (2) synclines, in which the strata on opposite flanks dip to-

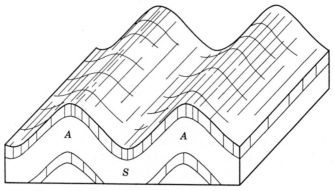

Fig. 35. Symmetrical anticlines (*A*) and syncline (*S*).

ward the axis (Fig. 35). In other words, an anticline is convex upward, whereas a syncline is convex downward.

If the dip and strike remained constant, folds would of necessity encircle the earth. This is not known to occur, and is explained in most instances by a dying out of the fold along the axis, giving rise to what is known as plunge. Plunge may be visualized by imagining the folds as rotated around a line perpendicular to the axes. As shown in Fig. 37, the intersection of a plunging anticline with a horizontal plane gives rise to a convex curve the crest of which points in the direction of plunge. A plunging syncline gives rise to a similar curve, but in this instance the trough points in the opposite direction. In a perfectly circular-shaped anticline, or dome, there are an infinite number of axes disposed radially around the center; and, strictly speaking, the strata should be said to plunge, rather than dip, away from this point.

A fold may also be inclined in a direction at right angles to the strike of the axial plane. This condition is easily visualized

Fig. 36. Anticline near Watts Bar, Tennessee. (Stewart M. Jones.)

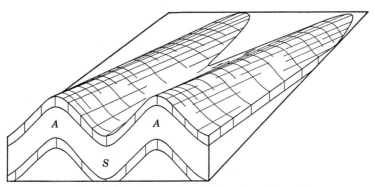

Fig. 37. Plunging anticlines (*A*) and syncline (*S*).

by imagining the axis to have been revolved around a line included in the axial plane (Fig. 38). Thus folds may be either

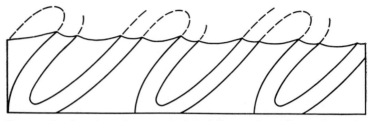

Fig. 38. Section through overturned anticlines and synclines. Note that on right-hand limbs of anticlines the strata are bottom side up.

normal or overturned. When inclination of the strata passes beyond the vertical the fold is said to be overturned, and when the axis is horizontal, or nearly so, the fold is said to be recumbent. Folds may be so tightly compressed that the strata on opposite flanks are parallel. Such folds are said to be isoclinal. A series of eroded isoclinal anticlines and synclines cannot be distinguished from a series of uniformly dipping strata on purely geometric grounds (see Fig. 39). Under certain conditions, folds may be fan-shaped (Fig. 40).

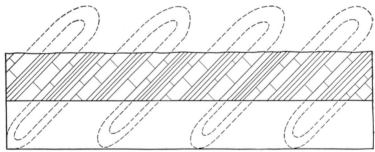

Fig. 39. Section through isoclinal anticlines and synclines. Note that unless the missing upper portion is restored the section might represent nothing more than a series of uniformly dipping beds.

Nearly all naturally occurring flexures exhibit some degree of distortion, a condition known as asymmetric folding. Asym-

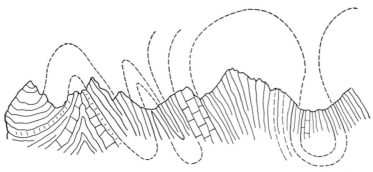

Fig. 40. Section through fan-shaped folds. (After A. Heim, *Mechanismus der Gebirgsbildung.*)

metry results in displacement of the axis from a central position. Asymmetric anticlines and synclines are illustrated in Fig. 41.

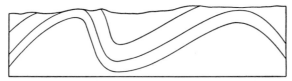

Fig. 41. Section through asymmetric anticlines and syncline.

Complex Folds. Simple, uncomplicated anticlines and synclines are rather uncommon. Generally, they are more or less distorted by faults, thickening and thinning of the strata, minor flexuring, etc. A series of alternating anticlines and synclines the over-all structure of which is anticlinal is known as an anticlinorium, whereas a series of alternating synclines and anticlines the over-all structure of which is synclinal is termed a synclinorium (Fig. 42).

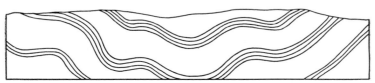

Fig. 42. Idealized section through a synclinorium.

When sufficiently compressed, recumbent folds may pass into faults. Figure 43 illustrates an extreme case of compressional

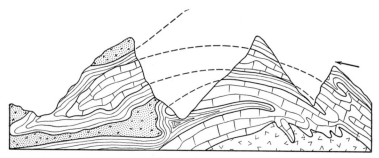

Fig. 43. Idealized section through a nappe. Arrow indicates direction in which forces that caused the folding were applied.

folding and overturning found in the Alps. European geologists call this type of structure a nappe.

GEOLOGICAL CONSIDERATIONS

Based on origin, two classes of folds may be distinguished: (1) depositional, and (2) tectonic. The former may result from a variety of conditions, a common example being deposition on a sloping surface, giving rise to what is known as initial dip (Fig. 44). Sediments may also be deposited over buried hills or ridges, resulting in an original anticlinal structure, which in fine-grained sediments may be further accentuated by greater consolidation of the thick strata on the flanks than of the relatively thin deposits on the crest. Extremely intricate small-scale folds may result from slumping of sediments deposited on very steep initial slopes.

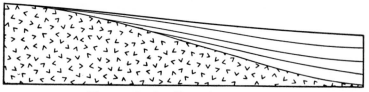

Fig. 44. Section illustrating initial dip.

Tectonic folds may originate through crustal movements which may be caused by forces acting in either a horizontal or vertical direction, or any component thereof. Vertical forces may originate from volcanic intrusions, deep-seated crustal movements, or collapse of substrata. The origin of horizontally directed forces is obscure, but they have been active since earliest geological time. When deforming forces act on alternately hard and soft materials, the soft materials tend to yield plastically, and may move into areas of lower pressure. In this manner, deeply buried salt beds which have been subjected to horizontal compression may be squeezed upward into the crests of anticlines, giving rise to what are known as salt domes (Fig. 45). Salt domes are common in

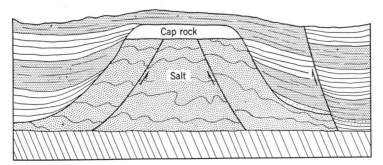

Fig. 45. Section through a salt dome.

the Near East, the Carpathian Mountain region, North Germany, Central America, and the Gulf Coast region of the United States. Under other conditions, the softer strata may be merely distorted, or squeezed into low-pressure areas, with consequent thickening and thinning of the beds. Folds that show this condition are known as similar, whereas those that do not are referred to as parallel folds. Similar folds are usually tectonic in origin, and they are more frequently associated with faults than are the parallel type.

Folding and Age Relationships of Strata. Folds are often too large and complicated to be detected by direct observation, and frequently it is impossible to distinguish between certain types on purely geometric grounds. Generally, folding cannot be worked out satisfactorily without geological mapping of a fairly large

area. Because of the structural complexities encountered in many
regions, it is often not possible to arrive at a complete solution
merely by plotting dips and strikes; the relative ages of the strata
involved and the influence of topography must also be taken into
consideration. In any series of sedimentary beds that have not
been overturned, the oldest stratum will lie on the bottom and
will be overlain by successively younger deposits. If the topography is reduced to a relatively flat surface, it follows that in anticlinal folds the oldest beds will occupy an axial position, and in
synclines the opposite relationship will occur. If the normal
order of succession is reversed, the fold has been overturned.
Were it not for the age relationships of the strata involved, a
civil engineer could doubtless map structure as well as, if not
better than, a geologist. It should be noted that only rarely
will the topography and underlying rock structure correspond.
For example, synclinal ridges and anticlinal valleys are very
common.

Dimensions. The size of folds may vary from a fraction of an
inch to many miles; flexures 100 miles or more in length and 10
to 100 miles in width are relatively common in many parts of the
world. In engineering operations small folds a few to 1000 feet
or so across are usually more important than larger regional
features.

Fold Problems

The chief problems presented by dipping strata are determination of the thickness of beds (measured at right angles to the
stratification) and calculation of depth to a stratum. They can
be solved by trigonometry, or graphically.[7,8] When outcrops are
rare or lacking, the dip and strike can always be determined from
the elevations and locations of three wells or drill holes, provided that the depth to a given stratum, or horizon, is known at
all three points. Calculation of dip and strike from these data is
known as solving the three-point problem; it gives very accurate
results provided that the horizon selected is truly a plane and not
a warped surface. This problem is readily solved by construction
and simple proportions.[1] It is often desirable to reconstruct the
geometric pattern of folds from a large number of dip and strike
observations. The various methods useful for this purpose have
been described by Busk [3] and need not be repeated here.

FAULTS

Faults are fractures along which formerly continuous beds have been dislocated in a direction parallel to the fault surface. They are more commonly associated with hard and brittle strata than with more plastic materials. The displacement may vary from a few inches or less to many miles. In some instances the fault is a single even fracture approaching very nearly the ideal condition of a plane, but usually the dislocation is rather uneven. In such cases the term "surface" rather than plane should be used. It is also very common for dislocation to occur along a number of closely spaced more or less parallel fractures, giving rise to what is known as a fault zone. Like folds, faults die out gradually at both ends.

GEOMETRY OF FAULTS

The geometry of faults is a rather complex subject, and so many conflicting nomenclatures have been used from time to time that the Geological Society of America undertook to standardize the terminology.[10] A fault may be regarded as a plane the dip and strike of which are analogous to the dip and strike of a stratum. The inclination of the fault plane to the vertical (complement of the dip) is known as the hade. The block above the fault is known as the hanging wall; the underlying block is termed the foot wall. Faults never offer direct evidence of the actual direction of the movements involved, the apparent movement indicated being purely relative. Thus, in the example illustrated in Fig. 46 the right-hand block may have moved down, the left-hand block may have moved up, both may have moved down but at different rates, both may have moved up unequal distances, or the left-hand block may have moved up and the right-hand block down. Furthermore, the direction of displacement may have changed from time to time, and only the net displacement and the direction of the last movement can ordinarily be determined. When the hanging wall has been displaced downward with respect to the foot wall, the fault is said to be of the normal, or gravity, type (Fig. 46). When the direction of displacement is reversed, a reverse or thrust fault results. If the displacement is entirely in a direction parallel to the fault trace, the fault is said to be of the horizontal slip variety (Fig. 48). Displacement of faults may also be oblique or rotational. The more important

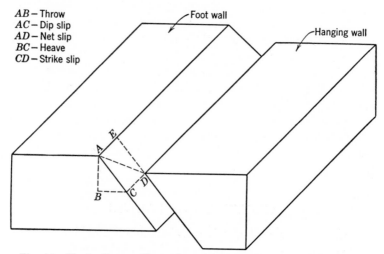

Fig. 46. Block diagram illustrating nomenclature of fault movements.

terms used in describing the components of fault movements are illustrated in Fig. 46. Problems arising from faulting are best solved by the methods of projection.[1]

GEOLOGICAL CONSIDERATIONS

Although faulting may exist in almost any type of rock, it is usually only in sediments that the displacement and other details can be worked out with any degree of precision. The committee of the Geological Society of America [10] approved the following nomenclature for faults in stratified rocks:

A strike fault is one whose strike is parallel to the strike of the strata.

A bedding fault is a special form of strike fault whose surface is parallel with the bedding of stratified rocks.

A dip fault is one whose strike is approximately at right angles to the strike of the strata.

Fig. 47. Thrust faults in tuff beds exposed in a cut on the Inter-American Highway near Colon, El Salvador. (T. F. Thompson.)

An oblique fault is one whose strike is oblique to the strike of the strata.

A longitudinal fault is one whose strike is parallel with the general structure of the region.

A transverse fault is one whose strike is transverse to the general structure of the region.

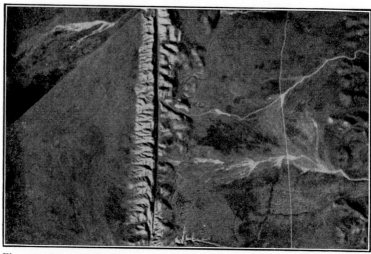

Fig. 48. San Andreas fault in San Luis Obispo County, California. The fault is of the horizontal slip variety, and the stream in upper left-hand corner has been displaced upward (north) relative to its extension in the upper right-hand corner. Bulging along fault trace has been caused by compressional components. (Fairchild Aerial Surveys, Inc., Los Angeles, California.)

Groups of Faults. Faults usually occur in groups and may form systematic surface patterns, or they may show an almost completely random arrangement. In faults in groups, terms like concentric, radial, rectilinear, etc., are often used to describe the fault pattern. A block that has been elevated along faults relative to the surrounding area is known as a horst; when the block has been depressed relative to the surroundings it is called a graben. A fault block is a mass completely or partially bounded by faults. A mass of rock that has been broken from the sides and caught between the walls of a fault is known as a horse (Fig. 49).

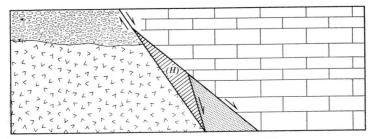

Fig. 49. Section through a horse (*H*).

Faulting and Age Relationships of Strata. Faulting is always accompanied and followed by erosion. The side which moves up relative to the other is subjected to more active erosion than the downthrown side. In time the overlying younger strata will be entirely removed from this side of the fault, and underlying older beds will be exposed at the surface; hence the rule that older rocks are always exposed on the upthrown side of a fault. If the fault surface dips away from the upthrown side, the displacement is of the normal type; if the opposite is the case, it is known as a reverse fault. An overthrust is a low-angle reverse fault in which the hanging wall has been pushed forward over the

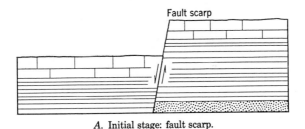

A. Initial stage: fault scarp.

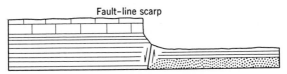

B. Later stage: fault-line scarp.

Fig. 50. Sections illustrating development of a fault-line scarp.

foot wall. In some instances the displacement along overthrust faults may be scores of miles. Isolated areas of overthrust rocks resting on underlying strata are known as outliers or *klippen;* they may represent all that remains of a formerly very extensive over-thrust sheet. Small patches of the underlying rocks surrounded by overthrust strata which have been exposed by erosion of an overthrust mass are known variously as inliers, windows, or *fensters.* A scarp caused by erosion along a line of faulting is known as a fault-line scarp, in order to distinguish it from a similar topographic feature formed directly by faulting known simply as a fault scarp (Fig. 50).

Deformation and Disruption along Faults. Only rarely will a fault exhibit a clean, uncomplicated break. Usually, the rocks will be found to have experienced more or less folding, fracturing, crushing, and grinding. Very often the walls are polished and traversed by more or less parallel striations forming what

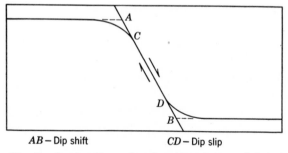

AB – Dip shift   CD – Dip slip

Fig. 51. Section illustrating drag along a normal fault.

are known as slickensides. Slickenside orientation gives an indication of the direction of movement along a fault, but it should be noted that similar evidences of earlier movements may have been entirely obliterated, and these may or may not have been identical in direction with later movements. Folding caused by frictional resistance of the rocks on either side of the fault surface is known as drag and usually indicates the direction of the last important displacement (Fig. 51). In cases where the frictional resistance has been great and the dislocating forces severe, the rocks on opposite sides of the fault surface may be broken into angular fragments known as fault breccia. Breccia zones vary greatly in width, and in extreme cases may be measured in hundreds or even thousands of feet. When the distorting forces have been exceptionally severe, as is often but by no means exclusively the case with thrust faults, the rock may be ground to a fine, clay-like powder known as fault gouge. Gouge zones vary greatly in width and range from less than an inch to over 100 feet measured perpendicularly to the fault plane. In addition to purely mechanical effects, faults may furnish channels for circulating water which may react with the wall rocks to produce various secondary minerals, the most important of which from an engineering standpoint are sericite and clays. The "heavy" ground encountered in the Moffat tunnel was caused by alteration of feldspars to montmorillonite by water circulating along an undetected fault.[6]

Nature and Extent of Fault Movements. Although faults of thousands of feet, or even miles, in displacement are not uncommon, it should not be supposed that single movements of comparable magnitude have occurred. From a study of faults known to be active at the present time, it is more reasonable to assume that such impressive movements represent the summation of a very large number of separate dislocations none of which was more than a few feet or inches in magnitude. The largest single displacement known took place at Yakutat Bay, Alaska, in 1899 and amounted to 47 feet.[11] However, even minor movements may cause severe local earthquakes capable of damaging engineering structures.

Criteria for Distinguishing between Active and Inactive Faults

Fractures that are known to have experienced dislocation within historic time are known as active faults. Because they present hazards to construction, the differentiation of active and inactive faults is a matter of considerable engineering importance, and it is unfortunate that frequently no very reliable decision can be made. The most direct and best evidence of activity is that furnished by seismographs and bench marks. If the seismograph records show that earthquakes occur along a fault, it should, of course, be regarded as active. Similarly, if accurately located bench marks exhibit horizontal or vertical displacements, any faults known to exist in the area should be regarded as active.

If a fault is known to be overlain by younger strata that are not displaced, it is permissible to regard it as inactive (Fig. 52). In faults that displace even the youngest rocks known in the area, some idea of potential future activity may be gained by a study of land forms, but it should be remembered that such evidence is indirect and subject to dispute. The physiographic evidences of active faulting are (1) bold escarpments, (2) sag ponds, (3) offset streams, and (4) shutterridges.[4]

Bold escarpments, especially when occurring in alluvium or other unconsolidated materials, are a good criterion of active

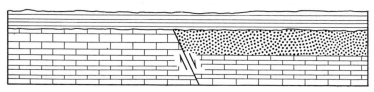

Fig. 52. Section through an inactive fault overlain by younger strata.

faulting, but care must be exercised not to confuse bold promontories formed by erosion with fault scarps. It should also be pointed out that it is possible for steep escarpments to develop by erosion along a line of faulting (fault-line scarp) rather than by the direct action of faulting alone. Fault-line scarps owe their existence to the presence of an erosion-resistant layer and can be detected easily when the downthrown side of the fault stands higher topographically than the upthrown side. Under other circumstances differentiation of fault and fault-line scarps is a much more difficult matter. Fault-line scarps may be regarded as proof that the fault is inactive.

Sag ponds are small undrained depressions caused by differential movement and subsidence of a block, or blocks, adjacent to a fault. Because they are subject to rapid destruction by headward erosion of adjacent streams and filling in by wash from the surrounding slopes, it is doubtful that sag ponds could exist in other than areas of active faulting.

If stream courses are found to be offset (Fig. 48), it is obvious that erosion has not been able to keep pace with crustal movements, and the fault should be regarded as active. In the same fashion, the interstream divides may also be offset, giving rise to what are known as shutterridges. Like the former, they are good indications of recent faulting.

Finally, it should be noted that in areas of deep soil cover, marked relief, and heavy rainfall, such as are commonly encountered in the tropics, erosion may be so rapid as to destroy physiographic evidences of faulting almost as fast as they are formed. Consequently, their absence must be interpreted with caution.

Use of Aerial Photographs in Structural Problems

In many regions aerial photographs taken at the proper altitude may furnish valuable clues to structure, and when faults are present such photographs may also aid in classifying them as "active" or "inactive." For best results, individual photographs should overlap sufficiently to be usable with the stereoscope. Although of great reconnaissance value, they should never be used as a substitute for studies on the ground. However, occasionally faults that could be detected with difficulty, if at all, by field mapping are readily detected from aerial photographs. Aerial photographs are very useful in engineering geology and are discussed more fully in Chapter 16.

JOINTS

Joints are planes or surfaces which intersect rocks, but along which there has been no appreciable displacement parallel to the joint surface. When displacement parallel to the fracture is measurable, the feature is known as a fault.

GEOMETRY OF JOINTS

Dip and strike of joints have the same significance as when dip and strike are applied to strata or folds and faults. On the basis of their relationship to the bedding or some similar feature of the beds that they intersect, joints may be classified as follows: (1) strike joints in which the strike of the joints parallels the strike of the bedding of sediments, the schistosity of a schist, or the banding of a gneiss; and (2) dip joints which strike parallel to the direction of dip of any of the above features. Oblique or diagonal joints are those that strike at a considerable angle to the strike or dip of the bedding of sediments, etc. A group of joints which are more or less parallel to each other is known as a set. A joint system consists of two or more sets, or a group of joints with a characteristic pattern. A conjugate joint system is formed by two mutually perpendicular joint sets.

Diagrammatic Representation. Joints are usually too numerous to be indicated on field maps in more than a very general and highly conventional manner, and various types of diagrams have been developed for the purpose of more detailed representation. The simplest type of diagram is a map on which the data derived from a study of many joints are combined in a single dip-and-strike symbol. Length of the strike lines indicates the direction and relative importance of the various joint groups. Another frequently used convention is to plot the strikes of individual joints on one semicircle and the dips on another. The semicircles are divided into sectors (usually 10° apart), and the number of joints in each sector indicated according to a suitable scale. From the two diagrams the percentages of joints striking and dipping in any specified direction can be read directly, but it is impossible to correlate the number striking in any specified direction with their dip. More precise methods involving the use of equal area projections have been described by Billings.[1]

GEOLOGICAL CONSIDERATIONS

In geology, joints are often classified according to their origin as tension joints and shear joints. It should be noted that the latter may be formed by either compressional or by shearing forces. For example, a joint system formed by two sets intersecting at 90°, one set of which strikes north-south, could have been formed by compressional stresses acting along a northeast-southwest line, or by a shearing couple acting in an east-west direction. Tension joints are usually easier to recognize; they may be formed in a number of ways. Compression and shear joints are not so easily recognized; they are difficult to interpret due to the fact that shearing stresses always have compressional components and vice versa.

Tension Joints. The most common cause of tension fractures is contraction during cooling of lava flows, which leads eventually to rupture along planes making angles of about 120°. Intersection of such planes forms prisms of roughly hexagonal shape. Because the mass is usually free to contract vertically, jointing tends to develop chiefly at right angles to the cooling surface, but horizontal fractures may also be formed by tension acting in a vertical direction. Fractures formed in this manner are

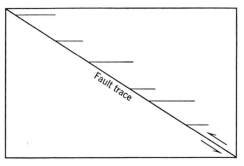

Fig. 53. Section illustrating development of tension joints along a thrust fault. (Adapted from Marland P. Billings, *Structural Geology*, copyrighted by Prentice-Hall, 1942.)

known as columnar joints; they are very common in basalt flows. Tension cracks, or joints, may also form during the drying out of water-soaked sediments. A common example of this type of joint is found in mud cracks which occur in rocks of nearly all geological ages. Tension joints may also result from friction along the walls of a fault (Fig. 53). Joints formed by tensional stresses set up around rising intrusive masses have been discussed elsewhere [1] and need not be repeated here. Folding always results in an extension of the mass in directions normal and parallel to the axis of distortion. Joints formed by the former process are known as extension joints; the latter are spoken of as release joints. These are especially well developed in brittle rocks occurring on the crests of anticlines. Figure 54 shows a cross section of the joint systems frequently encountered on the crests of anticlines.

Shear Joints. Conjugate systems, slickensided surfaces, and fractures cutting indiscriminately across both pebbles and matrix of conglomerates are usually interpreted to be shear joints, but it must be admitted that no very reliable criteria for recognition of these features exist. However, when the joint systems are correlated with folds, it is sometimes possible to demonstrate by analysis of the forces causing the folding that the joints occupy positions of resultant shear.

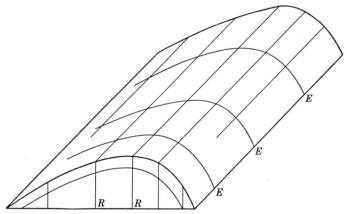

Fig. 54. Diagram illustrating development of extension (*E*) and release (*R*) joints along the crest of an anticline.

Sheet Joints. Inspection of many quarry faces and outcrop ledges often discloses the presence of numerous more or less parallel fractures called sheet joints, which tend to be spaced farther apart as the distance from the surface increases (Fig. 55). The trend of these fractures is always roughly parallel to the topographic surface. At depth, sheet jointing may not be visible, but its presence is indicated by planes of weakness that are utilized in quarrying. In addition to sheeting, large intrusive bodies usually possess two mutually perpendicular directions of parting which are oriented roughly normal to the surface. One of these is called the rift and the other the grain. Usually the former is better developed than the latter. Several hypotheses have been proposed for the origin of sheet jointing, the most acceptable of which is erosion and removal of the overlying materials, leaving the underlying rocks free to expand vertically while confined laterally. Sheet joints have been observed in sediments, but they are most prevalent in intrusive rocks.

Fig. 55. Sheet jointing in granite, Yosemite National Park, California. (S. F. Turner, U. S. Geological Survey.)

ENGINEERING SIGNIFICANCE

Because of their almost universal presence, joints are of considerable engineering importance, especially in excavation operations. It is desirable for joints to be spaced closely enough to reduce secondary plugging and blasting requirements to a minimum, but not so closely as to impair stability of excavation slopes or increase overbreakage in tunnels. Needless to say, the ideal condition is seldom encountered. Joints oriented approximately at right angles to the working face present the most unfavorable condition, whereas joints oriented approximately parallel to the working face greatly facilitate blasting operations and insure a fairly even and smooth break parallel to the face. Joints offer channels for ground water circulation, and in workings below the ground water table they may greatly increase water problems. They also may exert an important influence on weathering.

REFERENCES

1. Marland P. Billings, *Structural Geology,* Prentice-Hall, New York, 1942.

2. W. H. Bucher, The Mechanical Interpretation of Joints, *J. Geol.*, Vol. 28, 1920.

3. H. G. Busk, *Earth Flexures*, Cambridge Univ. Press, 1929.

4. J. P. Buwalda, Shutterridges, Characteristic Physiographic Features of Active Faults, *Proc. Geol. Soc. Amer.*, 1936.

5. F. H. Lahee, *Field Geology*, 4th Ed., McGraw-Hill Book Co., New York, 1941.

6. T. S. Lovering, Geology of the Moffat Tunnel, Colorado, *Trans. Am. Soc. Mining Met. Engrs.*, Vol. 76, 1928.

7. J. B. Mertie, Jr., Graphic and Mechanical Computation of Thickness of Strata and Distance to a Stratum, *U. S. Geol. Survey, Prof. Paper* 129, 1922.

8. J. B. Mertie, Jr., Stratigraphic Measurements in Parallel Folds, *Bull. Geol. Soc. Amer.*, Vol. 51, 1940.

9. H. F. Reid, Geometry of Faults, *Bull. Geol. Soc. Amer.*, Vol. 20, 1909.

10. H. F. Reid, W. M. Davis, A. C. Lawson, and F. L. Ransome, Report of the Committee on the Nomenclature of Faults, *Bull. Geol. Soc. Amer.*, Vol. 24, 1913.

11. Ralph A. Tarr, and Lawrence Martin, The Earthquakes at Yakutat Bay, Alaska, in September, 1899, *U. S. Geol. Survey, Prof. Paper 69*, 1912.

12. Karl Terzaghi, Effect of Minor Geologic Details on Safety of Dams, *Amer. Inst. Mining Met. Engrs., Tech. Publ.* 215, 1929.

C H A P T E R 5

Subsurface Water

If one were to select the branch of geology most important to civil engineering, it would probably be subsurface water. Few, if any, engineering projects are entirely unconcerned with this subject. Its importance is very apparent with respect to tunnels, dams, reservoirs, water supply, irrigation, and excavation. In foundation engineering, position of the water table is an important factor governing the selection and placing of piling. Stability of cuts and embankments is also influenced by underground water, as may also be the bearing capacities of highway and airfield subgrades. Also of great engineering importance are the chemical and physical changes in rocks produced by subsurface water, the leaching of limestone to form large underground caverns, to mention only one example. To treat in detail all situations where the subsurface water affects engineering works is beyond the scope of this book, and discussion is necessarily confined to general principles. Once the principles are understood, it is relatively easy to apply them to specific problems.

The usual geological approach to the study of subsurface water is mainly confined to examination of the physical characteristics of the water-bearing materials and their spatial relationships or structure. These considerations are of only secondary interest to the civil engineer who is primarily concerned with the amount of water present below the surface. Consequently, engineers and hydrologists have developed methods of making quantitative measurements based on more or less direct observations. These methods are largely outside the purpose of this chapter, however. It is fairly obvious that the comparatively indirect geological approach gains much from the engineer's quantitative data, and that proper interpretation of the engineer's measurements involves an understanding of subsurface geology.

GEOLOGIC AND HYDRAULIC CONSIDERATIONS

All water occurring beneath the earth's surface is known as subsurface water. Its immediate origin is rainfall. The portion of the rainfall that sinks below the surface has been estimated to range from about 50 per cent in humid regions to nearly 100 per cent in arid countries,[11] minus, of course, the very substantial amount lost by evaporation. In other words, the more humid the climate, other factors being equal, the greater is the proportion of runoff. Similar to the surface water, the subsurface water is always in motion, but at a much slower rate. In order to reduce confusion arising from multiplicity of terms, all definitions in this discussion will follow the usage of Meinzer.[6]

THE ZONE OF INTERSTITIAL WATER

Subsurface water always occurs in open spaces contained in the rocks forming the earth's crust. Near the surface all rocks possess a varying proportion of voids, or interstices, the size, shape, and abundance of which depend on the type of rock and the forces to which it has been subjected. As the depth below the surface increases the pressure becomes so great that even the strongest rocks cannot sustain the deforming stresses and through plastic flow tend to assume the densest possible form. As a result of flowage the interstices become closed, and any water occurring in the material can be held only in chemical combination. The top of the zone of flowage consequently determines the lower theoretical limit of the zone of interstitial water, which extends from there to the surface of the earth. As thus defined, depth of the zone of interstitial water is largely a function of rock strength. It has been estimated that for very strong rocks the zone of flowage ranges from 30,000 to 40,000 feet below the surface. However, it is probable that subsurface water does not extend to depths as great as those suggested by these figures. It is thought that on the average only 37 per cent of the openings in stratified

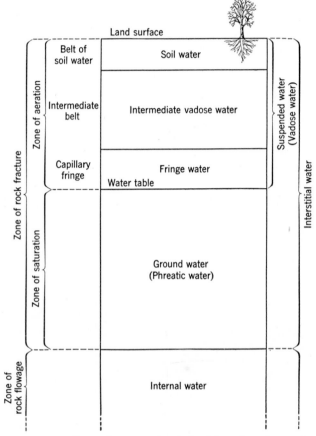

Fig. 56. Diagram illustrating mutual relationships of various subsurface water zones. (After O. E. Meinzer, U. S. Geological Survey.)

rocks and about 50 per cent of the voids in igneous materials are actually taken up by water.[11]

Zones of Aeration, Saturation, and the Water Table. The zone of interstitial water is classifiable into two major and three minor subdivisions. The major subdivisions are the zone of aeration and the zone of saturation. Meinzer defines these zones as follows: [6]

The *zone of saturation* is the zone in which the functional permeable rocks are saturated with water under hydrostatic pressure. The *zone of aeration* is the zone in which the interstices of the functional permeable rocks are not (except temporarily) filled with water under hydrostatic pressure; the interstices are either not filled with water or are filled with water that is held by capillarity. Impermeable bodies may be within the zone of saturation or within the zone of aeration, but they are in a sense not functional parts of either zone. If an impermeable body lies between a saturated and an unsaturated body it is rather arbitrarily considered to be in the zone of aeration ···.

A *water table* is the upper surface of the zone of saturation except where that surface is formed by an impermeable body ···. No water table exists where the upper surface of the zone of saturation is formed by an impermeable body.

The ground water table tends to parallel the topographic surface but is more subdued (see Fig. 57). At any given point depth

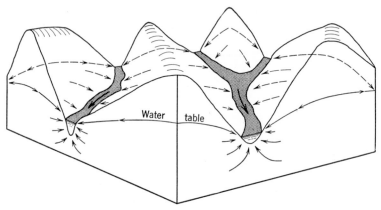

Fig. 57. Block diagram illustrating movement of the ground water and relationship of water table to the ground surface. (Robert I. Kaufman.)

to the water table is not constant; it rises and falls with variations in rainfall, atmospheric pressure, and along sea coasts with fluctuations of the tides. Under various conditions the water table may reach the surface giving rise to springs, rivers, lakes, and swamps.

No climate is so dry as to prevent the formation of a zone of saturation, but in very arid regions the water table may be deep.   In arid climates wind erosion may produce depressions that intersect the water table, thus giving rise to moist conditions on the surface.   The water in the zone of saturation is variously known as the ground water or phreatic water.   The terms "underground water," "subterranean water," and even "subsurface water" have been used as synonyms for the ground water, but their use is to be discouraged.   The ground water follows the laws of hydrostatics; water in the zone of aeration is suspended water and obeys the laws of gravity and capillarity.   The zone of aeration may be divided into the belt of soil water, the intermediate belt, and the capillary fringe.

Belt of Soil Water.   The belt of soil water is that part of the earth's crust immediately below the surface from which water is discharged into the atmosphere in perceptible quantities by the action of plants and soil evaporation.   Water in this belt is known as soil water.   Hygroscopic water is soil water that is in equilibrium with atmospheric vapor.

Intermediate Belt.   Below the belt of soil water and above the capillary fringe is the belt of intermediate water.   According to Meinzer,[6]

Both the belt of soil water and the capillary fringe are limited in thickness by definite local conditions, such as character of vegetation and texture of rock or soil, but the intermediate belt is not thus limited.   It is the residual part of the zone of aeration.   It may be entirely absent or may attain a thickness of several hundred feet.   Where the thickness of the zone of aeration is equal only to the aggregate thickness of the belt of soil water and the capillary fringe these two belts come into contact with each other, the intermediate belt wedging out, and discharge of the ground water into the atmosphere begins.   Where the zone of saturation approaches still nearer to the surface and the zone of aeration becomes still thinner the belt of soil water includes part or all of the capillary fringe and may even extend into the zone of saturation.   Where the height that water will rise by capillarity is equal to or greater than the thickness of the zone of aeration the capillary fringe comes as near the surface as is permitted by discharge by plants and soil into the atmosphere.   Where the belt of soil water extends into the zone of saturation or includes a temporary zone of saturation the water is regarded as both soil water and ground water ⋯.

The Capillary Belt. The capillary belt, or fringe, includes the zone of interstitial water coextensive with the belt of ground water, but in which the water is raised against the force of gravity by capillary action. Since the force of capillary action is a function of the diameter of the pores and the surface tension of the liquid, only materials that possess openings of the proper size (0.5 to 0.0001 mm in diameter) exhibit a capillary fringe. However, nearly all natural materials contain at least some proportion of openings of suitable size, and complete absence of a capillary fringe is very rare. Thickness of the capillary fringe depends on local conditions; in silts it may be as much as 8 feet or more. In finer materials it may be thicker. Where the capillary fringe approaches sufficiently close to the surface it may furnish water to plants, giving rise to what is often called natural subirrigation. It should be noted that capillary movement may take place horizontally as well as vertically. Identification of the capillary fringe is a matter of some practical importance, for water brought up by capillarity may greatly reduce the strength of silty and clayey highway and airfield subgrades as well as that of building foundations.

The Water Table. As previously defined, the water table is the upper surface of the zone of saturation, except where that surface is formed by an impermeable body. Configuration of this surface may be shown by contour lines, in which case all points on a given contour will represent an equal elevation of the water table. A gradient of the water table at any specified locality is the rate of change of altitude of the water table per unit distance. If the direction is not specified it is understood to be that of the maximum gradient. A vertical section of the water table is known as the water table profile. A line on the land surface all points of which are the same vertical distance above the water table is known as an isobath. A ground water divide is a line on the water table on each side of which the water table slopes downward in a direction away from the line (Fig. 57). It is analogous to a topographic divide. Generally, ground water divides tend to be associated with surface drainage divides, but in some localities there is little or no relationship between the two.

GROUND WATER MOVEMENT

Although the ground water is in motion, the rate is usually so slow that the pressure may be regarded as hydrostatic. Usually the pressure varies from point to point and always decreases in the direction of movement. Loss of pressure is due to internal and external friction, the resulting loss of pressure divided by the distance over which it occurred being known as the hydraulic gradient. The over-all direction of ground water movement is always in the direction in which the hydraulic gradient is a maximum. The velocity of movement is expressed by the relationship $V = Ki$, where i is the hydraulic gradient and K is a constant generally referred to as the coefficient of permeability. Permeability may thus be defined as the property of solids which permits passage of gases or liquids through or into the mass without impairment of structure or displacement of parts. It should be clearly distinguished from porosity, which is the amount of void space, usually expressed as the percentage of voids with respect to the total volume. A body may be porous without being permeable, but it cannot be permeable without being porous.

The following types of ground water movement may be distinguished: percolation, defined as movement of water through soil or rock under hydrostatic pressure, except where the movement takes place along large openings, such as caves; capillary migration, which occurs principally in the capillary fringe and in the zone of aeration. Percolation may be regarded as capillary percolation (through capillary interstices) and supercapillary percolation (through supercapillary interstices, or openings larger than capillary interstices). A subterranean stream is a body of water flowing through a very large interstice, such as a cave or group of intercommunicating channels.

Geologic Structure and Ground Water Movement. If the earth's crust were of equal permeability throughout, configuration of the water table would be solely a function of topography and rainfall. Movement would take place with equal ease in all horizontal directions, and there would be but a single saturated zone. Thick deposits of uniform unstratified materials probably approximate this condition, but they rarely occur in nature. Because the materials composing the earth's crust are usually not homogeneous with respect to permeability, movements of the ground water are complex, and configuration of the water table is highly variable. Different rocks and soils exhibit wide ranges in permeability, and in stratified deposits permeability parallel to the stratification may be hundreds of times greater than at right angles thereto. Hence the importance of correlating the principles of hydraulics with geological structure. In tracing the movements of the ground water it is necessary to distinguish between only two types of materials: relatively permeable rocks known as aquifers, and relatively impermeable beds referred to as aquicludes.[10] The horizontal and vertical arrangement of aquifers and aquicludes and their inclination with respect to a horizontal datum, usually taken as sea level, furnish the basis of a geological analysis of ground water movements.

Perched Water Tables. If, where the overlying head of water is sufficient to produce a downward movement, an aquiclude

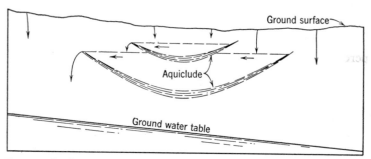

Fig. 58. Section illustrating development of two perched water tables by basin-shaped aquicludes intercalated in pervious materials. Direction of subsurface water movement indicated by arrows.

tends to channel the flow along its upper surface it is known as a negative confining bed. The water table above such a bed is said to be perched and should not be confused with the main water table which lies below. It is possible to have more than one perched water body, provided that the proper arrangement and number of aquicludes and aquifers exist (see Fig. 58). Perched water bodies may be either temporary or permanent. An example of the former may be seen in the formation of surface water during periods of rain and melting snow. Ground water is said to be semiperched if it has a greater pressure head than an underlying water body from which it is not separated by an unsaturated bed. In this instance, however, there is only one water table which is described as semiperched. In semiperched water bodies the underlying water is under subnormal head.

Artesian Water. If, where the hydrostatic head is sufficient to cause a resultant upward pressure, an aquiclude tends to channel the flow of water along its under surface it is known as a positive confining bed. If the confining bed were removed, the water

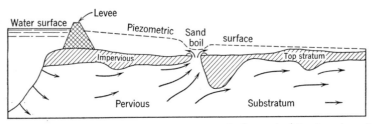

Fig. 59. Section through a levee at flood stage, illustrating development of sand boils caused by artesian pressure in the pervious substratum. Direction of ground water movement indicated by arrows.

would tend to rise to a height determined by the hydrostatic pressure, and an artesian flow would result in instances where the height of rise is greater than the distance to the ground surface. The surface to which the confined water would rise if conditions permitted is known as the piezometric surface, and an isopiestic line of an aquifer is the term applied to a contour of the piezometric surface. The hydraulic profile of an aquifer is the vertical section of the piezometric surface. Ground water, therefore, may be under artesian, normal, or subnormal head, depending on whether its static level is above or below the surface of the zone of saturation. In order for artesian conditions to exist, an aquifer must be overlain by a positive confining bed and receive a supply of water the free surface of which stands at a higher elevation (Fig. 59). These conditions are frequently encountered along the flanks of dipping strata.

Ground Water Dams. A ground water dam is a body of relatively impermeable material occurring in such a position as to impede the horizontal movement of ground water.[6] A water

Fig. 60. Sand boil near Baton Rouge, Louisiana. Height of sand bags gives a rough measure of the difference between the ground and piezometric surfaces. (Corps of Engineers, U. S. Army.)

table that shows a pronounced descent along such a dam may be termed an interrupted water table. In some cases ground water may spill over a dam to form a ground water cascade.

SPRINGS

If the water table intersects the surface at any point springs may result. Springs have been grouped into various classes on the basis of: (1) character of the openings furnishing the water, (2) rock structure and force bringing the water to the surface, (3) lithological character of the aquifer, (4) geological horizon of the aquifer, (5) "sphere" into which the water is discharged, (6) quantity of water discharged, (7) uniformity in the rate of discharge, (8) permanence of the discharge, and (9) features produced by the springs or otherwise associated with them.[6] It is unnecessary to consider all these points, but some of them may have engineering significance.

Rock Structure and Force Bringing the Water to the Surface. Springs may be either artesian or gravity in type. The former occur where an aquifer under artesian head has been exposed to the surface by natural agencies. Such springs may be identified by their high hydrostatic head and regularity and permanence of flow. A gravity spring is one which is formed by an outcrop of the water table. The water in such springs percolates from the water-bearing strata, or flows from large openings in the rock under the action of gravity. Gravity springs may be divided into depression springs, contact springs, and fracture and tubular springs. A depression spring is one that exists simply because the land surface extends down to the water table. A contact spring is one in which the water flows to the surface from a permeable bed over the outcrop of a less permeable material that retards or prevents the downward percolation of water and thus deflects it to the surface.

Quantity of Discharge. Discharge from springs is usually measured in gallons per minute, or in cubic feet per second. On the basis of discharge, springs have been grouped into eight orders of magnitude ranging from 100 cubic feet per second for those of the first order to less than 1 pint per minute for those of the eighth order.

Uniformity and Permanence of Discharge. The variability of discharge from a spring may be expressed quantitatively by the ratio of its fluctuation to its average discharge. No measure of variability is reliable unless measurements have been made over a considerable number of years, but it is permissible to speak of the variability over a specified period. With respect to permanence of discharge, springs may be classified as perennial (discharge continuously), and intermittent (discharge at only certain periods and at other times are dry). All intermittent springs are variable, but perennial springs may be constant, subvariable, or variable.

Quality and Temperature of the Water. The quality of water in springs has received much attention because of the supposed therapeutic value. However, the character of dissolved salts may furnish valuable supplementary data on composition of the rocks at depth. Relative to temperature of water, springs are classified as thermal and non-thermal. A thermal spring is one which supplies water at a temperature appreciably above the mean annual temperature of the surrounding atmosphere. Thermal springs are subdivided into hot and warm springs. A hot spring is one supplying water at a temperature above 98° Fahrenheit; a warm spring supplies water at a temperature lower than that of the human body.   Geysers are hot springs that intermittently erupt jets of hot water and steam.[1]

JUVENILE AND CONNATE WATERS

The question as to what proportion of the water contained in the earth's crust may be attributed to magmatic origin deserves a brief mention.   Although volcanologists seem to be in general agreement that water is an original constituent of molten rock, the amount of water derived from this source is probably very insignificant when measured over any short period of time.   From a geological point of view, however, magmatic, or juvenile, water is of great interest, and probably it should be considered as the ultimate source of all the water on our planet.   In some regions, especially in oil fields, salt water having a chemical composition very similar to sea water is often encountered.   The generally accepted view is that these fluids represent ocean water trapped in interstices of the sediments at the time they were laid down.   Such fluids are referred to as connate and may be regarded as "fossil" sea water.

GEOLOGICAL WORK OF THE SUBSURFACE WATER

In contrast with the surface water, the geological work of the subsurface water is chiefly chemical in character.   However, where the volume and velocity of the subsurface water reach sufficient magnitude, as in caves and subterranean rivers, it may erode and deposit mechanically in much the same manner as the surface water.   Civil engineers are only too familiar with this phenomenon as exhibited in the piping of foundations of earth dams.   On the basis of geological work accomplished, it is convenient to distinguish between the belts of weathering and cementation.   The belt of weathering includes the interval between the surface of the earth and the top of the water table, and therefore corresponds to the zone of aeration.   The belt of cementation is coextensive with the zone of ground water.   Ordinarily, the effects produced in the belt of weathering are those most frequently encountered in engineering work.   However, the position of the water table is not constant throughout geological time, and the zone of weathering today may have been in the zone of cementation a relatively short time in the geologic past and vice versa. Failure to consider the effects of fluctuations in the water table has been the cause of many errors in estimating construction costs.

The Belt of Weathering. This belt extends from the ground surface to the top of the zone of saturation. Near the ground level the surface and subsurface waters merge, and together constitute one of the principal agencies of soil formation. Surface weathering and soil formation are described in the following chapter.

Solution is the chief process occurring in the belt of weathering. The solvent action of liquids depends on their temperature, pressure, and chemical composition. Natural waters are usually either alkaline or acidic, the latter being the more common of the two. Silicates are soluble in basic solutions, but as a rule acidic waters have little dissolving effect on such materials. However, acidic waters react readily with carbonates to form soluble bicarbonates; the reaction of carbon dioxide-charged waters with limestone is shown by the following equations:

$$H_2O + CO_2 = H_2CO_3$$

$$H_2CO_3 + CaCO_3 = Ca(HCO_3)_2 \text{ (soluble)}$$

These reactions furnish the explanation of the formation of subsurface cavities and sink holes. The latter are deep depressions with vertical walls formed by the collapse of the roofs of underground caverns (Fig. 61). When subjected to prolonged weathering, regions underlain by limestone may develop an integrated subsurface drainage accompanied by the partial to almost complete extinction of the surface drainage. In such areas large surface streams often drain into sink holes and other solution features to become part of the subsurface drainage, where continuation of their courses can be traced with difficulty, if at all. Topography characterized by numerous sink holes and subterranean drainage occurs in many parts of the world; it is especially well developed in the region of Karst in Dalmatia; hence the term "karst topography" is often applied to such landscapes. It has been estimated that 96 tons of material are removed annually by solution from each square mile of the earth's surface distributed as follows: calcium carbonate, 50 tons; calcium sulphate, 20 tons; sodium chloride, 8 tons; silica, 7 tons; alkaline carbonates and sulphates, 6 tons; magnesium carbonate, 4 tons; and iron oxide, 1 ton.[11] These figures are doubtless somewhat incorrect and at best are only averages, but they give some idea of the relative abundance and solubilities of the materials involved.

Fig. 61. Sink hole near Meade, Kansas. (W. D. Johnson, U. S. Geological Survey.)

The Belt of Cementation. The upper limit of this belt is defined by the water table, but its lower limits are indefinite. Similar to the zone of weathering, the belt of cementation is characterized by both solution and deposition. It is probable that the amount of moisture entering the belt of cementation over any given period of time is approximately equal to the quantity withdrawn by springs and seepage.[11] If the relative amounts of dissolved material entering and leaving the belt of cementation were known accurately, it would be possible to state whether solution or deposition predominates. This question cannot be answered at present, but it seems probable that in the belt of cementation solution and deposition are nearly balanced. As compared with the belt of weathering, however, deposition is the dominant process.

Reactions in the belt of cementation differ from those in the belt of weathering in degree rather than in kind. Oxidation, hydration, carbonation, and solution occur, but their relative importance differs considerably from that observed in the belt of weathering. The source of oxygen in subsurface water is the

Fig. 62. Room in Carlsbad Caverns, New Mexico. (W. T. Lee, U. S. Geological Survey.)

atmosphere, and by the time subsurface water reaches the water table nearly all the oxygen may be lost. In the deeper zones of the earth's crust carbonation and hydration are the chief chemical reactions taking place. Precipitation of dissolved materials may result from mingling of solutions of different compositions, reaction with the wall rocks, and changes of temperature and pressure.

Cementation is the term applied to the binding together of rock particles by deposition of mineral matter in the interstices. As a result of this process gravels and sands are transformed into conglomerates and sandstones, whereas clays are changed to shales. The principal agent of cementation is the ground water, but gases and solutions expelled from cooling igneous bodies may be important locally. The more important cementing materials are oxides, carbonates, silicates, and sulphides. Of the oxides, silica, iron oxide, and aluminum oxide are the most important. Of the silicates hydrous varieties are by far the most common. Calcite, dolomite, and siderite are quantitatively the most important carbonates, whereas marcasite and pyrite are the most common sulphides. Silica derived from colloidal silicic acid released in the belt of weathering is the most common cementing substance, followed in order of relative abundance by calcium carbonate and dolomite. Cementation may also occur locally in the belt of weathering.

In addition to depositing cementing material the ground water may dissolve and reprecipitate mineral matter already present in the belt of cementation. Hence finely disseminated silica occurring as an original constituent of limestones and shales may be rearranged into definite nodules, bands of nodules, or bands. Soluble matter may also be segregated into irregularly shaped masses known as concretions (Fig. 63). Growth of concretions

Fig. 63.   Iron carbonate concretions in shale, Northumberland County, Pennsylvania.   (Pennsylvania Geological Survey.)

usually starts from a nucleus of foreign material around which are deposited successive layers of concretionary material. Growth of masses such as concretions may exert considerable pressure on the enclosing rocks as the result of the force of crystallization.

Fissures and fractures that are filled with mineral matter are known as veins; they may often be the source of valuable mineral deposits. However, it is generally believed that in most mining

Fig. 64. Calcite veins (white) in dolomite, Lancaster County, Pennsylvania. (Pennsylvania Geological Survey.)

areas the vein-filling materials were deposited by ascending solutions expelled from deep-seated igneous bodies rather than by meteoric waters. It is noteworthy that in some mines veins exceeding 40 feet in thickness are found. It seems unreasonable to assume that open fissures of this magnitude ever existed at depth, and almost equally unreasonable to attribute their thickness to pushing aside of the wall rocks by the force of growing crystals. The most reasonable explanation seems to be that the veins were repeatedly opened by crustal movements, each opening being followed by deposition of vein material.

Metasomatism. This may be defined as "the process by which original minerals are partly or wholly altered into other minerals, or are replaced by other minerals, or are recrystallized with or without mineral changes, or one or more of these together." [11]

In the type of metasomatism known as replacement there is subtraction atom by atom, or molecule by molecule, of the original substance accompanied by the almost simultaneous deposition atom by atom, or molecule by molecule, of another material. Thus in the case of silicified wood there is the removal of the original plant material and substitution of silica without loss of the most delicate microscopic plant structures. Metasomatism may be of some engineering importance because of the chemical and physical changes produced in rocks that have undergone this process. For example, solutions circulating along faults or other openings may induce profound changes in the wall rocks. Feldspars in granite may be altered to montmorillonite, a very weak and plastic clay mineral, with resulting loss of compressive and shearing strength of a formerly very competent rock. Metasomatism may be caused by both meteoric and magmatic waters, and is often associated with hot solutions. Metasomatism is usually difficult to detect from cursory examination of field exposures or hand specimens; it can usually be discovered by microscopic examination of thin sections.

FACTORS INFLUENCING SUBSURFACE WATER SUPPLIES

The principal factors influencing subsurface water supplies with respect to quantity are: (1) permeability, (2) porosity, (3) degree of saturation, (4) volume of the aquifer, (5) geologic structure of the aquifer, (6) water-yielding capacity of the aquifer, and (7) precipitation and runoff. Quality of subsurface water is mainly dependent on mineralogical character of the aquifer and overlying rocks, and on the culture of the surrounding area. Near the ocean the quality of the subsurface water may also be influenced by diffusion of salt water under the land surface. As the following discussion indicates, location of well fields is a rather complex geological and engineering problem which generally requires fairly extensive and detailed studies.

Permeability, Porosity, and Saturation. For a rock to contain water it must be porous, but in order to yield water to a well it must also be permeable. Porosity may result from various causes, the most important of which are solution, voids between the constituent particles of clastic sediments, and the presence of joints or other types of fissures. When combined with sufficient permeability, the higher the porosity of an aquifer the greater the amount of water it will yield. Uniformly porous beds present few problems in locating and spacing of wells, but in areas where porosity has resulted from solution and/or fracture openings, well location is usually rather difficult. Detailed mapping of geologic structure is ordinarily required, for only those wells that penetrate fractures or cavities may be expected to contain water in quantity. Gravel offers the best possibility for a high yield of water to a well, followed in order of relative favorability by sand, sandstone, limestone, and basalt.[5] Limestones are very unpredictable, some rivaling gravels in productivity, others being as unfavorable as clays and shales. Fine-grained unconsolidated materials such as loess, fine sand, and glacial tills may produce enough water to supply successful wells. Dense, consolidated rocks such as quartzite, granite, and gneiss may produce water when sufficiently fractured and jointed to possess openings. The most unproductive of all materials are clays and fine silts, the interstices of which are too minute to yield the high percentage of water that they may contain. The degree of saturation of any rock depends on its position with respect to the water table. Above this surface the saturation may range from zero to nearly 100 per cent, whereas below it saturation is nearly always complete.

Change of Permeability with Draw-Down. Under certain conditions withdrawal of water may result in consolidation of the material forming the aquifer.[10] In extreme cases of overpumping a depression of the land surface has actually been observed. At Mexico City, for example, subsidences of over 15 feet have followed excessive pumping with consequent lowering of artesian pressures in the stratified materials forming the aquifer.

Permeability and Specific Yield. Permeability of the aquifer is not the only factor influencing the amount of water furnished to a well. Part of the moisture in an aquifer is held by molecular attraction and other more or less obscure forces and cannot be recovered. The portion of the water that an aquifer will yield under the direct action of the force of gravity is called gravity water. The specific yield of a rock or soil is the ratio of the volume that the saturated material will yield by gravity to its own volume. The term effective porosity has been used in approximately the same sense as specific yield, but usually it refers to a somewhat more general, though closely related, condition. According to Meinzer,[6] the effective porosity of a material is the ratio of the volume of fluid that the saturated material will yield under any specified hydraulic condition to its own volume. The specific retention of a rock or soil is the volume of water that the saturated material will retain against the force of gravity to its own volume. However, the specific yield and specific retention of a small and of a large body of the same material are not comparable. In a small sample the percentage of water yielded will be less than the specific yield. This is caused by difference in lengths of the capillary tubes involved; a short capillary tube, such as predominates in a small sample, may hold all the water whereas a longer tube may contain an excess of water over the amount it will hold by molecular attraction. This excess of moisture will be drained by gravity and accounts for the higher specific yield that occurs under natural conditions.

Relationship of Water-Yielding Capacity to Rock Texture. The influence of rock texture on specific yield has been expressed as follows: [5]

> The relative amounts of ground water yielded and retained differ in different kinds of rocks and soils. The retaining force is chiefly adhesion, which increases with the aggregate area of the rock surfaces in contact with the water. Therefore, in rocks of uniform porosity the yield is least in those which have the smallest interstices. A clean gravel—that is, one which is not mixed with fine-grained materials—may have no higher porosity than a bed of silt or clay, yet it may be an excellent source of water, whereas the silt or clay may be worthless. The specific yield of the gravel may be nearly equal to its porosity, but that of the clay may be almost or quite zero, all or nearly all of its water being held against gravity. Dense rocks, such as limestone or lava containing good-sized solution channels or joints, may have low porosity and yet be excellent sources of water because the interstices which they contain are large and hence yield freely nearly all their water. Unassorted clayey material, such as the more dense boulder clay deposited by glaciers, has a low porosity, and, moreover, most of its interstices are small. It does not contain very much water even when saturated, and it holds against gravity most of that which it does contain. Its specific yield is still lower than its porosity.

Relationship of Yield to Period of Draining. The greater portion of the gravity water is yielded promptly, but there is apparently no limit to the period during which progressively slower draining will occur. Experiments may be cited in which a coarse sand yielded 26.88 per cent of its volume in $2\frac{1}{2}$ years distributed as follows: 10.68 per cent in the first half hour, 15.56 in the first hour, 24.28 per cent in the first 9 days, leaving only 2.60 per cent for the entire remaining period.[5] It was still yielding minute quantities of water at the end of the $2\frac{1}{2}$ years during which the experiment was conducted. Fine-grained materials not only yield less water but yield it more slowly. For most materials no serious error is involved if, after several days' draining, the water content remaining is regarded as the specific retention.

Determination of Specific Yield. Because the specific yield determined from measuring the water retention of small samples is not the same as that for a larger volume of the same material, it is advisable to determine the specific yield directly from field measurements. The following methods may be used for determination of specific yield: (1) draining of high columns of saturated materials in the laboratory, (2) saturating in the field a considerable body of material situated above the water table and above the capillary fringe and allowing it to drain downward naturally, (3) collecting samples immediately above the capillary fringe after the water table has gone down an appreciable distance, as it commonly does in summer and autumn, (4) ascertaining the volume of sediments drained by heavy pumping, a record being kept of the amount of water pumped, (5) ascertaining the volume of sediments saturated by a measured amount of seepage from one or more streams, (6) making indirect determinations in the laboratory with small samples by the application of centrifugal force, and (7) making mechanical analyses and estimating therefrom the specific retention and the specific yield.[5]

METHODS OF ESTIMATING GROUND WATER SUPPLIES

Although the amount of subsurface water available is usually large, if water is withdrawn from an underground source in greater quantities than are furnished to it depletion will eventually result. The rate at which water may be withdrawn from an aquifer without depleting the supply to the extent that withdrawal at this rate is no longer economically feasible is known as the safe yield. The balancing of additions of water from every source (ground water increment) to the area under consideration against withdrawals of every kind (ground water decrement) is known as the ground water inventory. It is obvious that the ground water increment must equal the precipitation minus evaporation and transpiration losses (absorption of water by plants) minus the runoff. Since these factors vary from year to year and even from day to day, the longer the period of observation, the more reliable will be the results. Two methods of making ground water inventories are in use: one is based on the analogy between surface and subsurface reservoirs, and is applicable to formations having a water table; the other is based on the analogy between subsurface water supplies and conduits, and is chiefly applicable to artesian basins. The factors involved in making ground water inventories are so numerous and complex that complete treatment would require a book about the same size as the present volume. Consequently, only a brief mention of the chief items can be included. For more detailed treatment of the subject the works of Meinzer,[8] Barrows,[2] and Tolman [10] should be consulted.

Precipitation. This is the general term covering all sources of moisture such as rain, snow, and hail. Precipitation may be measured by several types of instruments, the chief factors affecting accuracy of the results being density and location of recording stations and the period of observations. Estimates of the period required for accurate determination of the mean annual precipitation range from about 30 to 40 years, but a 20-year record may be expected to show an error of somewhat less than 4 per cent. For most purposes considerably more than 4 per cent error in measurement is acceptable.

Evaporation. The chief factors influencing evaporation are temperature of the soil and air, relative humidity, and wind velocity. Because these factors vary for soil and water areas, it is customary in practice to determine evaporation from soil and water areas separately. In measuring evaporation from soil areas an instrument known as a lysimeter is used. A common type of lysimeter consists of an open-topped watertight box which is placed in the ground and filled with a sample of the soil to be tested. The rim of the box projects slightly above the ground surface and prevents seepage into the lysimeter from adjacent areas. Rain which falls on the lysimeter must either evaporate or pass through it into the underlying soil. When sand is the type of soil tested, the difference between rainfall and the amount of seepage through the lysimeter represents the amount of evaporation, but in the case of humus the water losses include some transpiration and interception losses as well. Several formulae expressing evaporation losses from a water area as a function of wind velocity, relative humidity, temperature of the air, and temperature of the water have been developed and give satisfactory results under ordinary conditions. However, they all fail under certain limiting conditions, such as maximum wind velocity, very dry air, or high relative humidity. Because there is usually not so great a range in evaporation over any specified period as there is for precipitation, a shorter time is required to obtain average monthly or yearly evaporation losses.

Transpiration. This is a very complex phenomenon which depends on the amount and distribution of the rainfall, evaporation, and even the amount of light available. Transpiration losses tend to vary with temperature. The water requirements for certain types of crops have been estimated, and if data on crop yields from an area are known with sufficient accuracy this information may be used to determine transpiration losses.

Runoff. In general, surface flow plus percolation equals precipitation minus evaporation, transpiration, and water added to the underground supply. Surface flow can be estimated by measuring the area of a watershed and determining the flow of all streams draining the area. Percolation can be estimated on the assumption that minimum flow divided by the mean discharge is equal to percolation, or it may be measured directly with a lysimeter. It is obvious that if the mean annual precipitation is known the ground water increment must equal precipitation minus evaporation, transpiration, runoff, and percolation. Methods of ground water inventories involving such procedure, or variations thereof, may be defined as inventories based on the concept of ground water supplies as underground reservoirs. The

amount of water added to the ground water may be estimated
directly by measuring seepage into wells or tunnels, but such
methods usually involve considerable disturbance of natural conditions.   Furthermore, it is usually not possible to secure a
representative distribution of measuring stations.

Methods Based on the Concept of Aquifers as Conduits.   As
previously mentioned, these methods are chiefly applicable in
artesian basins, especially in cases where the wells are located
at considerable distance from the area of replenishment.   It is
apparent that if the velocity of ground water movement, the
cross-sectional area of the aquifer, and rate of recharge are
known, the rate of replenishment to the zone of withdrawal can
be readily calculated.   Velocity of ground water movement can
be measured in the field, or may be calculated from laboratory
determination of permeability.   In some instances, the water
yielded by consolidation of the overlying and reservoir rocks
following relief of artesian pressure caused by the draw-down
must also be taken into consideration.   Cases are on record in
which the amount of water furnished from this source was very
considerable.   There are reasons for believing that a large part
of the water furnished by consolidation is permanently lost to the
reservoir and cannot be replenished.

QUALITY OF SUBSURFACE WATER SUPPLIES

The chief factor influencing chemical composition, and hence
the quality of subsurface water supplies, is the mineralogical character of the reservoir rocks.   For example, in limestone areas the
underground water usually contains a considerable proportion of
dissolved carbonates.   Beds of rock salt furnish a ready source of
chlorides, and gypsum and anhydrite tend to supply quantities of
sulphates and sulphur.   Waters containing humic and other acids,
dissolved sulphates, chlorides, and similar chemicals may act cor-
rosively on steel and iron and may be deleterious to concrete.
Bacteriological contamination is almost always the result of human occupation of a region, and it tends to vary with density of
population.   Chemical composition of the ground water in any
given area may vary with depth below the surface.   Thus in Saudi
Arabia salt water frequently occurs near the surface, but at depth
almost chemically pure artesian water may be found.   When

the upper saline waters are cased off, these wells furnish an excellent source of drinking water.

Effect of Sea Water on Coastal Water Supplies. The effect of sea water on subsurface water supplies has been studied in northern Europe, the Hawaiian Islands, and in the United States.[3] Sea water is more dense than fresh water, and as a result along sea coasts a wedge of fresh water may float on top of saline water. As shown in Fig. 65, the height of the overlying fresh water with re-

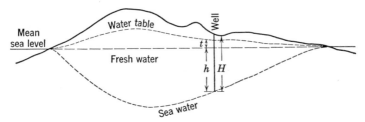

Fig. 65. Diagram illustrating formation of a fresh-water wedge overlying salt water. (After John S. Brown, U. S. Geological Survey.)

spect to sea level determines the depth of the salt- and fresh-water boundary below sea level. If H equals the total thickness of fresh water, h the depth of fresh water below sea level, t the height of fresh water above mean sea level, then

$$H = h + t$$

Let g equal the specific gravity of sea water, and assume that of fresh water as equal to 1. Then

$$h = \frac{t}{g - 1}$$

In all cases $g - 1$ will be the difference between the specific gravities of salt and fresh water. In the North Sea the specific gravity is estimated to be 1.027; hence

$$h = 37t$$

Application of this formula to conditions in northern Europe is reported to furnish fairly accurate results. It should be noted that, if a well supplying water from a body of fresh water floating on sea water is pumped, in addition to the normal cone of depression a cone of salt water intrusion will be formed If the water table is drawn down to sea level, salt water will rise to the same level. Furthermore, if such a body of fresh water is allowed to produce in excess of replenishment, salt water will gradually rise toward the bottom of the well. Consequently, great care must be exercised not to withdraw more fresh water over a given period of time than is furnished to the area from all natural sources. It should also be observed that the boundary between the salt and fresh water will fluctuate in elevation with the tides, provided that the reservoir rocks are sufficiently permeable.

REFERENCES

1. Eugene T. Allen and A. L. Day, Hot Springs of the Yellowstone National Park, *Carnegie Inst. Wash. Publ.* 466, 1935.
2. H. K. Barrows, *Water Power Engineering*, McGraw-Hill Book Co., New York, 1934.
3. John S. Brown, A Study of Coastal Ground Water with Special Reference to Connecticut, *U. S. Geol. Survey, Water Supply Paper* 537, 1925.
4. M. L. Fuller, Total Amount of Free Water in the Earth's Crust, *U. S. Geol. Survey, Water Supply Paper* 160, 1906.
5. O. E. Meinzer, The Occurrence of Ground Water in the United States with a Discussion of Principles, *U. S. Geol. Survey, Water Supply Paper* 489, 1923.
6. O. E. Meinzer, Outline of Ground-Water Hydrology with Definitions, *U. S. Geol. Survey, Water Supply Paper* 494, 1923.
7. O. E. Meinzer, Compressibility and Elasticity of Artesian Aquifers, *Econ. Geol.*, Vol. 23, 1928.
8. O. E. Meinzer, Outline of Methods for Estimating Ground-Water Supplies, *U. S. Geol. Survey, Water Supply Paper* 638-C, 1932.
9. O. E. Meinzer, Editor, *Hydrology*, Physics of the Earth, IX, McGraw-Hill Book Co., New York, 1942. Reprinted by Dover Publications, Inc., New York 19, N. Y., 1949.
10. C. F. Tolman, *Ground Water*, McGraw-Hill Book Co., New York, 1937.
11. C. R. Van Hise, A Treatise on Metamorphism, *U. S. Geol. Survey, Mono.* 47, 1904.
12. Gerald A. Waring and O. E. Meinzer, Bibliography and Index of Publications Relating to Ground Water Prepared by the Geological Survey and Cooperating Agencies, *U. S. Geol. Survey, Water Supply Paper* 992, 1947.
13. Leland K. Wenzel, The Thiem Method for Determining Permeability of Water-Bearing Materials, *U. S. Geol. Survey, Water Supply Paper* 679-A, 1936.
14. Leland K. Wenzel, Methods for Determining Permeability of Water-Bearing Materials, *U. S. Geol. Survey, Water Supply Paper* 887, 1942.

C H A P T E R 6

Rock Weathering and Soils

The term weathering includes the changes which occur essentially in place when rocks are in contact with the atmosphere, surface water, and organisms. If transportation of the material is involved, the process is known as erosion. Weathering and erosion are closely related, each aiding and preparing the way for the other, and any distinction between the two is more or less an arbitrary one. The changes brought about by weathering are important in soil formation, and from the soil scientist's point of view soil is essentially an end product of weathering. The subject is too involved to be treated thoroughly in a book devoted to engineering geology, and those wishing a more comprehensive knowledge should consult the references listed at the end of the chapter. The paper by Reiche [18] is a good review of weathering processes and has furnished much of the material in this discussion.

The civil engineer should know something about weathering, for many engineering operations are concerned with weathered rock rather than with completely fresh materials, and engineering projects have come to grief because of failure to appreciate the influence of weathering on construction problems. In other instances, failure to anticipate deep or irregular weathering has led to greatly increased costs. The engineer should also know something about soils and the soil-forming processes in the sense that these terms are understood by the pedologist, as the scientific student of soils is called. The soil classification used in nearly all published soil maps follows the established pedological practice and thus differs markedly from that used by soil mechanics specialists. If engineers, and more particularly those specializing in highway and airport construction, are to use these maps they must be able to translate the pedological terminology into engineering language. Much preliminary exploration may be saved by this procedure. To correlate the pedological and engineering usages is beyond the scope of this book, and those interested in pursuing the subject more thoroughly should consult references 1, 4, and 21 at the end of the chapter.

WEATHERING

Weathering agencies are of two types: (1) physical, and (2) chemical. The former includes such items as temperature changes, crystal growth, pressure, and the action of plant roots and burrowing animals. The second involves numerous and often complex reactions. Physical weathering results in disintegration and chemical weathering in decomposition of the parent material.

MECHANICAL AGENCIES

Geologic literature contains numerous references to rock spalling and disintegration thought to have resulted from differential expansion and contraction caused by variations in solar heating. It is probable, however, that temperature variations produced in this manner are not sufficient to cause all the effects attributed to them.[3, 6] Only in the case of forest fires and the hotter desert regions do temperature changes seem sufficient to cause significant spalling and disintegration of rocks.

Unloading and Exfoliation. In the discussion of sheet jointing in Chapter 4 it was mentioned that when overburden pressures are released by uplift and erosion rocks are free to expand vertically leading to the development of stresses that find relief in the formation of more or less closely spaced parallel joints. Thin concentric slabs may separate and fall away from such jointed rock masses, causing originally angular-shaped blocks to assume rounded outlines (Fig. 66). Exfoliation, as the process is called, leads eventually to the formation of large rounded rock masses known as exfoliation domes. Half Dome in Yosemite Valley [16] and Stone Mountain in Georgia are two well-known examples. Exfoliation may also occur in small isolated boulders, and it is

Fig. 66. Exfoliation of granite, Royal Arch Lake, Yosemite National Park, California. (S. F. Turner, U. S. Geological Survey.)

apparent that in these instances unloading cannot be the explanation. Hydration is the cause usually attributed to this comparatively small-scale effect.

Crystal Growth. It has long been known that under certain conditions growing crystals may exert forces sufficient to disrupt rocks. In cold or temperate climates growing ice crystals are an effective agent, and in arid regions formation of saline crystals may be of considerable importance. The disruption of rocks by growing ice crystals should not be confused with frost heaving described in Chapter 11.

Colloidal Plucking. It has been suggested that it is possible for contraction of colloids to exert sufficient tension to loosen or remove flakes from rocks with which they are in contact.[18]   It is known that a film of drying gel may pull flakes of glass from a tumbler, but little or no study of the significance of this process in weathering has been made.

Organic Activity.   The ability of growing plants to rupture and dislodge sizeable blocks of rock is well known, as are the activities of earthworms and other burrowing organisms.   Although mechanical activities of organisms are important in facilitating weathering processes, the contention that the chief role of organisms is chemical seems to be well founded.

Importance of Mechanical Weathering.   The geological importance of mechanical weathering is twofold: (1) it aids and prepares the way for erosion; and (2) it accelerates chemical weathering.   Erosion is discussed in the following chapter; hence

Fig. 67.   Weathered granite, Arbuckle Mountains, Oklahoma.   Disintegration has widened joints and led to formation of the loose surface rubble visible in foreground.   (Corps of Engineers, U. S. Army.)

here it is only necessary to consider the second item. Chemicals attack solids at and along their surfaces. When other factors remain constant, the greater the surface exposed, the more effective is the chemical action of liquids in contact with solids. When the size of particles is reduced, the surface area per unit volume is tremendously increased with a corresponding increase in chemical activity.

CHEMICAL AGENCIES

All chemical weathering takes place in the presence of water, which is often called the universal solvent because all materials are at least to some degree soluble therein. The chief reactions are hydration, hydrolysis, oxidation, reduction, carbonation, and exchange. The chemical reactions associated with weathering are generally accompanied by an increase in volume and liberation of heat. It is estimated that the change of granite to soil involves a volume increase of about 88 per cent.

The Colloidal State. When a certain degree of fineness is reached, properties of solids dispersed in water or other suitable media are influenced by electrical charges on the surfaces of the particles and associated phenomena, resulting in what is known as the colloidal state. The dimensions at which this state is reached in soils range from about 0.002 to 0.0001 millimeter. At the lower end of the scale every known soil constituent is in the colloidal state, whereas clay may exhibit the colloidal properties at the upper limit. True colloidal particles exhibit the Brownian movement, and do not, therefore, settle out of suspension unless their electrical charges are neutralized. The coagulation, or precipitation, of colloidal particles is known as flocculation. The chemical reactions taking place during weathering were once thought to be entirely ionic in character, but it is now realized that colloidal phenomena play a very important part.

Hydration and Hydrolysis. Under natural conditions, the reactions of hydration and hydrolysis are usually closely associated, each preparing the way for the other. Simple hydration, or adsorption, depends on the formation of electromagnetic bonds between the crystalline particles and the surrounding water. Adsorbed water is more or less firmly held, but it can be driven off at fairly low to medium temperatures. In some instances part of the adsorbed water is liberated at one temperature and the remainder at a higher one. The liberation of adsorbed water plotted against temperature is often a characteristic property of a mineral, and thermal dehydration curves have been developed as a useful method of identifying the minerals in clays.[17] Although the alteration of anhydrite to gypsum is probably the most frequently cited example of hydration, it is by no means the most important; the most significant hydrations occur in the aluminosilicates, especially when in the colloidal state. In this instance, hydration and hydrolysis take place simultaneously with the release of colloidal silicic acid, hydrogen and hydroxyl ions, and ions of the various metallic elements. The volume increase accompanying hydration is an important factor in the disintegration of granular igneous and metamorphic rocks. Hydration is also important in preparing the way for other chemical reactions such as oxidation and reduction.

Oxidation. Evidence of this process is seen in the red color of many soils, where it is caused by the presence of ferric iron in the form of hematite (Fe_2O_3). It may take place inorganically or by the action of various forms of bacteria of which the iron and manganese types are probably the most important. These organisms have the ability of producing iron and manganese hydroxides from lower valence compounds such as carbonates. Sulphur and phosphorus may also be oxidized by bacterial action. The oxidation of sulphide minerals releases sulphuric acid, which is an important agent in rock decay.

Reduction. According to Reiche, reduction reactions may occur in a number of ways, among which are photosynthesis, fermentation of cellulose, and other bacterial actions. Fermentation of cellulose liberates hydrogen, methane (CH_4), and oxygen. The oxygen may combine with carbon to form CO_2, or it may enter into the composition of organic acids. Soil nitrates are reduced to nitrites and gaseous nitrogen which may form ammonium salts or ammonia, depending on the species of bacteria involved. Ammonia is a common constituent of poorly drained soils. Ferric iron may be reduced to ferrous, or even the metallic form. Sulphates are bacterially reduced to sulphides. Hydrogen sulphide is formed in organic muds by the putrefaction of albuminous matter.

Carbonation. The formation of carbonates is one of the more important processes of chemical weathering. Carbon dioxide from the atmosphere unites with water to form carbonic acid which in turn reacts with metallic ions to form carbonates. The metallic ions are formed chiefly from the breakdown of silicates and to a lesser extent the sulphides and other ore minerals. Iron sulphide may be transformed into iron carbonate, and calcium feldspars may break down to form clay and calcium carbonate. Unknown millions of tons of oxygen have thus been removed from the atmosphere since geologic time began and have been fixed in the carbonate and oxide rocks. There is no evidence that the oxygen content of the atmosphere has changed materially during geologic time, and it is apparent that the oxygen fixed in the rocks must eventually be returned to the air in some manner. Huge quantities of carbon dioxide are given off by volcanic eruptions, and it is thought that in the deeper layers of the earth's crust carbonates are transformed into silicates with the liberation of carbon dioxide, forming what has been called the metamorphic cycle. The oxygen in the atmosphere is largely derived through the vital processes of plants, which take in carbon dioxide and liberate oxygen.

Exchange Reactions. When solutions are in contact with solids, ions from the solution may replace atoms in the crystal lattice of the solid. In order to maintain the electric neutrality of the system, an ion is either given off by the solid in exchange for the one adsorbed, or is obtained by a secondary reaction in the solution, resulting in what is known as ion-exchange reaction. The adsorption of calcium ions and the liberation of sodium ions by zeolites, employed in "water softening," is an example of ion exchange between solids and solutions known as base exchange. This process can be reversed by placing calcium zeolites in concentrated sodium brine. Hauser [10] lists the following replacement series in which substances to the right tend to displace those to the left in either series.

Anions: SO_4 F NO_3 Cl Br I CNS OH
Cations: Li Na; K Mg Ca Sr Ba Al H

According to Reiche, in the chemical weathering of the non-aluminous silicates the chief effect of exchange is to liberate colloidal silicic acid and various soluble compounds of which ferrous carbonate is the most important. With the aluminosilicates the reactions are more complicated. The chief end product is what may be called solid aluminosilicic acid ($H_4Al_2Si_2O_8$). In tropical climates clay minerals break down with leaching of silica, a reaction which is thought to be the result of acid radical exchange. Although in no sense dependent on it, exchange reactions are greatly accelerated by the presence of CO_2 derived from carbonic acid.

End Products. The products of weathering include materials moved in solution, or colloidal suspension, and residual substances. The former tend to be transported from the scene of activity and belong to the domain of sedimentation rather than of weathering. The residual products remaining after leaching of soluble constituents merit a brief mention. The chief residual product of weathering is clay. The clay minerals may be divided into three groups: illites (also called hydrous micas and bravaisites), montmorillonites, and kaolinites.[7] In the first two, part of the aluminum is usually replaced by iron, magnesium, or both. In the illites, potassium is an essential constituent. It is believed that the illites are usually the result of alteration taking place on the ocean bottom rather than on the surface of the earth. Montmorillonites and kaolinites are formed by subaerial weathering and also by the action of hot solutions. Formation of montmorillonites appears to be promoted by deficient leaching by alkaline or neutral ground water, whereas the kaolinites seem to form by thorough leaching by acidic waters. Mixtures of the three groups are thought to result from leaching of marine sediments by carbonated waters. Montmorillonites may be transformed into kaolinites by weathering, but the reverse process is doubtful. Oxides of iron and aluminum are also residual products of weathering and are formed usually, but not exclusively, in tropical climates. Other end products are minerals such as quartz, zircon, and rutile, which are not appreciably soluble under ordinary conditions of weathering.

Depth of Weathering. Depth to which weathering may penetrate depends mainly on depth of the water table, time, climate, permeability, and type of rock. In recently glaciated areas weathering may be practically non-existent, whereas in the humid tropics it may extend to depths of 100 feet or more. In the Transvaal, granites are reported to be weathered to a depth of nearly 200 feet. In calcareous rocks the actual depth of weathering is difficult to estimate, owing to removal of overlying materials by solution. In some areas the residual clays formed by leaching and destruction of carbonate rocks may reach a thickness of 10 feet, corresponding to an original thickness of about 400 feet of limestone and shale. Weathering may be unusually deep along joints and faults (see Fig. 68). For example,

Fig. 68. Joints in limestone widened by chemical weathering and later filled with silt and clay. (Missouri Pacific Railroad.)

in one of the Los Angeles-Colorado River Aqueduct tunnels the presence of weathering along joints in sandstone and quartzite permitted large blocks to slide into the tunnel. Where no weathering was present along the joints, the blocks supported themselves by natural arch action. In this instance, the presence of a mere film of weathered materials created an important engineering problem.

WEATHERING OF CONSTRUCTION MATERIALS

Like the bedrock, construction materials are exposed to weathering, the effects of which may be very important. In large cities the air may contain appreciable quantities of acid, thus accelerating the normal rate of weathering. Moreover, concrete and building stones may be in contact with surface waters containing dissolved sulphates, or other salts, which penetrate pores and crystallize there. The resulting pressure of crystallization may aid in rapid disintegration of the material.

Building Stones. The rate of weathering of building stones depends not only on their composition, but also on climate. Limestones and marbles slowly dissolve and smooth-finished stone may eventually take on a rough appearance. Solution is especially rapid if the rock contains appreciable amounts of pyrite or other readily oxidized sulphide minerals, but this is not often the case. Stones containing calcareous cements are subject to slow disintegration caused by solution of the cementing material. Sandstones are not soluble, but because of their porosity they are subject to mechanical disintegration which may be especially rapid in cold climates. The life (period elapsing before serious deterioration occurs) of coarse sandstone (brownstone) in buildings in New York City is estimated to be as low as 5 years, whereas compact sandstone may last up to 200 years. The difference is almost entirely the result of the high porosity of the brownstone and the intensive frost action resulting therefrom. Serpentines and some varieties of marble may have a life of only 2 to 3 years.

Factors influencing the rate of weathering of building stones are, perhaps, best appreciated from the history of the obelisk which stands in Central Park, New York City.[13] This monument consists of a very coarse, porphyritic granite containing abundant quartz and microcline (a potassium feldspar), rare oligoclase, hornblende and mica, and scarce magnetite and pyrite. It was quarried at Syene, Egypt, sometime before 1600 B.C. About this time it was erected at Heliopolis, near the present city of Cairo. Here it stood until 525 B.C., when Cambyses took the city and an attempt was made to efface the inscriptions on the base by fire. It is possible that the obelisk was thrown down by the Persians;

and, if so, it remained on the ground for the next 513 years. After nearly 1600 years exposure to the dry desert climate of Heliopolis, it was re-erected at Alexandria in 12 B.C. Here it stood exposed to the damp and hot climate of the Mediterranean coast for almost 1900 years until it was transported to New York in 1881. Examination of the rock by competent petrographers disclosed that at this time the obelisk had experienced very little chemical alteration and only very slight mechanical disintegration despite 3480 years of extreme vicissitudes.

By 1884, only a little over 3 years after the arrival of the obelisk in New York, scaling of fragments from the monument became so pronounced that the Park Commission appointed a committee to find ways and means for its preservation. Careful study disclosed that hydration had widened the original pore spaces producing incipient disintegration. Widening of the pores promoted and aided frost action which served as the chief agent of destruction. It was recommended that the surface be sealed with paraffin, and it is interesting to note that 220 square yards of surface absorbed 67.75 pounds of paraffin, despite the fact that the fresh rock had a porosity of only ½ of 1 per cent. An equal area of brownstone took 40 to 50 pounds. This treatment appears to have been effective. The obelisk in the Place de la Concorde in Paris is also reported to have experienced notable weathering since its transportation to Europe. Surely no more convincing evidence of the importance of climate in weathering could be imagined than that furnished by this exceptionally well-documented slab of rock.

SOILS

There is at present no generally accepted definition of soil. Consequently, it is necessary to know what the user has in mind when he employs the term. To a civil engineer soil is any material that is unlithified and can be excavated without blasting. To a farmer "soil" is the upper humic layer that supports plant life, but to the pedologist the term has a somewhat different meaning. To him, soil is any material exhibiting the soil profile described below. In the remaining portion of this chapter the pedologist's definition is followed, but in other parts of the book the civil engineer's usage is generally applied.

The Soil Profile *

Horizon A: Eluvial, or Leached, Horizon. Of the three soil horizons horizon A is characterized by mineral decomposition, solution and abstraction of soluble compounds, and the development of a pervious aggregate structure. It rarely exceeds 18 inches in thickness and is sometimes absent. Occasionally, it may reach a thickness of several feet. Following partial drying, the materials may become powdery or granular. Clayey soils may shrink and crack upon drying. The A horizon frequently has a prismatic structure and occasionally a columnar structure. As a rule, it is both pervious and absorbent and delivers water to the underlying horizon.

Horizon B: Illuvial Horizon, or Subsoil. This is the zone of precipitation and accretion of materials dissolved from the horizon above. Calcium carbonate and clayey material containing aluminum and ferric hydroxides may accumulate to form hardpans. Such accumulations cement the subsoil firmly and limit the downward movement of roots and water. Claypan is an accumulation of stiff, compact, relatively impervious clay. It is not cemented, and if immersed in water is readily worked into a soft mass. The B horizon may be pervious because of: (1) aggregate grain structure, (2) vertical parting, or columnar structure, which may be even more pronounced than in horizon A. The clayey material in horizon B may give it a large absorptive capacity (field capacity). It is usually less than 3 feet thick.

Horizon C: Parent Material, or the Substratum from Which the Soil Originated. The parent material may be decomposed rock in place or freshly deposited alluvium, etc. Weathered material in place is known as the regolith. This zone is usually not more than 6 feet thick, but may reach as much as 50 feet or more.

Variations within horizons are designated by subscripts such as A_1, A_2, B_1, B_2, etc. Organic materials on top of the mineral

---

* This term has long been used by pedologists to designate the changes brought about in the uppermost layers of the earth's crust by the soil-forming processes. Soil engineers apply the term to all changes observable in materials overlying the bedrock regardless of the nature or origin of these changes. Thus the term as now used in engineering has a considerably broader meaning than in pedology and geology.

zone are labeled A_0, F, H, etc.   Calcium carbonate accumulations in the upper part of the C horizon of grassland (chernozem) soils are termed the C_{ca} horizon.   Because of overlapping or gradational boundaries, it is not always easy to differentiate between horizons.   The soil profile is illustrated in Fig. 69.

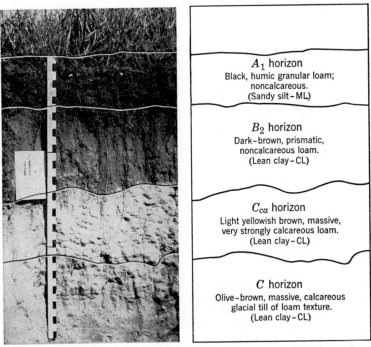

A_1 horizon
Black, humic granular loam;
noncalcareous.
(Sandy silt – ML)

B_2 horizon
Dark-brown, prismatic,
noncalcareous loam.
(Lean clay – CL)

C_{ca} horizon
Light yellowish brown, massive,
very strongly calcareous loam.
(Lean clay – CL)

C horizon
Olive-brown, massive, calcareous
glacial till of loam texture.
(Lean clay – CL)

Fig. 69.   The three soil horizons as developed in a North Dakota chernozem.   (W. M. Johnson, Soil Survey, Soil Conservation Service, U. S. Department of Agriculture.)

The development of the soil horizons may be either weak or strong, depending on the rate of aggradation or degradation. When the removal of soil material by wind, rainwash, and running water is slow, horizons A and B are thick, especially B.   If degradation is fast, horizon A may be removed as rapidly as formed, exposing horizon B at the surface.   At times, both horizons may be removed.   A deep water table favors leaching and redeposition.   Alternate wetting and drying also favor the development of thick soil horizons. Horizon *A* may be diluted by incorporation of wind-borne, rain-washed, or water-borne materials, and hence may be only slightly leached. Horizon *B* may be "drowned" by rapid accumulation of material at the surface. In

Fig. 70. Sandy clay resting on calcareous conglomerate (caliche) near Fillmore, Utah. Sharp, irregular upper surface of the caliche indicates that the overlying materials were deposited after formation of the caliche. (Soil Conservation Service, U. S. Department of Agriculture.)

general, soils of aggradation have poorly developed horizons. Youthful soils usually show weak zonation, whereas mature soils usually are characterized by pronounced zoning. Senile soils are commonly featured by the presence of hardpans. Flood plain soils are built up rapidly and often "drown out" deep root systems. They are usually pervious to semipervious, but they may also be nearly impervious.

Soil as a Physical System

Soil is a physical system consisting of various properties that are functionally related.[11] This relationship may be expressed as follows:

$$\text{Soil} = \text{function (climate, organisms, topography, parent material, time)}$$

This equation states that the magnitude of any one of the properties of a soil such as clay content, porosity, or density is determined by the variables listed within the parentheses. When the variables are evaluated, the soil type is fixed. Thus, for a specified climate, group of organisms, topography, and parent material exposed to the soil-forming agencies for a definite period of time, one, and only one, type of soil can result. Soils that have been exposed for a sufficient length of time for the soil-forming agencies to have run their course are described as mature. With maturity the process of transformation has been completed, and a state of equilibrium has been reached. In view of the delicate nature of the equilibrium and the numerous factors involved, however, it is seldom that soils reach full maturity. Early workers were disposed to regard the parent material as the most important of the soil-forming factors, but it is now generally agreed that a given parent material may develop different types of soil, depending on environmental conditions, chiefly climate and vegetation. Other factors remaining constant, at the stage of maturity there is also a tendency for different parent materials to develop the same type of soil. In India, for example, laterite (a soil high in iron and aluminum oxides) is known to have formed on such diverse parent materials as basalt and limestone.

Major Soil Types

It was formerly thought that two types of mature soils could be distinguished: (1) pedocals, and (2) pedalfers.[15] The first was considered to comprise those soils that possess an horizon in which lime and magnesium carbonates are more abundant than in zones above and below, and was thought to be characteristic of regions of low rainfall. Pedocal soils were considered to range from the black earths, or chernozems, containing abundant organic matter, to gray desert soils with little or no organic matter. In the United States, the secondary calcium carbonate horizon is often referred to as caliche (Fig. 71). Soils

Fig. 71. Caliche (white layer indicated by shovel) in a loam near Meadow, Utah. The underlying materials are calcareous clays deposited in extinct glacial Lake Bonneville. (Soil Conservation Service, U. S. Department of Agriculture.)

marked by a zone of iron oxide accumulation, grouped as pedalfers, were considered to be characteristic of humid climates. The laterites of tropical regions, described in a following paragraph, were considered to be an extreme example of the pedalfer group.

It is now known that the pedocal and pedalfer classification does not serve the purposes intended, and the United States Department of Agriculture [20] classifies soils under the following broad groups: (1) azonal, (2) intrazonal, and (3) zonal. Azonal soils lack well-developed profile characteristics, owing to their youth or other conditions. This group includes alluvial and skeletal soils, dry sands, etc. Soils of this type are often spoken of as "immature," although immaturity in the sense of inadequate length of exposure to the soil-forming processes may not be the governing factor. Intrazonal soils reflect some local factor of relief, parent material, or age more than the normal effect of climate and vegetation. This group includes bog and meadow soils, as well as types like rendzina (a dark, organic, clay soil developed by weathering of limestone). Zonal soils have well-developed soil characteristics reflecting the activity of the soil-forming agencies, climate, organisms, and vegetation. They include a wide variety of materials, ranging from tundra to laterite soils. Podzols, chernozems, and red desert soils are also included.

Geologic Processes Active during Soil Formation

From a geological point of view, soils may be classed as either residual or transported. Residual soils consist of more or less insoluble materials that have remained essentially in the place in which they are formed. They cover a larger portion of the earth's surface than transported soils. Residual soils may reach full maturity or old age. Foreign materials are not added as the soil forms and are not incorporated in the soil horizons. In transported soils, on the other hand, materials are added during soil formation. Soil horizons may be suppressed by rapid addition of new material. Such soils are always immature. The following types of transported soils may be distinguished: (1) Colluvial soils containing materials contributed by rainwash and concentrated at the base of declivities. (2) Eolian soils consisting of wind-blown materials which are deposited either in moist places or in localities covered by vegetation. The building up of such soils is usually rapid. (3) Alluvial soils, the best examples of which are found in flood plains of large rivers such as the lower Mississippi. Deposition in times of flood may be very rapid, and soil zoning may be completely suppressed. If the water table is deep, roots may penetrate to great depths. Permeability of portions of the alluvium penetrated by roots may be greater than that of portions not so penetrated, but it is usually considerably lower than that of the river sands and gravels which underlie the soil.

Distribution of Soils According to Climate

In frigid zones soil, strictly speaking, seldom exists; it is replaced by so-called skeleton soils consisting of physically disintegrated rocks which have undergone little or no chemical weathering. Tundra soils are characterized by shallow profiles, accumulation of undecomposed plant material, often in the form of peat; permanent ice may be present in the lower horizons. In consequence of the slow rate of chemical weathering, clays are very scarce although silts are often very abundant.

Desert soils are marked by the predominance of physical over chemical weathering, are very low in organic matter, and have a neutral to alkaline reaction. These soils are often characterized by a hard surface crust. The so-called lime crusts occur in Texas, New Mexico, Palestine, and North Africa, where the French geologists refer to them either as tufa or as the *croûte calcaire*. In the United States, the term "caliche" is often used to designate materials of this type. Depth of the crust ranges from a few inches to about 3 feet. In Algeria and parts of Morocco, the natives excavate and live below the crust. The lime crusts of North Africa are generally found in and above fine-grained soils and have probably been formed by evaporation of lime-charged waters brought to the surface by capillary action. Gypsum crusts occur in poorly drained areas of Egypt and the central Australian deserts. Crusts found south of the lime-crust belt of Africa have a red to brown color and are often alluded to as "desert sunburn." [1]

In temperate regions of humid climate a type of soil known as podzol is very common. Podzols are characterized by pronounced A, B, and C horizons with a surface rich in organic matter overlying a white or ash-gray leached zone, which overlies in turn a brown zone formed by accumulation of aluminum and iron.

Humid tropical areas are noted for the occurrence of soils

known as laterites. They are characterized by accumulation of iron and aluminum oxides near the surface and the leaching of silica. Lateritic soils of Cuba and Central America contain as much as 90 per cent clay-sized particles, but are nevertheless well drained. Such soils are ideal for highway subgrades. Iron concretions formed in the upper horizons of lateritic soils are used in India and Thailand as a source of road material. In both podzols and laterites, silica may be redeposited in lower horizons to

Fig. 72. Aluminum oxide nodules in laterite, Central Africa.

form the "hardpans" so familiar to builders. Formation of laterites does not seem to be well understood, and the literature discussing them is contradictory. The zone of iron and aluminum oxide accumulation has been used for centuries as an iron ore and building material. Blocks cut from this material become hard and resistant when dried; they were used in construction of the famous deserted city of Angkor, Cambodia.

Time as a Soil-Forming Factor

This is a somewhat debatable subject. It is often stated that soil once gone cannot be replaced and that hundreds or thousands of years are required for soil formation. Several well-documented examples of rapid to fairly rapid soil formation may be cited in opposition to this view. About 1 foot of soil has been formed on the limestone roof of a fortress at Kamenetz-Podolsk in the Ukraine during the 230 years since it was abandoned. About 14 inches of soil have formed on volcanic ash deposited during the eruption of Krakatoa, in the East Indies, in 1883; and investigators in the West Indies have concluded that "within 10 to 20 years, sterile volcanic ash may give rise to fertile soil." [11]

Weathering of Granite as an Example of Soil Formation

Processes involved in soil formation can be summarized and made more realistic by consideration of a concrete illustration: the weathering of granite in a temperate humid climate, for example. Granite, it will be recalled, is a rock consisting of quartz, orthoclase and plagioclase feldspars, mica, and varying amounts of ferro-magnesium minerals such as hornblende. It is one of the strongest and most durable of rocks, but, when exposed to the atmosphere, frost action and other physical agencies of weathering eventually reduce it to a mass of more or less discrete fragments like those shown in Fig. 67.

As the size of fragments becomes smaller and smaller, chemical weathering becomes progressively more pronounced. Among the first minerals to be attacked chemically are those high in iron and magnesium. Hornblende, a complex silicate of calcium, magnesium, iron, sodium, and aluminum, is transformed into a mass of calcium, magnesium, and sodium carbonates, iron and aluminum oxides, and colloidal silica. The oxides of iron and aluminum are always more or less hydrated and tend to be concentrated residually. The remaining decomposition products tend to be transported in solution. Brown and yellow streaks of limonite staining disintegrated rock fragments are often an early sign of chemical weathering. Orthoclase feldspar breaks down according to the following equation:

$$2KAlSi_3O_8 + 2H_2O + CO_2$$
$$= H_4Al_2Si_2O_9 \text{ (clay)} + K_2CO_3 + 4SiO_2$$

The potassium carbonate is largely carried away in solution; that remaining behind is an essential plant food which promotes the growth of vegetation. The silica is released in the colloidal form and is largely carried off in suspension. Weathering of feldspars proceeds along surfaces and cleavage and fracture planes, and feldspar crystals streaked with thin veins of clay mark an early stage of the process. In time, the entire crystal is reduced to a shapeless mass of clay. Micas generally do not alter as readily as feldspars, but in time they too may be reduced to clay. Quartz, not appreciably soluble, is concentrated residually in the form of sand. The end result of chemical weathering is thus a mass of sandy clay with varying amounts of calcium carbonate and iron and aluminum oxides.

Plant growth becomes established early in the process, certain types of trees and bushes becoming rooted long before extensive chemical weathering has occurred. Their roots help to break up the material mechanically, and humic acids formed by decaying vegetable matter are important elements in rock decay and soil formation. In acid environments calcium carbonate is leached from the soil, the end result being a mixture of clay and iron and aluminum hydroxides, forming a soil of the podzol type. In humid tropical regions, humus does not accumulate on the ground surface, humic acids are not formed, and under these conditions clays are leached from the soil, causing iron and aluminum oxides to accumulate residually. The end result is laterite, an extreme example of the zonal type of soil.

It should not be forgotten that soil formation is a dynamic process involving numerous and complex agencies. Soils are not formed once and for all, but are continually changing. Materials are removed from the surface by erosion, but the humic layer and lower soil horizons respond by moving downward into previously unaltered materials, provided the rate of erosion does not exceed that of the soil-forming processes. Changes in vegetation, slope, or climate, also bring about corresponding changes in the soil, but a soil does not necessarily reflect the climate and vegetation prevailing in the region today. During the Ice Age the climate of western United States was much more humid than now, and deep soil profiles were developed in areas where only shallow profiles are forming at the present time.

Mineral Composition of Soils in Relation to Physical Properties

The mineral composition of coarse-grained soils has comparatively little influence on their engineering properties, but in silts and clays it may have an important bearing on plasticity, water content, etc. This phase of soil mechanics has not been thoroughly explored, but studies by Grim [8, 9] have demonstrated its potentialities. It is known, for example, that sodium montmorillonite increases greatly in volume when exposed to an excess of water, whereas calcium montmorillonite swells but little. The swelling properties of montmorillonite can be destroyed only by complete drying out of the adsorbed water. In nature, this process is very slow, and montmorillonite clays usually continue to increase in volume when exposed to water after partial desiccation. In the presence of limited amounts of water, calcium montmorillonite reaches equilibrium at a higher moisture content than the sodium type. Consequently, soils containing calcium montmorillonite may have higher natural water contents than the sodium varieties. Montmorillonite has a high base-exchange capacity, and if concrete is placed directly on a sodium montmorillonite clay it is possible that a change in physical properties of the soil may result.

Colloidal organic material generally increases the plasticity of clays. According to Grim,[8] the explanation seems to be that organic colloids are adsorbed by montmorillonite in much the same manner as water, with consequent decrease in its adsorption capacity of water. In the presence of excess organic material, montmorillonite carrying adsorbed organic material has a marked capacity for forming complexes that do not swell in water. However, these complexes react with certain organic liquids to form gels that greatly increase in volume when exposed to water. These gels are very stable, exhibit little or no air drying, and have very low bearing power. The properties of soils formed in bogs and swamps seem to be often influenced by the above-mentioned reactions.

REFERENCES

1. D. J. Belcher, L. E. Gregg, and K. B. Woods, The Formation, Distribution, and Engineering Characteristics of Soils, *Eng. Bull. Purdue Univ., Research Ser.,* No. 87, 1943.
2. D. J. Belcher, L. E. Gregg, D. S. Jenkins, and K. B. Woods, Origin and Distribution of United States Soils, Prepared Co-operatively by *Tech. Develop. Serv., CAA and Eng. Exp. Sta., Purdue Univ.,* 1946. Contains a very useful soil map of the United States.
3. Eliot Blackwelder, The Insolation Hypothesis of Rock Weathering, *Am. J. Sci.,* Vol. 26, 1933.
4. K. D. Glinka, *The Great Soil Groups of the World and their Development* (trans. by C. F. Marbut), Edwards Brothers, Ann Arbor, 1927.
5. E. R. Graham, Acid Clay, an Agent in Chemical Weathering, *J. Geol.,* Vol. 49, 1941.
6. David Griggs, The Factor of Fatigue in Rock Exfoliation, *J. Geol.,* Vol. 44, 1936.
7. R. E. Grim, Modern Concepts of Clay Materials, *J. Geol.,* Vol. 50, 1942.
8. R. E. Grim, Some Fundamental Factors Influencing the Properties of Soil Materials, *Proc. 2nd Intern. Conf. Soil Mech. and Foundation Eng.,* Vol. 3, Rotterdam, 1948.
9. R. E. Grim, Mineralogical Composition in Relation to the Properties of Certain Soils, *Géotechnique,* Vol. 1, No. 3, June, 1949.
10. E. A. Hauser, *Colloidal Phenomena,* McGraw-Hill Book Co., New York, 1939.
11. Hans Jenny, *Factors in Soil Formation—A System of Quantitative Pedology,* McGraw-Hill Book Co., New York, 1941.
12. J. S. Joffe, *Pedology,* Rutgers Univ. Press, New Brunswick, 1936.
13. Alexis A. Julien, A Study of the New York Obelisk As a Decayed Boulder, *Ann. N. Y. Acad. Sci.,* Vol. 8, 1893–1895.
14. W. P. Kelley, Hans Jenny, and S. M. Brown, Hydration of Minerals and Soil Colloids in Relation to Crystal Structure, *Soil Sci.,* Vol. 41, 1936.
15. C. F. Marbut, A Scheme for Soil Classification, *Proc. 1st Intern. Congr. Soil Sci.,* Vol. 4, 1928.
16. F. E. Matthes, Geologic History of Yosemite Valley, *U. S. Geol. Survey, Prof. Paper* 160, 1930.

17. P. G. Nutting, Some Standard Thermal Dehydration Curves of Minerals, *U. S. Geol. Survey, Prof. Paper* 197-E, 1943.

18. Parry Reiche, A Survey of Weathering Processes and Products, *Univ. New Mexico Publ. Geol.* 1, 1945.

19. C. S. Ross, Clays and Soils in Relation to Geologic Processes, *J. Wash. Acad. Sci.*, Vol. 33, 1943.

20. U. S. Department of Agriculture, *Soils and Men*, Yearbook, 1938.

21. Hans F. Winterkorn, Engineering Uses and Limitations of Pedology for Regional Exploration of Soils, *Proc. 2nd Intern. Conf. Soil Mech. and Foundation Eng.*, Vol. 1, Rotterdam, 1948.

Erosion and Deposition
by Streams

Erosion and deposition by streams are of great importance to engineers. Engineering structures have been destroyed by stream erosion, and others have been made totally useless after a comparatively short period by stream deposits. All types of river engineering, especially flood control and navigation work, require a detailed knowledge of stream action. The number of variables influencing stream action is very large; and, although the effect of any one is fairly well known, each river seems to combine the separate variables in a rather unique manner. Consequently, it is difficult to generalize concerning stream behavior, and prediction of stream action progresses slowly. The geologist's interest in stream action often stops after he has fitted his observations into the classical concept of the erosion cycle, which views all streams as essentially alike in being fated to flow ultimately on very low slopes on a nearly flat surface, or peneplain, of their own making. From the engineer's viewpoint, however, no two rivers are alike. Even in the improbable event of two streams having the same hydraulic characteristics, slight differences in soils of the drainage basins may cause marked differences in their sedimentary loads. For this and other reasons, it is seldom safe to proceed by analogy, and experience gained from construction of engineering works on one river is not always an infallible guide to the planning of similar works on another.

The present practice is to supplement accurate and detailed observations of the past behavior of the stream with model tests. However, such tests have not always proved satisfactory where transportation and deposition of sediment must be taken into account.

The complete investigation of a river for engineering purposes involves the application of many sciences: hydraulics, hydrology, pedology, economics, geology, and botany. The engineer is, of course, keenly interested in river hydraulics which enables him to estimate discharges and other flow characteristics. Hydrology is concerned with determination of runoff, precipitation, humidity, temperature, and evaporation, and is in turn somewhat dependent on geology. Pedology and botany furnish important information concerning the runoff, and economics determines the feasibility of the proposed engineering works. Geology is thus only one of the elements entering into potamology, as the scientific study of rivers is called.

RIVER MECHANICS

The flow of water through stream channels and the accompanying erosional and depositional action of the flowing water constitute what may be referred to as the field of river mechanics. Both engineers and geologists are greatly concerned with erosional and depositional action, which in turn depend in a large measure on hydraulic factors. Consequently, for a better understanding of stream dynamics the geologist must turn to the hydraulic engineer; for assistance in problems involving erosion and deposition the engineer may profit from the studies of geologists.

HYDRAULIC CONSIDERATIONS

The basic law of fluid flow in open channels was announced in 1775 by Antoine de Chézy. This law is given in the following equation, which states that the velocity of flow is proportional to the square root of the slope and the hydraulic radius (cross-sectional area divided by the wetted perimeter).

$$V = C \sqrt{RS}$$

In this equation V equals the velocity, R the hydraulic radius, S the slope, and C the so-called coefficient of roughness. The Chézy formula is not very useful in practical work because of the difficulty of evaluating C.

The hydraulic slope of a river changes with stage, and the velocity is therefore higher during flood stages than at times of low water. Although the total drop over a long stretch does not change very much, local variations may be very pronounced. The effect of variation in stage is further accentuated by variations in depth of the channel. At low water stages the shallow sections between deep pools act as submerged dams and thus tend to lower the slope upstream and to steepen it downstream, but at flood stages variations in slope tend to be smoothed out. Because of variations in slope, stages, width, roughness of the channel, and other factors, the velocity is seldom constant, and there are normally great variations in the rate of flow past successive points on a river channel. Variations are often even greater within a single cross section. This variation in rate of flow is responsible for much of the erosional and depositional action of streams.

In a straight channel of uniform cross section, the thread of greatest velocity of flow is near the middle. In the vicinity of bends, however, the thread of greatest velocity impinges on the outside of the loop causing the water to pile up in a manner analogous to superelevation on the outsides of highway curves. The resulting variation in height of the water surface has been thought to cause a transverse circulation to take place away from the concave and toward the convex bank. This transverse flow superimposed on the mean forward velocity has been assumed to result in a helical spiral considerably longer than the width of the river. Helical flow has been thought to be responsible for moving the bedload from one convex bank to another, but it is probable that its importance has been greatly overestimated. It is even doubtful that helical flow takes place to any considerable extent or the bedload actually moves from one convex bank to another.

Matthes [14] has pointed out that during rising flood stages the thread of maximum velocity tends to shift from the concave bank, and during bankfull stage it is usually a considerable distance away. During floods the velocity near the concave bank is relatively low, and the thread of greatest velocity approaches the convex bank. The river scours its bed near the convex bank and tends to straighten its channel. During falling stages the convex banks receive sedimentary deposits and are built out, while the thread of greatest velocity hugs the opposite bank. Erosion resulting in steepening of the concave bank is therefore principally a falling-stage phenomenon.

The flow of rivers is seldom streamlined, for at even moderate velocities turbulence appears. Turbulence is superimposed on the mean forward velocity, and in this type of flow the water particles may be visualized as moving very haphazardly in random directions and at varying speeds. These variations can be handled only by vectorial methods, and a complete mathematical theory of turbulence involves difficulties beyond the competence of the average geologist and engineer. There is no doubt, however, that turbulence is of great importance in both erosion and deposition by running water.

The most impressive departures from streamlined flow, leading to formation of various types of eddies, are caused by irregularities in shape of banks. In the vicinity of projecting points the current is deflected obliquely away from the bank, causing an upstream current to form on the bank below. The circulation thus formed is known as a suction eddy. On the Mississippi River, suction eddies 800 feet long and 200 feet wide have been observed.[2] If at the same time a vertical component is imparted to the current, a pronounced vortex may result. The velocity of flow in vortices is often much greater than the normal river current. When the current is swift a vortex may be torn from the place where it formed and carried downriver. When the bottom of such a vortex fills with water, the vortex appears to be overflowing. This phenomenon is known as a "boil," and an eddy in a swift current may throw off a succession of boils that travel downstream. Vortices often scour deep holes in the stream bed, and Matthes[15] has suggested that channel deepening caused by vortex action be designated by the Dutch word *kolk*. Bank caving resulting from suction eddies is usually greatest just below and on the lower side of the points that generated the eddies. Thus, although the material causing an eddy may be resistant to scour on its upstream side, it may not be able to resist the erosive

force of the reverse current forming the eddy. As a result, the
point and eddy tend to move upstream. At times several points
are formed which move slowly upstream while maintaining about
the same spacing and shape.

When the main thread of the current crosses from the upper
side of a sharp point to the opposite bank, a part of the flow is

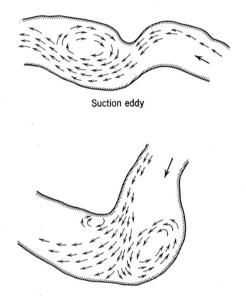

Suction eddy

Suction and pressure eddies

Fig. 73. Diagrams illustrating formation of suction and pressure eddies.
(After D. O. Elliot, Mississippi River Commission, Corps of Engineers,
U. S. Army.)

directed upstream forming what is known as a pressure eddy.
In contrast to suction eddies, pressure eddies usually move
slowly and may cause deposition of silt. The flow in pressure
eddies is not uniform, and they may often reach gigantic dimensions. The formation of a pressure eddy on the bank of a
stream is often accompanied by formation of a suction eddy on
the opposite bank. Deposition of sediment by a pressure eddy
is well illustrated by the following example described by Elliott.[2]
In 1886 a pressure eddy formed in the Mississippi River which extended from the foot of Market Street to Wolf River above the
city of Memphis, Tennessee. The eddy took 35 minutes to fill

and empty, and by 1888 had a width of nearly 2000 feet. While filling, the flow of the current was upstream, but this direction was reversed during the emptying portion of the cycle. This eddy formed a bar which eventually became tied to the river bank. By 1915 the bar extended for nearly a half mile and at low water covered an area of 100 acres.

STREAM CHANNELS

Stream channels may be either cut into bedrock or built up by the combined erosional and depositional action of the stream. From an engineering point of view this distinction is important. Bedrock channels are usually comparatively stable; while those formed in alluvium are normally very unstable. The greatest engineering difficulties are encountered in channels of the latter type. Brief descriptions of the salient characteristics of both types of channels follow.

Bedrock Channels

Such channels are formed by erosional activity of the stream. Consequently, they are generally characterized by rather steep slopes, and are usually rather narrow. Because of inequalities in hardness of the bedrock, the longitudinal profile of the channel is seldom uniform. Shoals, rapids, and waterfalls are common. Bedrock channels are usually fairly straight, but under certain conditions fantastic loops and windings may occur. They often follow faults, joint systems, or beds of weak rock, and foundations for large structures may thus contain important flaws. Rock channels are generally found in the upper reaches of streams, but many African rivers flow in bedrock to within a few miles of the ocean. Bars and other features formed by depositional activity are rare and generally are found only in the vicinity of local obstructions. With continued erosion, bedrock channels tend to become alluvial, and some rivers exhibit repeated alternations of bedrock and alluvial channels.

Alluvial Channels

In their lower courses streams tend to deposit part of their sedimentary load, forming what are known as alluvial valleys.

Such valleys may be formed by streams that have no marked tendency to build up or erode their valleys ("graded" or "poised" streams), or by streams that are actively building up their valleys. Channels in alluvium are characterized by low slopes, general absence of rock ledges, and are constantly shifting both vertically and laterally. The path of strongest flow, known as the channel line, usually coincides with the thalweg, or line connecting the points of greatest depth. Both the channel line and thalweg shift from one bank to the other, the comparatively straight reaches between bends where the channel line shifts from one bank to the other being known as crossings (see Fig. 74). Such channels present the appearance of a succession of bends connected by relatively straight reaches. Exceptionally deep pools tend to form on the outside of bends, whereas crossings are comparatively shallow. The bottoms of the deeper pools may be considerably below sea level. The valley of the Mississippi River from Cape Girardeau, Missouri, to the Gulf of Mexico is a fair example of an alluvial valley the surface features of which have been formed by a poised stream.

Braided Streams. When rivers carrying large amounts of sediment debouch onto flat plains the check in velocity resulting from the lowered slope causes them to deposit much of their load and actively build up their valleys. Because of the large quantities of sediment deposited, such streams are unable to maintain a definite channel, and their courses become a succession of bars with numerous shallow channels between. Such streams, described as braided, are well illustrated by the North Platte and other rivers in stretches east of the Rocky Mountain front. An example of stream braiding caused by overloading with glacial materials is shown by Fig. 110.

Meandering Streams. Meandering streams are characterized by S-shaped courses, fashioned in alluvial materials, which are free to shift their locations and adjust their shape as the channel moves downvalley. As pointed out by Matthes,[14] the term is strictly applicable only when channel migration affects both the bends and intervening straight reaches. Merely tortuous, or crooked, channels should not be classed as meanders. The behavior of meandering streams is of great engineering importance, and thus deserves close examination.

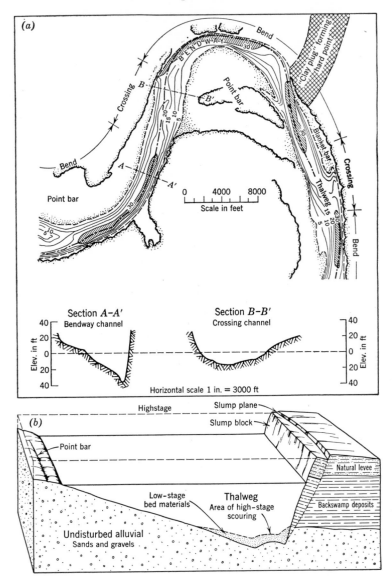

Fig. 74. Diagrams illustrating features of Mississippi River bends. (After H. N. Fisk, Mississippi River Commission, Corps of Engineers, U. S. Army.)

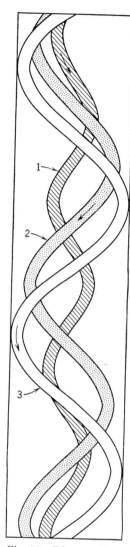

Fig. 75. Diagram illustrating downstream migration of meander loops in homogeneous materials. (After J. F. Friedkin, Corps of Engineers, U. S. Army.)

Meanders have been attributed to local obstructions, rotation of the earth, changes in river stages, etc. Laboratory work [6] indicates the primary cause to be erosion of the banks with resulting overloading of the stream and attendant deposition farther downstream of the materials removed from the upstream bank. Meandering can take place in degrading (eroding), aggrading (depositing), or poised streams; the only requirement being that the stream be able to erode its banks. Although most common in the lower reaches of rivers, meandering may occur wherever conditions are favorable. It is found at high altitudes where stream beds consist of coarse gravel and small boulders or in low-water channels in dry beds of streams in arid regions.

Laboratory studies mentioned above indicate that in uniform materials lying on a uniform slope a series of symmetrical and uniform bends develops. The material eroded from the concave bank upstream travels only a short distance and is largely deposited on the convex bank of the loop immediately downstream. Relatively little material crosses over to the opposite side of the channel. Meandering is thus essentially a trading process in which the material eroded from the caving bank is deposited downstream in the form of a bar. The final result is to cause individual meander loops to shift downstream in the manner illustrated in Fig. 75, and it has been well said that meandering streams are continually making detours around their own deposits. The rate of migration depends on many factors, the most important of which seems to be resistance of the banks. Clays and silts are relatively resistant to erosion, whereas sands and gravels are readily worn away. During a period of 33 years, sandy materials in a bend of the Mississippi River near Rosedale, Mississippi, were cut back at an average annual rate of 150 feet, but in 1948 recession amounted to 1800

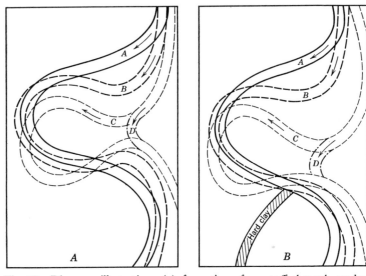

Fig. 76. Diagrams illustrating: (*a*) formation of a cutoff through gradual approach of opposite arms of a meander loop; (*b*) formation of a cutoff through influence of a hard point in the banks.

feet per year. In contrast to this extremely rapid recession, during the period from 1765 to 1820 clay deposits occurring in this area were cut away at an average rate of 25 feet per year.

Figure 76*a* shows the classical concept of widening of meander loops leading eventually to cutting through of the narrow neck thus formed, with attendant formation of oxbow lakes and cutoffs. However, laboratory work does not support this widely held view. In model tests of 50 laboratory streams in uniform materials, not a single cutoff of this type developed.[6] It is likely that, as illustrated by Fig. 76*b*, most natural cutoffs are caused by local differences in erosional resistance of the materials forming the banks. There are equally good reasons for believing that departments from the normal size and shape of meander loops are caused by variations in bank resistance and more or less obscure hydraulic factors. Geologic studies leading to the location of "hard points" are thus an essential step in engineering studies concerned with the behavior of meandering streams.

The results of laboratory studies are well substantiated by observations on the lower Mississippi River where the effects of bank materials are amply demonstrated. For example, meandering is best developed in stretches composed of loose, easily eroded materials. It is practically absent below Donaldsonville, Louisiana, where the banks consist mainly of deltaic deposits of clays and silts.[3] Furthermore, the relation of "hard points" to cutoffs and meander shapes can often be demonstrated in detail in various segments of the river.

River Outlets or Distributaries

An important characteristic of alluvial rivers is that the banks of the main streams are higher than surrounding portions of the flood plains. Consequently, drainage of the flood plains is normally away from the trunk channels, and it is common for the lower courses of alluvial rivers to branch into numerous outlets, or distributaries. Natural outlets may be either flood outlets or all-stage outlets. The former function only during periods of high water; the latter are operative at all stages. All-stage outlets are common features of river deltas. Not all overflow drainage lines are true outlets, for many of them merely return their discharge to the main river at some point downstream. Under certain conditions, the main river may be permanently diverted into an outlet channel which follows a shorter route to the sea. Outlets may form in many ways. They usually are old channels of the trunk stream, or they may in exceptional cases be formed by crevassing of natural levees forming the banks. At times, crevassing of natural levees may permanently divert most of the river discharge into a new channel. A spectacular example of this occurred on the Hwang Ho River of China in 1852, when the river broke out of the old channel 300 miles above the mouth and formed a new channel with its mouth nearly 300 miles north of the old one. The former mouth near Nanking is at present of only minor importance. Less spectacular, but of no less importance, is the threatened capture of the Mississippi River by the Atchafalaya River, its only remaining natural outlet. The Atchafalaya River leaves the Mississippi some distance above Baton Rouge, Louisiana, and enters the Gulf of Mexico by a course some 160 miles shorter than the main channel past New

Fig. 77. Junction of the Atchafalaya (foreground), Red (upper left), and Mississippi Rivers (upper right). The Atchafalaya and Mississippi are connected by a channel 6½ miles long known as Old River which is seen just below the words "point bars." (Mississippi River Commission, Corps of Engineers, U. S. Army.)

Orleans. The rapid enlargement of the Atchafalaya channel since 1927 has raised justifiable fears that it soon may become the main channel, and studies of ways and means to correct the situation are underway.[5, 9]

TRANSPORTATION AND DEPOSITION OF SEDIMENT BY STREAMS

The transportation and deposition of debris by running water are very complex phenomena that have baffled students of potamology for many decades. Theoretical studies are encumbered with mathematical difficulties; and, owing to the difficulty of selecting materials which on a reduced scale possess physical similarity to natural debris, model tests have not proved as valuable as in other phases of hydraulics. Geological observations of streams are usually lacking in the quantitative data necessary to solve the problem. Stream transportation and deposition usually occur in a more or less alternating manner; to say that a stream is depositing indicates only that deposition is the dominant process. The balance between transportation and deposition depends on river stages. Deposition is usually most marked during falling and low-water stages; during rising and flood stages transportation may predominate. Except for colloidal particles and material carried in solution, deposition of transported materials is brought about by lowering the velocity below a certain critical point. Deposition builds up the channel and increases the velocity of flow and is thus a self-limiting process. In consequence of erosion in reaches of high velocity and deposition in reaches of lower velocity, steep slopes are flattened and flat slopes are steepened. The over-all tendency is for the stream to grade its channel so that the velocity is just sufficient to carry the load. In geologic literature such a stream is said to be "graded" or "at grade." Engineers prefer to use the word "poised" to describe this condition. As the process continues, the progressive decrease in volume of material carried as load permits continued lowering of the slope. This last effect must be very slow, and thus has relatively little engineering significance.

Derivation of Sedimentary Load

The sedimentary load of rivers is derived from many sources: particles displaced by alternate freezing and thawing, glacier action, burrowing animals, soil creep and landslides, wind action, and the dynamic force of falling raindrops. Particles loosened by these various agencies are transported by surface runoff. During periods of extremely high runoff detritus avalanches may take place in which immense quantities of material are moved toward the stream. Boulders are rolled long distances on a medium of sands, silts, or clays. Although in most instances the bulk of the sedimentary load has been prepared by weathering, streams carrying suspended matter act as rather effective rasps and chip fresh rock particles from the channel walls, a process known as corrasion. In channels composed of loose, easily eroded materials, the force of moving water alone may be sufficient to dislodge particles.

Movement of Detritus

Streams may transport material in the following ways: (1) as bedload, or material rolled or pushed along the bottom of the channel; (2) material held in suspension by upward currents caused by turbulence; (3) material lifted from the bottom by eddies but almost immediately redeposited, a process known as saltation; (4) extremely fine material held in suspension by the Brownian movement; and (5) material in chemical solution. The last is very important in ground water erosion discussed in Chapter 5. In stream transportation the first three methods are generally considered the most important.

The Bedload

Except for sediment transported by saltation, which temporarily constitutes a part of the bedload, all material in this category is transported entirely by being rolled or pushed along the bottom, either by impact or the frictional drag of the water. The size range of particles constituting the bedload is a function of velocity, and the ability of a stream to move materials in terms of their particle size is known as the competence. It is often stated that the competence of a stream varies with the square of the velocity. However, average velocity may have little meaning, owing to the effects of turbulence, and the velocity near the stream bed is very difficult to measure. For this and other reasons, it is extremely difficult to find a quantitative expression of competence in terms of velocity. In some rivers transportation of the bedload takes place at all stages, but in most instances uninterrupted movement occurs only during periods of high water. Character of the bedload also tends to vary systematically from the source to the mouth of a stream. In the source region velocities are high and the bedload may consist principally of boulders. Owing to abrasion these become progressively smaller and rounder as they move downstream. At the same time a process of sorting takes place in which fine particles are moved farther downstream than coarser ones. As a result, the bedload tends to range from large boulders in the headwaters to mainly silt and clay near the mouth. In most rivers the bedload constitutes only a small fraction of the total load, but movements of the bedload may have a very important influence on channel characteristics.

Because the bedload is rolled or pushed along the bottom, deposition of these materials entails merely a cessation of forward movement caused by lowering of stream velocity. Deposition of the bedload occurs in artificial and natural reservoirs, on the upstream sides of obstructions such as rock ledges, on crossings, and on the convex shores of bends. Deposition at crossings results from a check in velocity caused by a spreading out of the flow as the stream emerges from the deep pool immediately upstream. It takes place mainly during high-water stages. During falling and low-water stages, the crossings may be scoured, and the material may be carried into the deep pool immediately downstream. Scouring of crossings at low water may be a selective process in which only fine materials are transported downstream and the coarser particles are left in place, causing the crossing to become "armored" with a layer of hard gravel. As has previously been mentioned, the chief site of bedload deposition is thought to be on the convex banks of meander loops.

In situations where the banks cave into the stream the caving material is temporarily added to the sedimentary load. If the stream is in a poised condition, the effect of a sudden addition of material is to cause it to readjust its suspended and bedloads. Part of the caving material is taken into suspension while an equivalent amount of coarser material already in suspension settles out. The remainder temporarily constitutes a part of the bedload. Most of the caved material is eventually carried to a bend downstream, but the total quantity of sedimentary load some distance farther downstream remains unchanged. This apparent paradox is well illustrated in the lower Mississippi River where 900,000,000 cubic yards of bank material are estimated to be furnished to the river each year. Despite this addition of sediment, very little of it finds its way to the Gulf, for the amount of sediment passing Cairo, Illinois, is approximately the same as that delivered at the Head of Passes. The material furnished by the caving banks is merely shifted from the concave banks to the convex banks immediately downstream from the points at which the cavings occurred.

The Suspended Load

Particles making up the suspended load are transported chiefly by turbulence. Before turbulence can have an effect, the upward velocities of eddies must be sufficient to lift particles from the stream bed. At any point, however, the upward and downward currents must balance each other, but the concentration of sediment is greater near the bottom than near the top. Consequently, downward currents carry less sediment than upward ones, and the over-all result is for material to be held in suspension. Particles transported in this manner range in size from very coarse sand to clay. It should be noted that in turbulent transportation the coarser particles will drop out as points of lesser turbulence are approached, the over-all movement being from points of higher to lower turbulence. The general law governing deposition of both the suspended and bedloads is that deposition takes place mainly in relatively slack and less turbulent areas. In large meandering streams, such as the Mississippi River, these are found mainly on the insides of bends. Straub [19] believes that turbulence may account for the rapid lateral shifting of channels in stream beds made up of fine materials. This investigator has observed a lowering of the bed of the Missouri River in a single cross section of as much as 15 feet in a single day, accompanied by a corresponding rise at another point in the same cross section. It may also account for the variation from coarse to fine sediment observed in a section extending from the deep pool near the concave bank to the edge of the convex bank, an effect formerly attributed to helical flow.

Deposition of Materials in Solution

Deposition of the portion of the load carried in chemical solution is brought about by chemical means. Thus in many arid regions evaporation may result in deposition of calcium carbonate caused by loss of carbon dioxide from calcium bicarbonate or by simple supersaturation. Mingling of waters carrying different materials in solution may also cause deposition of dissolved substances. Colloidal material may be flocculated by contact with sea water, and tidal and deltaic deposits may carry high percentages of colloidal matter. Chemical deposition in harbors and near river mouths annually necessitates dredging costing millions of dollars.

ALLUVIAL VALLEYS

Valleys formed in unconsolidated stream deposits are known as alluvial valleys; because of the rapid changes in materials normally encountered they present engineers with many difficult and important problems which cannot be appreciated without some understanding of alluvial deposition. The water level in streams is subject to considerable seasonal variation; and, as has been previously mentioned, changes in the stream regimen are always accompanied by delicate adjustments of the channel. Floods usually cause accelerated erosion of the channel whereas low stages are accompanied by deposition. This process is known as "scour and fill," and cases are known in which as much as 100 feet of cutting have been followed by an equal amount of deposition. As a result of cut and fill action, the lower courses of most stream valleys are characterized by alluvium-mantled floors known as flood plains. In instances where stream action has proceeded uninterruptedly, the alluvial mantle consists of a relatively thin layer overlying an eroded floor of bedrock. In the lower Mississippi River, however, thickness of the alluvium approaches 400 feet and the base often extends far below sea level. Geologic studies conducted for the Mississippi River Commission by Fisk [3] indicate that the great thickness of alluvium in the lower Mississippi Valley has resulted from filling of a gorge formed by stream action during the Glacial Epoch, when sea level was approximately 400 feet lower than at present. Studies by Dutch engineers have disclosed similar conditions on the lower valley of the Rhine River, but there depth of the alluvium is not so great. It is probable that similar studies of other great rivers would lead to comparable results.

Natural Levees

When a river overflows its banks the water spreads out with consequent loss of velocity, resulting in deposition of materials in long ridges roughly paralleling portions of the river banks. These features are known as natural levees. They are found on the outsides of bends and along both banks of straight reaches. Materials composing natural levees consist mainly of sands, silts, and clays, with the finer materials concentrated a considerable distance away from the banks. On the lower Mississippi River natural levees seldom exceed 10 feet in height, and in the delta region they may extend to a greater depth below than above the average ground surface. Weight of the materials has apparently caused settlement of the underlying soft clays. Natural levee widths range from ¼ to over 4 miles and are usually greatest on the outsides of meander loops. In flood plain areas the points of highest elevation are normally formed by natural levees.

Crevasses. At times of flood a river may break through its natural levees to form crevasses (see Fig. 78). Deposits laid down in crevasses are usually thin, but have many irregularities resulting from the branching character of crevasse channels. Firm-looking silts filling old crevasses may form veneers over troublesome plastic clays and prove misleading in levee or highway work unless very careful borings are made.

Point Bars

As has already been mentioned, at times of high water the thread of greatest velocity of flow is on the insides of river bends; at low water it tends to shift to the outsides. As a result, during low water sands and silts are deposited in the slackwaters on the insides of bends, forming features which Mississippi River men call point bars. As the bend expands outward, successive ridges may form, the height of which may be as much as 10 feet above mean low-water level. Their shape tends to conform to the curvature of the channel in which they were laid down, but in response to downstream migration of meanders they often truncate each other in a complicated manner.

Swale Fillings. The low areas, or swales, between successive ridges develop dense willow growths that trap fine sediment. Consequently, the ridges are usually separated by belts of silts and clays containing more or less organic matter. These features are called swale fillings, and they may exert a marked influence on foundation conditions, and seepage under artificial levees. Under other conditions, the river may scour across a swale, which it gradually enlarges to form a chute cutoff. In contrast to the more widely known variety of cutoff formed by cutting through a meander neck, chute cutoffs develop slowly and

Fig. 78. Aerial view of typical meander belt deposits and adjacent backswamp, False River Cutoff, Louisiana. (Corps of Engineers, U. S. Army.)

only a short segment of the channel is abandoned. Like swales, chutes may become filled with fine-grained sediments forming what may be called chute fillings.

Channel Fillings

Natural cutoffs are caused by the gradual approach of opposite arms of a meander loop and the eventual cutting through of the narrow neck thus formed. They give rise to the familiar oxbow lakes which represent the old channel of the river. Channel segments abandoned in this manner are slowly filled with sediments that become progressively finer in grain toward the central portion of the former meander loop. Generally, the basal portion of the deposited material is compact plastic clay with less plastic clays and silts above. Maximum thickness of such deposits may equal the depth of the former river channel, which in the lower Mississippi Valley may be as much as 175 feet. River men refer to this type of deposit as a "clay plug." Because of their cohesive character, clay plugs tend to form hard points on the river banks and greatly to retard bank erosion. For this reason, the presence or absence of clay plugs may be an important consideration in selecting locations for structures near river banks.[4] Soft clay deposits filling cutoff river channels can form very troublesome foundations. Levees built on these materials often exhibit large settlements and may be subject to failure. Clay-filled abandoned channels and swales may also exercise an important influence on seepage under artificial levees (see Fig. 59).

Backswamp Deposits

The regions on the outsides of natural levees are lower topographically than those in the immediate vicinity of the river and thus are inundated during times of flood. Such areas are known as backswamps. Materials laid down in backswamps are mainly clays with a high organic content and are often calcareous. In the lower Mississippi Valley backswamp deposits range from 20 feet in thickness in the latitude of Memphis, Tennessee, to over 70 feet in central Louisiana. Backswamps can usually be identified by their low-lying wooded character and lack of directional surface drainage.

Deltaic Deposits

The more or less triangular-shaped areas of alluvial deposits found in the vicinity of river mouths are known as deltas. Owing to recent changes in geological conditions, some rivers apparently have not had time to build up a delta, but most large rivers have formed deltas comparable to their size. Deltas are formed by a gradual filling in of the ocean bottom by alluvial deposition, and they tend to extend themselves in a seaward direction. The

Fig. 79. Distributaries at the mouth of the Mississippi River with natural levees extending far out to sea. (New Orleans District, Corps of Engineers, U. S. Army.)

comparatively flat areas built up by deltaic deposition are often referred to as deltaic plains. According to Russell and Russell,[18] the Mississippi River delta has a structure resembling a pile of leaves. The principal vein of a leaf is marked by natural levee deposits bordering the channel of the trunk stream and its lateral veins by similar deposits bordering the distributary channels. Intervenous areas are represented by marshlands in which swamp deposits are laid down during floods. The serrate margins of a leaf are formed by re-entrant marine embayments, marginal brackish-water lakes, and similar features. Position of the trunk channel has varied greatly in the past, causing individual delta leaves to overlap in a complicated manner. Once removed from the zone of active sedimentation, a delta leaf tends to subside beneath the sea, and subsidence accompanied by marine erosion is actively destroying the older delta leaves. As in the case of alluvial deposition in general, the controlling features are the natural levees, which in this instance are extended out to sea. The low-lying areas between natural-levee ridges are occupied by marshes in which very soft deposits of organic clays are formed. Numerous borings in the city of New Orleans and vicinity amply substantiate the above interpretation of deltaic sedimentation and structure.

THE "POISED" OR "GRADED" STREAM

The concept of a "graded" or "poised" stream has already been introduced but a more detailed discussion is necessary in order to illustrate its practical importance. A poised stream is in a state of delicate balance, and if any change of the controlling factors is introduced the balance is shifted in a direction that tends to absorb the effect of the change. Such adjustments may be surprisingly rapid, a characteristic of poised streams being "the almost telegraphic rapidity with which the preliminary reactions are propagated upvalley and/or downvalley from the point where a change has been introduced ····." [12] It is often assumed that a poised stream must have a very low slope. Although most streams of low slope are at, or near, grade, others like the Shoshone River may be poised although the fall may be as much as 30 feet per mile, or more. The Shoshone is poised at this slope because of the abundance of coarse debris supplied by the rugged Absaroka Range. In other words, size and quantity of the load are important factors in determining the slope of poised streams. The following examples have been taken largely from a paper by Mackin. [12]

Effect of Increased Load

If a poised stream receives an increased load it will respond by building up the slope of its channel below the point of entry of the excess load. Steepening of this segment of the channel enables the stream to transport more load to the next segment downstream, resulting in increased deposition in that segment and consequent steepening of the slope. A wave of deposition is propagated downstream in this manner and continues until a sufficient length of channel has been built up to slopes that enable the stream to transport all the debris supplied to it. An outstanding example of this process is found in the effects of hydraulic mining on the west slopes of the Sierra Nevada range of California. After about 30 years of mining operations it was found that approximately 600,000,000 cubic yards of material had been deposited in the Yuba River, increasing the slope of the stream from about 5 feet per mile to as much as 20 feet per mile.[8]

Effect of Decreased Load

The effect of decreased load on a poised stream is to cause it to make up for the deficiency by picking up material from the channel bottom. The result is downcutting with lowering of the slope. This process continues until the slope has been lowered to a point where the reduced velocity is just sufficient to transport the reduced load. Reservoirs act as settling basins, and downcutting below the reservoir may result from their construction.

Effect of Changes in Discharge

If a stream receives most of its load near the upper reaches, any change in discharge below this region will cause either deposition or erosion. A decrease in discharge requires an increase in slope, because the load must be carried in a narrower cross section. The stream deposits part of its load, building up the slope to a point at which the increase in velocity resulting therefrom is just sufficient to enable it to transport the load. Should the discharge increase, the stream will respond by downcutting until the channel has been lowered to a point at which the decrease in velocity prevents further downcutting. An illuminating example of the effect of changes in discharge is found in the lower Mississippi River system. The threatened capture of the Mississippi by the Atchafalaya River mentioned earlier is thought to have been accelerated by downcutting of the Atchafalaya channel as the discharge increases and silting of the Mississippi channel as the discharge decreases.

Effect of Rise of Base Level

The term base level indicates the level of a body of standing water into which a stream discharges and below which it cannot cut appreciably. Since sea level represents the ultimate level of all the water on the earth, it may be considered as the ultimate base level and is usually referred to simply as base level. The base levels brought about by bodies of standing water at elevations above sea level are only temporary in the geologic sense, and from the geologic point of view they can persist for only a short time. The effect of a rise of base level is to cut off the lower reaches of a stream leading in most cases to formation of a delta. As the delta grows, the velocity of the stream is checked and deposition results. Continuation of this process results in a wave of deposition which is propagated upstream until the channel is built up to a level corresponding to the new and higher base level. This process can take place above reservoirs, and in the case of the Elephant Butte Dam the channel of the Rio Grande has been raised at least 10 feet since 1916. The site of the town of San Marcial, near the head of the reservoir, is now largely buried in silt. As much as 2 to 4 feet of silting has occurred at Albuquerque 100 airline miles upstream and 500 feet above the reservoir level. However it is probable that at least a part of this silting is attributable to increased load supplied by tributary streams and withdrawal of water from the main channel for irrigation, and thus it is not entirely the result of construction of the reservoir.

The tendency of poised streams to deposit material far upstream from the point of ponding must be taken into account in estimating the rate of silting in reservoirs. As Mackin has pointed out, if the rate of silting is estimated merely by dividing the yearly increment of debris carried by the streams entering the reservoir by the capacity of the reservoir, considerable error may result because of failure to allow for the progressive increase in proportion of the load which may be deposited in the stream beds.

Lowering of base level causes increase in velocity of flow resulting in downcutting which is propagated upstream until the slope is flattened to a point where no further downcutting can take place. The new profile thus established will tend to parallel the original one. Such changes are often observed with any change corresponding to lowering of a barrier in the stream's path. A man-made example is the shortening of meander loops by artificial cutoffs.[16]

LAND FORMS PRODUCED BY STREAM EROSION

The surface of the earth has been nearly everywhere profoundly modified by stream erosion. This alone would justify a discussion of the subject, but there are other and better reasons why the engineer should know something about stream erosion. The face of the earth as we see it today represents the combined effects of composition and structure of the underlying rocks, the surface agencies operating on them, and the length of time they have been in operation. The surface agencies may be flowing water, waves, ice, or wind, each of which modifies a given rock structure in its characteristic manner. The various surface features produced by the combined effects of the rock structure, erosional agency, and the stage to which it has progressed are known as land forms. The study of land forms is important to the engineer for, as explained above, the land form is generally closely related to the materials from which it has been fashioned, and to identify the land form is often equivalent to identifying the soils and/or rock of which it is composed. The aerial photographic interpretation of soils discussed in Chapter 16 depends largely on an understanding of land forms. Land forms caused by wind, waves, and ice are treated in following chapters.

DEVELOPMENT OF INTEGRATED DRAINAGE SYSTEMS

On a surface newly exposed to erosion, the initial drainage is concentrated in the original irregularities. Excavation of definite channels generally begins in the lower portions. These channels are gradually extended headward as progressively greater proportions of the runoff are concentrated into them. Lateral drainage into the primary channels tends to bring into being secondary drainage channels known as tributaries. A continuation of this process leads to the development of an increasingly complex net of branching tributaries. Eventually, the stream net extends across the newly exposed slope until the distance which runoff at any point must travel before reaching a channel is less than a certain critical distance required to institute erosion. At this time development of the drainage net is complete, and the divides separating it from adjacent basins are established. Examples of various stages of drainage net development may actually be seen in various parts of the world. It should be recognized, of course, that the effect of geologic structure may alter details of the drainage pattern.

Horton [10] has shown that the relative numbers and lengths of streams in a drainage basin are governed by a mathematical series. Provided that streams of a certain order or above are present and the number of streams of at least four orders and their average lengths are known, the number and average length of tributaries of other orders can be calculated. The order of a stream is determined by considering unbranched tributaries at the headwaters as constituting the first order; tributaries of the second order receive only tributaries of the first order; third-order tributaries receive first- and second-order tributaries; and so on until the main stream is reached. If Horton's methods are applied, streams of the fourth order, or above, must be present.

THE EROSION CYCLE

The changes occurring between the uplift and initial exposure to erosion of a region and the eventual wearing down of this highland area to an almost featureless plain similar to the original surface constitute what is known as the erosion cycle. The concept of the erosion cycle makes it possible to work out the later stages of the geologic history of a region in considerable detail; and the proper understanding of land forms developed by stream erosion is also somewhat dependent on this concept. The subject is a very large one, and space permits only the briefest possible outline. Additional considerations affecting the concept of the erosion cycle are given in the following chapter.

Stages of the Erosion Cycle

During early stages the streams are mainly occupied in extending and deepening their courses. As the courses are deepened the slope is lowered, and the transporting powers of the streams decline. From this point onward the streams are more active in widening than in deepening their valleys. Valley widening is aided by weathering, by mass movement of surficial materials lying on oversteepened valley walls, by slope and tributary wash, and when conditions are favorable by the development of meanders. As the valleys are widened and extended the interstream divides become progressively narrower. The formerly flat uplands progressively disappear and are replaced by narrow ridges. When this stage is reached the region is said to be mature as distinguished from the earlier, or youthful, stage. With continued erosion maturity gives way to old age. This stage is characterized by the cutting down of divides and continued broadening of the valley floors. In brief, during the erosion cycle an upraised area is first dissected into a network of streams and tributaries separated by residual flat uplands, which are eventually cut back until maximum relief, or the stage of maturity, is attained. With continued erosion relief dwindles, divides are lowered, and the area is reduced to a low-lying surface referred to as a peneplain.[1] It should be noticed that the terms "youth," "maturity," and "old age" are purely relative and have no absolute time significance. The alluvial valley of the lower Mississippi River, for example, is almost senile in terms of the erosion cycle, but in actual age is probably one of the youngest major features on the surface of the earth. Carbon 14 analyses indicate that it dates mainly from the last 8000 years or so.

The Erosion Cycle in Flat-Lying Sedimentary Rocks

If a region of nearly flat-lying sedimentary rocks of varying thickness and resistance to erosion is uplifted, the major streams will respond by cutting very deep, relatively straight and narrow valleys known as canyons. Erosion by tributaries, weathering, and slope wash gradually widen the original narrow trunk valleys, but the harder rock ledges tend to keep their original precipitous slopes. If a rock layer of greater hardness, or of the same hardness but greater thickness, lies under a softer or thinner ledge, the upper layer will retreat at a faster rate proportional to its relative thickness and/or hardness. In other words, the rate of valley widening and configuration of the valley walls

Fig. 80. Grand Canyon from Point Sublime, showing dissection of nearly horizontal strata into ridges and pyramid-shaped buttes. (Drawn by W. H. Holmes, U. S. Geological Survey. From *Outlines of Physical Geology*, by Longwell, Knopf, and Flint, John Wiley & Sons, New York, 1941.)

depends on the relative resistances of the rock layers, a deduction that is amply borne out by detailed studies of the Grand Canyon and similar features. Harder ledges tend to form nearly vertical cliffs; softer beds assume more rounded and flatter outlines. As the process continues, portions of the formerly continuous upland surface may become detached from the intervening plateaus. These detached plateau remnants are known as mesas when their summit areas are large and as buttes when the summit areas are small in comparison with their heights. At the stage of maturity, the region consists largely of mesas and buttes, and in old age nearly all traces of the former upland surface have been removed. It should be mentioned that relatively flat-lying

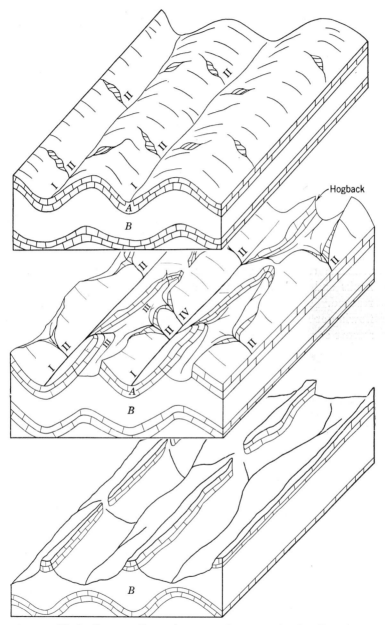

Fig. 81. Block diagrams illustrating successive stages in the dissection of folded strata.

sedimentary rocks possess nearly the same erosion resistance in all horizontal directions. As a result, stream patterns developed in such materials form intricately branching systems resembling the veins of a leaf, constituting the so-called dendritic drainage pattern. Dendritic drainage patterns are not confined to relatively horizontal sedimentary rocks; they may be formed in lava flows and in bodies of igneous rock such as batholiths, which also possess fairly uniform resistance to erosion.

The Erosion Cycle in Folded Sedimentary Rocks

In folded sedimentary rocks the erosion cycle produces many interesting features, but owing to complexity of the subject only the briefest sort of outline can be given. Figure 81 illustrates a region underlain by layers of unequal resistance to erosion that have been folded into alternate anticlines and synclines. If such a region is exposed to erosion, the trunk streams develop in the synclinal troughs. Tributaries then develop on the anticlinal flanks and extend their courses headward until the crests are reached. The anticline is subsequently "unroofed" by tributaries flowing at right angles to the streams labeled II. Because of its steeper slope, the stream marked IV is able to extend its valley headward until it intersects the upper reaches of the stream labeled III. The headwaters of this stream are thus diverted into the drainage system of stream IV, a phenomenon known as stream capture, or piracy. In time, the resistant layer *A* is removed from the anticlinal crests, leading to the formation of asymmetrical ridges called hogbacks. (If the dip of the strata is low, only a few degrees, the term cuesta is used.) If layer *B* is less resistant to erosion than *A*, the former topographically low synclinal valleys will become ridges, and the former anticlinal ridges will become valleys. With continued erosion the resistant layer responsible for the synclinal uplands will be removed and the surface will be reduced to a peneplain, the only vestiges of the former existence of the folds being their eroded roots lying below the surface of the peneplain. This greatly oversimplified example constitutes the most general case of stream erosion of folded rocks. All other types and configurations of folded and/or faulted rock structures may be regarded as special cases and analyzed in a similar manner.

It should be noted that the drainage pattern produced in folded rocks consists of more or less parallel trunk streams flowing approximately parallel to the axes of the larger folds with tributaries flowing nearly perpendicular to the trunk streams. This is the so-called trellis drainage pattern. In more or less circular-shaped features, such as domes and volcanoes, the trunk streams tend to radiate away from the uplift, and the tributaries tend to form concentric, semicircular patterns.

Interruptions of the Erosion Cycle

If the earth's crust were static, erosion would have long since reduced its surface to a plain lying only slightly above, or possibly even below, sea level. Obviously, the surface configuration of the land must be the result of interaction between two opposing forces: (1) those which operate on the surface and which tend to reduce it to sea level, and (2) those which originate within the earth and as a result of which areas are either elevated or depressed. Since there is no particular connection between these forces, the erosion cycle may be interrupted at almost any stage. Elevated peneplains that have entered a second cycle of erosion are very common; indeed, it is almost impossible to point out a mountain range that is definitely known to be in the first cycle of erosion. The even crests of the Appalachian Mountains, for example, are believed to be the remnants of a peneplain formed long ago and later uplifted. Present relief of these folded mountains is entirely the result of erosion and not of folding. A characteristic feature of areas in the second cycle of erosion, other than flat-topped ridges developed on folded rocks, is the presence of incised meanders. Incised meanders differ from the alluvial meanders previously described chiefly in their being fashioned largely in bedrock. Shape of incised meanders is thought to be inherited from former alluvial meanders which have been cut into bedrock in consequence of the increased velocities resulting from the uplift. Good examples of incised meanders are found in the tributaries of the Colorado River, the middle Rhine, and other streams too numerous to mention. The erosion cycle may also be interrupted by changes in climate leading to glaciation, or by desiccation leading to desert conditions.

STREAM TERRACES

Many stream valleys are bordered by rather flat, step-like side benches, which usually slope gently toward the central axis of the valley. Such features, known as terraces, are much favored by engineers in the location of highway and railway routes. They also furnish good sources of construction materials. For these reasons they are of considerable engineering importance.

Terraces may be formed in a number of ways, and space does not permit a thorough examination of the subject. The simplest method of origin is by intermittent uplift of the land surface. Between periods of uplift, a stream may deepen and widen its valley, and if this stage persists long enough a considerable thickness of alluvial materials may accumulate. An ensuing period of relatively rapid uplift may then cause the stream to cut down, leaving remnants of the former wide flood plain as terraces. Continuation of alternate periods of stability followed by uplift may result in the development of multiple terraces bordering the valley walls. Benches originating in this manner, often referred to as valley plain terraces, may be defined as surviving parts of formerly continuous valley floors that have been cut into during episodes of accelerated downcutting. It should be noted that intermittent uplift may produce terraces in bedrock as well as in alluvial deposits.

Changes in sea level are equivalent to changes in elevation of the land surface, and river terraces may be formed by changes in the level of the sea. The valleys of the lower Mississippi River and tributaries are bordered by four, often very broad, terraces. Fisk [3] believes these features to have been caused by changes in sea level, during the Glacial Epoch, accompanied by uplift of the land surface north of an axis between Baton Rouge and New Orleans, Louisiana, and subsidence of the land south of this line. As a result, to the north of this axis the terraces stand well above the present stream channels; to the south they have been buried under still later deltaic deposits. The best foundation conditions to be found in New Orleans and its environs are furnished by these buried terrace deposits.

River terraces may also develop through downcutting of a meandering stream. The process of meander development causes successive meander belts to shift downstream. At the same time the valley may be deepened, and remnants of the meander deposits may remain as terraces. Mackin [12] believes that the terraces of the Shoshone River (Fig. 82) have been formed by alternate cutting and deposition of a meandering stream without

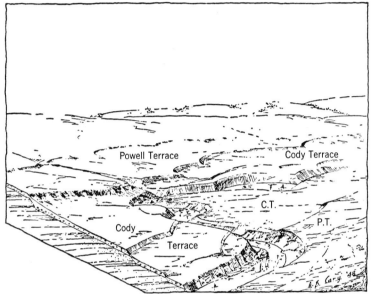

Fig. 82. Terraces along the Shoshone River near Cody, Wyoming. (After J. Hoover Mackin.[12])

intermittent uplift, or other interruptions of the normal cycle of erosion. As contrasted with valley plain terraces, benches of this origin do not correspond in elevation on opposite sides of the valley, or even at some distance upstream or downstream on the same bank.

Changes in climate may also lead to formation of river terraces. For example, if the climate becomes more humid, not only will the streams become larger, but increased growth of vegetation may serve to restrict the supply of rock waste available to them. Both of these factors will tend to bring about accelerated downcutting that may leave remnants of the former flood plains as terraces.

The subject of terrace origin might be almost indefinitely elaborated. Enough has been said, however, to indicate the relationship between mode of origin and composition of terraces. Thus, not all terraces will provide suitable and abundant sources of construction materials; on the other hand, the configuration of the bedrock below the terrace deposits will not always be a reasonably even one. In the case of meandering streams that were actively cutting down before the process was reversed by accumulation of alluvial fill, for example, the upper surface of the bedrock is likely to be an extremely irregular one. It is in these connections that the geologist can aid the engineer not only in the search for construction materials but also in the location and interpretation of borings.

REFERENCES

1. W. M. Davis, *Geographic Essays*, Ginn & Co., Boston, 1909. Reprinted by Dover Publications, Inc., New York 19, N. Y., 1954.
2. D. O. Elliott, *The Improvement of the Lower Mississippi River for Flood Control and Navigation*, U. S. Waterways Experiment Station, Vicksburg, Miss., 1932.
3. H. N. Fisk, *Geological Investigation of the Alluvial Valley of the Lower Mississippi River*, Miss. River Comm., Vicksburg, Miss., 1944.
4. H. N. Fisk, *Fine-Grained Alluvial Deposits and Their Effects on Mississippi River Activity*, U. S. Waterways Experiment Station, Vicksburg, Miss., 1947.
5. H. N. Fisk et al., *Geological Investigation of the Atchafalaya Basin and the Problem of Mississippi River Diversion*, Miss. River Comm., Vicksburg, Miss., April, 1952.
6. J. F. Friedkin, *A Laboratory Study of the Meandering of Alluvial Rivers*, U. S. Waterways Experiment Station, Vicksburg, Miss., 1945.
7. G. K. Gilbert, The Transportation of Debris by Running Water, *U. S. Geol. Survey, Prof. Paper* 86, 1914.
8. G. K. Gilbert, Hydraulic Mining Debris in the Sierra Nevada, *U. S. Geol. Survey, Prof. Paper* 105, 1917.
9. John R. Hardin, Mississippi-Atchafalaya Diversion Problem, *Military Engineer*, Vol. XLVI, No. 310, March-April, 1954.
10. Robert E. Horton, Erosional Development of Streams and Their Drainage Basins; Hydrophysical Approach to Quantitative Morphology, *Bull. Geol. Soc. Amer.*, Vol. 56, 1945.
11. John B. Leighly, Toward a Theory of the Morphologic Significance of Turbulence in the Flow of Water in Streams, *Univ. Calif. Publs. Geog.*, Vol. 6, 1932.

12. J. Hoover Mackin, Concept of the Graded River, *Bull. Geol. Soc. Amer.*, Vol. 59, 1948.

13. Gerard H. Matthes, Floods and Their Economic Importance, *Trans. Amer. Geophys. Union*, 1934.

14. Gerard H. Matthes, Basic Aspects of Stream Meanders, *Trans. Amer. Geophys. Union*, 1941.

15. Gerard H. Matthes, Macroturbulence in Natural Stream Flow, *Trans. Amer. Geophys. Union*, 1947.

16. Gerard H. Matthes, Mississippi River Cutoffs, *Trans. Amer. Soc. Civil Engrs.*, Vol. 113, 1948.

17. W. W. Rubey, The Force Required to Move Particles on a Stream Bed, *U. S. Geol. Survey, Prof. Paper* 189-E, 1938.

18. R. J. Russell and R. D. Russell, Mississippi River Delta Sedimentation, *Recent Marine Deposits*, American Association of Petroleum Geologists, Tulsa, 1939.

19. Lorenz G. Straub, Mechanics of Rivers, *Hydrology*, Physics of the Earth, Vol. IX, McGraw-Hill Book Co., New York, 1942. Reprinted by Dover Publications, Inc., New York 19, N. Y., 1949.

20. W. J. Turnbull, E. L. Krinitzsky, and S. J. Johnson, Sedimentary Geology of the Alluvial Valley of the Lower Mississippi River and Its Influence on Foundation Problems, *Applied Sedimentation*, John Wiley & Sons, New York, 1950.

21. A. O. Woodford, Stream Gradients and Monterey Sea Valley, *Bull. Geol. Soc. Amer.*, Vol. 62, 1951.

C H A P T E R 8

Desert Land Forms

A desert is any region where evaporation exceeds precipitation. The average annual rainfall is usually less than 10 inches. Although there are many localities where no rain falls for many years, there is probably no spot on earth that does not have at least some precipitation. Bahrein Island, in the Persian Gulf, has about 2 inches of rain per year and is one of the driest regions known. Here, as in other arid countries, rain may not fall for years and then 3 inches or more may be delivered in a single almost catastrophic downpour. Most deserts are located fairly near the tropics, and the temperatures may be high. As with the rainfall, however, they are subject to violent extremes. The maximum may reach 135° Fahrenheit, and the minimum may drop below the freezing point. Under such conditions, plant life is not abundant, but deserts do have a certain amount of vegetation. Animal life is also not abundant and shows various adaptations to climate.

Nearly all large deserts are bordered by steppes where the rainfall is somewhat higher and more regularly distributed. Vegetation and animal life are correspondingly more abundant. Such regions are essentially transitional between desert and humid areas.

Deserts may occur in the rain-shadow belt on the lee sides of mountain barriers that cause moisture-laden winds to drop their moisture on the windward slopes, or they may be found in broad belts both north and south of the equator where rising air currents of the tropics descend after having lost most of their moisture during the previous ascent. Such deserts are not dependent on topography and have been termed planetary deserts. The Sahara is an outstanding example of a desert of this type.

Deserts command attention for many reasons. About one quarter of the land surface of the earth is desert, and in the past the great desert regions acted mainly as barriers between the civilizations on their margins. The slow realization of the vast potential productivity of desert soils, aided by the discovery of valuable minerals and large quantities of petroleum, is beginning to alter the traditional role of deserts in human affairs. Highways, railways, airfields, and pipe lines span some of the world's largest deserts, and reclamation of arid regions is progressing rapidly in the United States, Mexico, North Africa, the Near East, India, and Soviet Russia. Increasingly large numbers of civil engineers will thus probably live and work in deserts, and for this if for no other reason they should know something about the processes at work in arid regions. Furthermore, deserts present many geologic-engineering problems, the chief of which are water supply, control of flash floods, delineation of the extent of debris-collecting basins, control of drifting sands, and the deterioration of concrete structures by alkaline waters.

STREAM EROSION AND DEPOSITION IN DESERTS

Geologic processes at work in arid regions have not been studied as thoroughly as those in humid climates, and there are wide differences of opinion concerning the relative importance of the various agencies. The early explorers of deserts were mainly from western Europe and eastern United States, and it is natural that on entering what appeared to be a completely new and strange world they were perhaps unduly impressed by the desert sun and winds. As a result, they tended to attribute most of the features of desert landscapes to the effects of the sun's heat and wind action. However, it appears that torrential rains, and not the wind, are the chief agency in the production of desert land forms. Indeed, as King [10] has pointed out, deserts are no more "abnormal" than are humid regions, and it is probable that desert landscapes do not differ fundamentally from those of humid areas. Deserts are, in fact, probably the best possible locations in which to study the work of flowing water.

Flowing water operates in deserts in much the same manner as in other regions, and where large rivers flow through deserts, the lower Colorado River, for example, they erode and deposit in much the same manner as described in the preceding chapter. The chief difference between most desert streams and those of humid regions is their intermittent character. This seemingly unimportant difference is responsible for many of the features of desert landscapes. Owing to the low rainfall and intermittent character of the runoff, desert regions generally lack drainage to the sea, and as a result no large integrated drainage systems normally develop. If the region is mountainous, the drainage collects in isolated intermontane valleys, giving rise to a series of comparatively flat alluviated basins at various elevations separated by rugged mountain ranges. The Basin and Range province of western United States, comprising the state of Nevada and parts of Oregon, California, Arizona, and Utah, is an excellent example of a region of interior drainage. Another very important difference is that torrential desert rains may give rise to sheet flows on a scale seldom, if ever, encountered in humid regions.

Rock Weathering and Erosion in Arid Regions

A traveler gifted with no more than ordinary powers of observation could hardly fail to notice that in desert regions long, comparatively straight mountain ranges often rise abruptly from relatively flat side slopes. Slopes of the surrounding plains range from about 1° to 6°; those of the mountain fronts range from about 15° to 90° and generally exceed 30°. Between the gentle slopes of the plains and the steep mountain fronts there is no transition, and it is difficult not to believe that two entirely unrelated processes are at work, one forming the mountain fronts and the other the surrounding plains. Studies by Lawson,[12] Bryan,[2] and others have shown this is not true. Both have been formed by weathering and stream action, and even if the alluvium on the lowlands were removed the mountain fronts would still rise just as abruptly from a comparatively flat rock-cut surface. A brief analysis of the processes involved in producing this type of landscape follows.

In regions of interior drainage, like the Basin and Range province, the area was originally broken into a series of steep-fronted mountain chains separated by intervening valleys. The originally very steep mountain slopes were attacked by weathering, which at first was most effective along their upper margins. Blocks loosened by weathering fell to the bases of the escarpments where they were further broken up by weathering and finally removed by running water. This process continued until the angle of slope was reduced to a point where weathering action was essentially uniform over the entire extent of the escarpments. From this stage onward, the slopes ceased to be lowered and the escarpments retreated parallel to themselves. The angle at which this point is reached depends mainly on joint spacing, which in turn is somewhat dependent on rock type. Massively jointed rocks, such as granites and gneisses, produce steep slopes (usually 35° or above), whereas closely jointed rocks like schists and felsites form comparatively gentle slopes. Stated briefly, the slopes of desert mountains represent an equilibrium between the various forces (weathering, gravity, sheet flow, etc.) tending to lower them and the resistance of the rocks to further lowering, whereas the slopes of the surfaces from which they rise are determined by the carrying capacities of the streams draining the mountain fronts. The sharp break in slope between the mountain fronts and adjacent plains is thought to be caused by erosion by sheets of water formed at the base of the escarpments as the velocity of runoff over the escarpment is abruptly checked on reaching the plains.[10] These sheets of water may flow in a laminar manner at first, but the increase in volume caused by rainfall on the plains soon converts this into turbulent flow, and under some conditions a hydraulic jump may be formed. Formation of a hydraulic jump and attendant cavitation erosion may account for the deep nicks that are sometimes found along the toes of the escarpments. In other words, flow conditions over the escarpments are analogous to those over spillways.

In closed basins, the debris shed from the retreating mountain fronts, after having been broken up still further by weathering, must be deposited eventually in lower parts of the basins. At an early stage of the process, the volume of the basins was relatively small and they filled very rapidly in comparison with retreat of the mountains that furnished the material. The rising masses of debris soon overlapped onto the platforms caused by retreat of the mountain fronts and protected the toes of the escarpments from further weathering. Consequently, the bases of the escarpments began to rise as they retreated, but as volume of the basins increased as the volume of debris furnished by retreat of the escarpments decreased this rise took place at an ever-decreasing rate, forming eventually a hyperbolic-shaped rock-cut surface that Lawson [12] called the "suballuvial bench" (see Fig. 83).

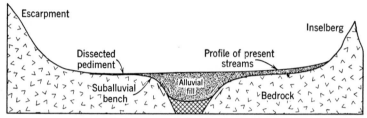

Fig. 83. Section across an alluviated desert valley. Vertical scale exaggerated. (Adapted from Kirk Bryan.[2])

The surface of the alluvium covering the suballuvial benches slopes downward toward the basins at an ever-decreasing rate. It also increases in thickness when followed in this direction and may often feather out completely near the mountain fronts, leaving smooth rock-cut surfaces covered by scattered patches of alluvium often referred to as "pediments." Pediments in this restricted sense are actually the exposed parts of Lawson's suballuvial benches. According to Davis,[5] during final stages of reduction of the mountain masses the detrital material feathers out completely and gives way to broadly convex rock-cut surfaces called "domes," which mark the last vestiges of former rugged mountains. There is some reason to doubt that these so-called domes actually exist, for careful study seems to indicate that most rock remnants left by erosion are concave in outline.[10]

In situations where the region is drained by through-flowing streams, recession of the mountain fronts takes place in the manner described above, but the debris is removed at about the same rate as it is formed and the rock-cut floors rise fairly uniformly toward the mountains, in the vicinity of which the angle of slope tends to increase. These rock-cut and/or alluviated surfaces are known as pediments. As time goes on pediments become wider and wider and may eventually coalesce to form an essentially flat surface known as a pediplain. As King [10] has pointed out, pediplains are analogous to peneplains, and most so-called peneplains should be thought of as having the characteristics of pediplains. Isolated concave-shaped masses of rock rising above

Fig. 84. Typical landscape in the Karroo region, South Africa. Wide pediment in foreground backed by steep escarpment developed on flat-lying sandstones and dolerites. (Photo by T. J. D. Fair, from L. C. King.[10])

a pediment or pediplain are known as *inselberge*, or "island mountains." They are entirely analogous to monadnocks, as similar features dotting a peneplain have long been called. Monadnocks have generally been thought of as having convex outlines caused by processes active during the so-called humid cycle of erosion. Convex residuals are found in nature, but their outlines may be attributed to exfoliation rather than stream work, and there seems to be no good reason for not considering the term "monadnock" as synonymous with *inselberg*. This should not be construed as meaning that monadnocks were necessarily formed under desert conditions, but rather as an illustration of the essential similarity of land forms developed in arid and humid regions. Pediplains, pediments, and associated *inselberge* cover vast areas of South and Central Africa, Mongolia, and Mexico. They also occur in the United States, although the term *inselberg* is seldom used here. The famous Cerro Azúcar at Rio de Janeiro is an *inselberg*. Marking as they do the end stages of processes that proceed at a very slow rate, pediplains and associated *inselberge* are far more impressive reminders of the length of geologic time than the Grand Canyon and similar spectacular features

Fig. 85.   Cerro Azúcar, a prominent landmark overlooking Rio de Janeiro, Brazil.   The alignment of this and smaller *inselberge* in background indicates the trend of a once-continuous mountain range.   (Pan American World Airways.)

which have been formed by geologic processes acting at nearly their maximum rates.

Stream Deposition in Arid Regions

Mountains cause the heated desert air to rise to elevations where it is cooled, and the heaviest precipitation in arid regions is usually in their vicinity.   Owing to the intermittent character of the rainfall, each rain finds an abundant supply of weathered rock ready for transportation.   Vast quantities of material ranging in size from clay to huge boulders are moved along stream channels, and on flat expanses sheet floods and mudflows may occur.[3]   Boulders many cubic yards in volume may be carried along in a thick suspension of silt and clay.   On reaching the foot of the mountains, stream velocities are abruptly checked, and the sedimentary load is dumped in the form of rudely stratified masses of poorly sorted debris ranging in size from clay to huge boulders. During early stages of reduction of the steep mountain fronts, this debris is deposited at canyon mouths to form steeply sloping conical-shaped masses known as alluvial cones.

Fig. 86. Coalescing alluvial fans on slopes of Calico Mountains, San Bernardino County, California. (Spence Air Photos.)

As reduction of the mountain masses proceeds, the stream beds are lowered. The upper portions of the alluvial cones are eroded and the material is deposited farther downslope. This process plus the debris supplied from the wasting mountain masses eventually transforms the originally steep alluvial cones into larger, more gently sloping deposits known as alluvial fans (Fig. 86). Individual fans may eventually coalesce to form vast plains that slope away from the mountain fronts. These features, known as piedmont alluvial fans, are common occurrences in desert regions. They are found in western United States, North Africa, and elsewhere. In Morocco, a piedmont alluvial fan, known as the Plain of Haouz, extends for a distance of nearly 300 miles along the north front of the High Atlas Mountains.

Aggrading alluvial cones and fans are characterized by branch-

ing networks of ephemeral stream channels that resemble blood
vessels. These channels branch from, rather than converge toward, the trunk streams (see Fig. 87). The reason for this behavior is simple. The water in the trunk channels carries so
much sediment that the slightest diminution in volume or loss of

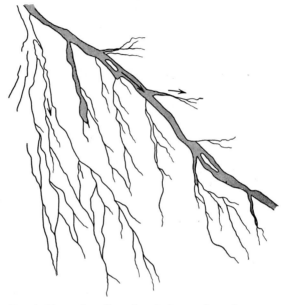

Fig. 87. Sketch illustrating diverging drainage channels on an aggrading
alluvial fan.

velocity is sufficient to cause deposition and blocking of the channels. Consequently, the discharge breaks through the sides of
the channels to form new channels which in turn are plugged by
deposition. This causes still further branching, and so on, until
the flow has either been dissipated or sunk into the porous alluvium. In brief, the upper levels of alluvial fans are subject to
flooding, whereas the lower levels are comparatively free from
floods.

The materials composing alluvial cones and fans generally reflect faithfully the rock types found in the basins of the streams
draining from the mountains. Regions of hard rocks, such as

granites and gneisses, generally furnish boulders, coarse gravels, and sands; shales, schists, and other soft rocks tend to produce silts and clays. The grain size of alluvial fan deposits generally tends to diminish away from the mountain fronts, but where the highlands contain large quantities of soft rocks this may not be true. Alluvial fan deposits are also subject to reworking by streams, and in closed basins, like those of the Basin and Range province, clays and silts may be laid down in ephemeral lakes, called playas, that form in the centers of the basins. Playa deposits may also contain soluble salts, some of which, like borax and potassium compounds, may have great commercial value.

Debris Control

The movement of debris along the upper slopes of alluvial cones and fans presents definite obstacles to development of these areas, and the problem is particularly acute in regions like southern California where urban and residential development of alluvial fans extends practically to the mountain fronts. It is estimated that the flood of 1934 at La Crescenta, north of Los Angeles, deposited 660,000 cubic yards of material.[6] It also killed 42 people and caused $5,000,000 of property damage. Remedial measures are thus well justified and are being carried out by the Corps of Engineers, U. S. Army, and other public agencies. The methods employed are: (1) terracing of slopes and establishment of an adequate cover of vegetation; (2) installation of drop structures; (3) building of check dams and barriers; and (4) construction of debris-catching basins.[6]

Watershed treatment by terracing and encouraging the growth of vegetation is perhaps the most permanent solution, but it is costly and considerable time is required for it to become effective. It is not applicable in rugged rocky regions where the soil cover is thin or lacking. Furthermore, plant growth may always be destroyed by fires, which in arid regions are very difficult to prevent.

Drop structures are used mainly in soil conservation, but they may also be useful where the debris comes mainly from canyons or gullies that have cut through otherwise stable drainage surfaces. They consist of a series of small check dams which cause the debris to be collected in a series of basins formed behind the barriers. The upstream basins fill first, and unless the deposits are removed the increased stream velocities in lower reaches may be detrimental.

Check dams consist of barriers erected across drainage channels to intercept debris. The storage is usually limited and their useful life short. Barriers are essentially large check dams.

Debris basins are hollows excavated near the mouths of canyons for the purpose of collecting and storing debris. They usually consist of bowl-shaped pits with U-shaped embankments on the downstream side. There is an inlet structure on the upper side and an overflow spillway on the downstream side. They are usually designed to hold only the debris from a single flood and must be cleaned out in order to function continuously. Owing to their accessibility, clearing the basins is much easier as a rule than in the case of check dams and barriers, which are usually located in rugged canyons.

Water Supplies in Arid Regions

Deserts are by no means lacking in water supplies, but the successful search for water entails a somewhat different approach than that used in more humid regions. The chief sources of supply are surface water, water in and along the lower boundary of alluvial deposits, and water in the bedrock.

There are few perennial streams in arid regions, and water from this source is necessarily limited. The ephemeral streams and gullies offer fairly large seasonal supplies, which can often be greatly increased by the construction of dams and reservoirs. Owing to the long periods that may elapse between rains and the relatively high costs, however, development of this source is often impractical. Small temporary surface water supplies can also be obtained from natural water holes. These are of two types, those in bedrock and those in alluvial materials. Water holes in bedrock, called rock tanks in southwestern United States, may be formed in various ways. They are often old plunge pools below rapids and waterfalls, or potholes formed by eddies. Depressions, called *charcos* in Mexico and southwestern United States, may be formed in alluvial materials by the concentration of sheet flow into gullies.[2] As the flow spreads out at the lower ends of the gullies, downward cutting ceases and elongated depressions are thus formed in the upper reaches. When floored with impervious materials, these depressions retain water for considerable periods after rains. Natural water holes are of importance to ranchers and absolutely vital to the survival of the nomadic peoples of the Old World desert regions. The Arab and Berber languages are, consequently, very rich in terms applying to the various types of water holes.

In places where pervious alluvial materials rest on impervious bedrock, large water supplies may be available, provided the configuration of the bedrock surface is such as to form adequate retention reservoirs. The rock-cut floors of the desert basins described in an earlier section often afford almost ideal catchment basins, and large water supplies can often be developed in the overlying alluvium. If such wells are to be used as a permanent source of supply, care must be taken not to exceed the safe yield. In North Africa, Persia, Afghanistan, and neighboring countries, inclined tunnels are driven into the upper levels of alluvial fans and sand dunes in order to intercept the water table.

Water in the bedrock of desert regions follows the same laws as those set forth in Chapter 5. Although the water table may be deep, subsurface water is available and often abundant. It is a remarkable fact that both the Sahara and the desert of Arabia are favored with unusually large amounts of pure artesian water, and it is the escape of such waters along joints and fissures that gives rise to many of the oases. The water stored in the artesian aquifers was probably derived from a former period of more humid climate, and the present rate of recharge must be very slow. Consequently, intelligent development of this source of water supply depends largely on maintaining a reasonable balance between the present recharge and the rate of withdrawal.

EROSION AND DEPOSITION BY THE WIND

Owing to their generally low water content and the scarcity of vegetation, desert soils have little cohesion. Particles 0.3 millimeter or less in diameter (medium-coarse sand and below) are moved readily by the wind, which in flat desert expanses has free play. With the exception of dust, materials carried by the wind seldom travel far and are generally deposited in more or less streamlined bodies known as dunes. The work of the wind is twofold: erosional and depositional, and, as in the case of stream action, the two generally go hand in hand. Wind action is responsible for many of the minor desert land forms, but, as previously explained, the major relief features are generally the result of crustal movements and stream action. The work of the wind is, nevertheless, of considerable importance to engineers, for it is involved in many practical problems, the chief of which is the control of drifting sands.

WIND TRANSPORTATION

Bagnold [1] has shown that materials may be carried by the wind in three ways: suspension, saltation, and surface creep. Suspended materials are held aloft by eddies and may thus travel great distances. Sand grains are generally too large to be held in suspension for more than very short periods, but dust particles may remain aloft for a considerable time. Dust from the Sahara has been known to fall in Paris. Sand grains whirled aloft by eddies drop along generally flat trajectories and on colliding with a hard surface may rebound to almost their original height. During rebound, they are carried forward by the wind and thus follow new trajectories similar to the original ones. It is thus apparent that sand particles moving over a hard surface may travel considerable distances, owing primarily to rebound. This process is known as saltation and is similar to, but not identical with, saltation along the bed of a stream. Grains rolled or pushed along the surface by impact of other grains moving by saltation are said to move by surface creep. Particles over 6 times greater in diameter than those transported by saltation can be moved in this fashion. It is probable that between one fourth and one fifth of the total weight of drifting sand moves by surface creep; the remainder is transported mainly by saltation. Sand particles seldom rise more than 6 feet above the ground during transit, and the bulk of the movement takes place within 18 inches of the ground surface.

To summarize, the transportation of sand by the wind takes place mainly along or near the surface of the ground, and the distance of travel during any particular storm is relatively short. Dust, on the other hand, may rise to great heights and travel long distances. The sand storms of popular story in which caravans are overwhelmed and buried are a considerable distortion of the facts. Dust storms may be dense enough to obscure the sun, but a sharp distinction should be made between the movement of dust particles held in suspension and the movement of sand grains along or close to the surface. Strictly speaking, sand storms do not occur.

WIND EROSION

In desert regions the wind winnows fine particles, chiefly silts and clays, from mixtures of fine and coarse materials leaving surface layers of gravels and coarse sands. In addition to sorting, the abrasive action of the wind flattens and smooths the upper surfaces of pebbles and boulders, forming gravel-strewn surfaces known as desert pavements. The constituent particles of desert pavements are often coated by a thin layer of black iron oxide caused by evaporation of waters charged with iron oxide contained in the interstices of the particles, forming what is known as "desert varnish." In the Sahara, extensive gravel-covered plains formed by the sorting action of the wind on river-deposited sediments are known as *reg*. Similar plains on the south bank of the Euphrates River, in southern Iraq and northern Saudi Arabia, are known as *dibdibba*. Desert pavements are readily crossed by wheeled vehicles and offer excellent sites for landing fields.

In areas where the soil consists of materials fine enough to be carried away by the wind, great bowl-shaped depressions may be formed. The lowering of a land surface by this process is known as deflation. It should be noted that deflation can lower a surface well below the level of stream channels, the only downward limit of the process being the water table. Depressions formed by wind erosion are widely distributed in North Africa and southern Asia. The "pans" of South Africa are of similar origin. The Qattara depression, about 100 miles west of Cairo, Egypt, is nearly 200 miles long and about 75 miles wide. The floor extends to 420 feet below sea level and is covered by salt marshes. Investigators seem to be agreed that this immense hollow has been excavated by the wind. The *sabkhas* of eastern Saudi Arabia bear certain, perhaps superficial, resemblances to deflation basins, but it is probable that those at low elevations along the coast represent old arms of the Persian Gulf and that those farther inland often occupy what are at least in part structural basins. These features may approach the dimensions of the Qattara depression but are generally much smaller. They are usually oval-shaped with the long axis paralleling the strike of the bedrock. Their floors consist mainly of silts and clays that probably were brought in by the wind and are held in place by the presence of moisture, for in many instances the silts and clays rest on layers of rock salt which in turn rest on highly concentrated brines. Moreover, *sabkha* surfaces often approximate very closely those of bodies of standing water.

Within a few feet of the ground surface, desert winds are charged with large quantities of sand. The natural sand blast thus formed acts in a highly selective manner. Weak rocks are rapidly cut away, causing harder materials to stand out in bold relief. The sharp edges of sand-blasted surfaces can be very destructive to automobile tires and about 2000 miles may often be very good mileage. Soft bands of alternating thin beds of hard and soft materials are cut away, causing the harder materials to stand out as long fluted ridges. If the strata are tilted, crenulated overhanging ridges may be formed, which in the deserts of central Asia are known as *yardangs*. The natural sand blast of the desert may also cut through the bases of telegraph poles and other wood or metal objects in contact with the ground, giving rise to a minor engineering problem.

DEPOSITION BY THE WIND

Deposits formed by the wind may be classified as sand shadows, sand drifts, and sand dunes. It is impossible in a brief chapter to give more than a generalized account of how these features are formed. Those interested in gaining detailed information should consult Bagnold's excellent treatment of the subject.[1]

Sand Shadows

Sand shadows are long tongue-like deposits formed on the lee sides of obstacles in the path of the wind. The air flowing past an obstacle is divided into streamlined and turbulent flow along a parabolic-shaped surface of discontinuity. Turbulence occurs inside the parabola, immediately behind the obstacle, but gradually fades away some distance downwind. Sand piled in front of the obstacle eventually reaches proportions where a part of it is swept around the sides. Grains passing around the sides are

Fig. 88. Sand shadows encroaching on a road in Algeria, French North Africa. (H. T. U. Smith.)

immediately swept into the turbulent zone behind the obstacle, where they accumulate. Two tongue-shaped wings on either side of the obstruction are formed first, but they eventually join to form a single elongated mound.

Sand Drifts

A sand drift may be formed in the lee of a gap between two obstacles or on the lee side of a cliff. The sand deposited in front of obstacles tends to drift toward any opening between them, and owing to the concentration of air currents through the opening it is readily swept through. After the air has passed through the gap, it is free to spread out. The velocity is checked, and a sand deposit results. After a time, this sand deposit itself becomes an obstacle to passage of the wind, thus accelerating growth of the deposit. The limit of its growth is reached when the frontage from which the sand can stream away as fast as it collects equals the frontage of the upwind collecting area from which the sand in the drift was derived.

Sand drifts below cliffs usually consist of a series of tongues projecting downwind from the mouths of re-entrants in the cliff face. The sand collects below these irregularities because they tend to concentrate the flow of air through them. The form of these deposits depends mainly on the direction of the wind relative to the cliff face. They eventually break away from the cliff and move across the plain, forming elongated subparallel mounds.

Sand Dunes

Sand dunes are deformable obstacles in the wind's path which, once formed, tend to perpetuate themselves. They differ from sand shadows and drifts in that no fixed obstacles are necessary to initiate their formation, and are found mainly in large, nearly flat desert expanses where as Bagnold[1] so aptly says: "Here, instead of finding chaos and disorder, the observer never fails to be amazed at a simplicity of form, an exactitude of repetition and a geometric order unknown in nature on a scale larger than that of crystalline structure. In places vast accumulations of sand weighing millions of tons move inexorably, in regular formation, over the surface of the country, growing, retaining their shape, even breeding, in a manner which, by its grotesque imitation of life, is vaguely disturbing to the imaginative mind." Sand dunes generally rise from a smooth rock or pebble-covered floor, and between the sand of the dunes and the materials of the intervening areas there is little or no transition. This almost uncanny ability of the wind to pile nearly every available grain of sand onto dunes is attributable to air currents set up by the dunes themselves.

Dunes may occur singly but are usually found in vast colonies that cover tens of thousands of square miles. About one seventh of the surface of the Sahara is covered by drifting sand dunes forming what is known as *erg*. The Grand Erg Occidental, the

Grand Erg Oriental, and the Libyan Erg each cover vast areas. The Egyptian Sand Sea and the Great Nefud of northern Saudi Arabia are other examples of large bodies of drifting sands. The dunes of the Rub Al Khali, or Empty Quarter, in the southern part of Saudi Arabia probably exceed 600 feet in height and have been known to move as much as 50 feet in a single storm. Sand

Fig. 89. Sand dunes on east side of Imperial Valley, Southern California. All-American Canal in foreground. (Spence Air Photos.)

dunes usually do not exceed 100 feet or so in height, however. They may move forward in a strong wind at a rate of 2 feet per hour. The material composing desert dunes is usually fine to very fine quartz sand. Dunes may also occur along coasts, where they tend to form elongated roughly parallel ridges that gradually move inland. Coastal dunes differ from desert dunes in many particulars and should not be confused with the latter.

Bagnold [1] has classified desert dunes into two chief types: one, an almost perfectly streamlined oval-shaped body whose symmetry is broken on the downwind side by a spoon-shaped indentation; the other, an elongated, more or less irregular ridge. Dunes of the first type are generally referred to as *barchans;* those of the second are often spoken of as *seifs,* from their fancied resemblance to an Arab sword.   Bagnold believes that *barchans* are formed by winds coming predominantly from one direction, whereas dunes of the *seif* type are formed when strong winds blow from a quarter other than that of the general drift of sand caused by the more persistent gentle winds.   *Barchans* may be transformed into *seifs* in the manner shown in Fig. 91.

Fig. 90.   Steep front of advancing sand dunes, White Sands National Monument, New Mexico.   These dunes are unusual in being composed mainly of gypsum rather than quartz sand.   (H. T. U. Smith.)

The spoon-shaped indentation on the downwind side of a *barchan*, mentioned in the preceding paragraph, is known as the slip face.   It is always present in *barchans* and may also be associated with *seifs*.   The slip face is formed by the tendency of the upper part of the lee face to advance faster than the lower, causing the sand to pile up on a slope that eventually exceeds the angle of repose, the maximum value of which is about 34°. As this critical angle is approached, the sand suddenly shears along a plane a few degrees flatter than the angle of deposition, and a miniature landslide results.   The surface along which the break occurred has the spoon-shaped outline of a typical shear failure and forms the slip face.   The sand involved in the failure falls to the base of the dune, where it forms a loose mass appropriately called "dry quicksand." As the dune migrates downwind, a succession of slip faces is formed, and the dry quicksands associated with them are more or less deeply buried under firmer accretion deposits formed on the upwind face. In places where the accretion deposits are thin, wheeled vehicles, or even a

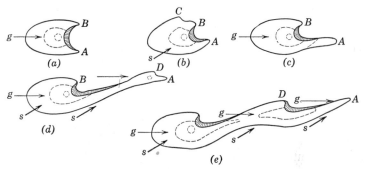

Fig. 91. Sketch illustrating formation of *seif*-type dunes from *barchans*. Arrow *g* indicates predominant wind direction; arrow *s* a strong cross wind alternating with *g*. When stage *D* (figure *d*) is reached, sand from wing tip *B* is added to that from wing tip *A* during both wind regimes to form a new dune. (After R. A. Bagnold, *The Physics of Blown Sand and Desert Dunes*, William Morrow & Co., New York. Copyrighted by Methuen & Co., Ltd., London, 1941.)

man on foot, may suddenly sink deep into the underlying loose materials. To put it briefly, sand dunes are very treacherous surfaces for the passage of traffic, wheeled vehicles particularly.

Control of Drifting Sands

The control of drifting sands formerly received little or no systematic study, and the methods employed were often of a "hit or miss" character. The over-all futility of such methods is well illustrated in certain localities of North Africa, where for centuries the natives have been carrying away in baskets the sands which threaten to engulf their oases. This direct and primitive method of sand control resolves itself into a labor of Sisyphus and should not be considered as other than a short-term expedient. Attempts to contain drifting sands by fixed solid barriers also generally belong in the category of futile expedients.

It is now fairly generally realized that effective control of drifting sands depends on an understanding of the aerodynamic principles governing the origin, growth, and migration of dunes. When these are properly understood, it is often possible to make the wind work toward, rather than against, the desired objective. Many methods of sand control are available; the choice between them often depends on the cost of the protective measure and the nature of the object to be protected. The methods employed to protect a long stretch of highway, for example, are not necessarily the same as those used to protect a single building. Often a combination of two or more methods is the best solution to the problem. Some of the more effective methods of sand control are planting, fencing, paving, and oiling. Brief descriptions follow.

Planting. In situations where sufficient water is available for the growth of plants, dunes can often be permanently stabilized by the planting of grasses, legumes, shrubs, and trees. In large desert areas, it is usually expensive and, therefore, applicable mainly to protection of areas where pleasing appearance is a factor. Planting is best adapted to coastal dune areas and has been used with success in the Cape Cod region [11] and similar localities. An outstanding example of the effectiveness of the method is found in the Landes region [14] of southwestern France, along the shores of the Bay of Biscay. Here thousands of square miles of formerly fever-ridden swamps and shifting sand dunes have been transformed into a growing productive area by the simple method of planting thick stands of maritime pines. The trees have not only stabilized the dunes, but their roots have broken up an iron oxide layer a foot or so below the surface. The improved drainage resulting has aided in the elimination of swamps.

Fencing. If the symmetry of a dune is disturbed, its migration is also affected. A simple open lath-type fence will accomplish this result by causing the sand to heap up on both the upwind and downwind sides. Eventually, the accumulations become so large that the fence ceases to be effective and it must be moved, or new ones placed.

Paving. Falling sand grains rebound more uniformly from a hard smooth surface than from a soft irregular one. Nearly all the grains rebound above the stagnant air bordering the ground surface and are swept along in the fast-moving upper air. Consequently, if a surface is paved, or oiled, drifting sands tend to migrate across it without forming thick or irregular accumulations. A good example of a practical application of this principle is seen in the prevention of irregular sand deposits on highways by oiling or paving a suitable expanse of the shoulder area. Paving may also be used in conjunction with other methods, where its chief function is to prevent the wind from becoming loaded with sand before arriving at a locality to be protected.

Oiling. In localities where oil is obtainable at reasonable cost, drifting sands may be stabilized by saturating the surface with oil. Selective oiling may also be employed in the removal of sand dunes. For example, if the symmetry of a dune of the *barchan* type is destroyed by oiling either the central portion or both wings, the dune is eventually blown away. Excellent descriptions of the experiences of the Arabian American Oil Company with this and other methods of sand control are contained in a paper by Kerr and Nigra,[9] which should be read by all those faced with actual problems of controlling drifting sands.

REFERENCES

1. R. A. Bagnold, *The Physics of Blown Sand and Desert Dunes,* William Morrow & Co., New York. Copyrighted by Methuen & Co., Ltd., London, 1941.
2. Kirk Bryan, The Papago Country, Arizona, *U. S. Geol. Survey, Water Supply Paper* 499, 1925.
3. W. D. Chawner, Alluvial-fan Flooding; the Montrose, California, Flood of 1934, *Geog. Rev.,* Vol. 25, 1935.
4. C. A. Cotton, *Geomorphology, an Introduction to the Study of Landscape,* John Wiley & Sons, New York, 1950.
5. W. M. Davis, Granitic Domes of the Mohave Desert, California, *Trans. San Diego Soc. Nat. Hist.,* Vol. 7, No. 20, 1933.
6. Burnham H. Dodge, Debris Control, in *Applied Sedimentation,* John Wiley & Sons, New York, 1950.
7. E. F. Gautier, *Sahara, the Great Desert,* Translated from the French by Dorothy F. Mayhew, Columbia University Press, New York, 1935.
8. Arthur Holmes, *Principles of Physical Geology,* Ronald Press Co., New York, 1945.
9. Richard C. Kerr and John O. Nigra, Eolian Sand Control, *Bull. Am. Assoc. Petroleum Geol.,* Vol. 36, No. 8, 1952.

10. Lester C. King, Canons of Landscape Evolution, *Bull. Geol. Soc. Amer.*, Vol. 64, No. 7, 1953.

11. Karol Kucinski and Walter S. Eisenmenger, Sand Dune Stabilization on Cape Cod, *Econ. Geog.*, Vol. 19, No. 2, 1943.

12. A. C. Lawson, The Epigene Profiles of the Desert, *Univ. Calif. Publs.*, *Bull. Dept. Geol.*, Vol. 9, No. 3, 1915.

13. Maurice Lelubre, Conditions structurales et formes de relief dans le Sahara, *Travaux de l'institut de recherches Sahariennes*, Tome VIII, 1952. Special number published for the 19th International Geological Congress in Algiers.

14. W. C. Lowdermilk, Les Landes, Where Over Three Quarters of a Century France Has Transformed Vast Mobile Sand Dunes and Waste Marshland into Rich Pine Producing Area, *Am. Forests*, Vol. 50, 1944.

15. A. M. Sears, *Deserts on the March*, Oklahoma University Press, Norman, 1935.

CHAPTER 9

Shore Lines and Beaches

Erosion and deposition along shore lines have been active from very early geologic time, but did not constitute an important engineering problem until economic development of shore lines reached a fairly advanced stage. It is estimated that in the United States alone the equivalent of a one-foot strip over the entire 52,000 miles of shore line is carried away each year, and some beach areas have a valuation of $20,000 per front foot.[9] The problem of shore line protection is complex, and it is not surprising that early attempts often failed, owing to inadequate appreciation of the varied factors involved. It is now fairly generally realized that the success of shore protection works depends largely on an understanding of the natural processes involved, and when adequately understood these forces can often be directed so as to work with rather than against the engineer. Establishment of the Beach Erosion Board in the early 1930's as a research, coordinating, and advisory body is tangible evidence of this broader point of view, and it marked a long step forward in the progress of coastal engineering in the United States.

Shore lines were one of the first subjects to be studied by geologists, but much geologic study was, and still is, concentrated on establishing and elaborating cycles of marine erosion and deposition and the refinement of criteria for distinguishing rising and sinking coast lines. Geologic-engineering studies of shore lines are less concerned with such matters than with the processes at work, and are especially concerned with the rates at which these processes operate. The results of such studies suggest that many early generalizations and dogmatisms should be abandoned. Much has been accomplished, but a great deal remains to be done. More extensive and accurate information concerning the transportation and deposition of materials by waves and currents is particularly needed.

OCEAN WAVES AND CURRENTS

Waves and currents are the forces involved in the evolution of shore lines, and at least a general understanding of their characteristics is essential to those concerned with coastal engineering.

OCEAN WAVES

Figure 92 presents the essential concepts and definitions involved in a discussion of wave motion. It should be noted that, although the wave shape advances, the water particles merely oscillate in a nearly circular orbit with very little forward move-

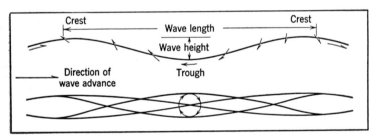

Fig. 92. Diagram illustrating oscillatory wave motion. (After F. P. Shepard, *Submarine Geology*, copyrighted by Harper & Brothers, 1948.)

ment. The very small net forward movement of water particles associated with the passage of a wave is known as the mass transport, but for the present purpose this may be neglected. In other words, in a wave motion of the type illustrated only energy is transferred from one point to another. Other definitions not shown on the illustration are: amplitude, equal to one half of the wave height; velocity, or the speed at which the wave advances; and period, or the time elapsing between the passage of successive crests or troughs. The rate of movement past a given point is equal to the wave length divided by the period. According to Mason,[9] the energy is proportional to the product of the period squared and the height squared. At a depth of one quarter of the wave length the diameter of the orbit of particle movement is only about one tenth of the wave height, and at a depth of one half of the wave length particle motion is essentially nil.[12] That some such relationship exists is indicated by the fact that a submerged submarine is not subjected to the same violent motion as are surface craft.

Height. Waves originate in the open ocean and are caused by the interaction of winds and the water surface. The height of ocean waves depends on the wind velocity and expanse of water across which steady winds travel (fetch). The higher the wind velocity and the longer the fetch the higher will be the wave, but at a height of about 50 feet a further increase in wind velocity serves mainly to blow away the wave crests. Consequently, about 50 feet is the limiting height of waves in the open ocean, but under exceptional storm conditions breaking waves may attain a height of about 100 feet. Waves in the open ocean are known as sea, and the continuation of these waves beyond the area affected by the wind is known as swell. Breaking waves or swell are known as surf. Seas and surf are generally irregular and confused in shape, but swells tend to be smooth and regular. The longest ocean waves ever observed were about 2600 feet long, moved at a rate of 68 knots, and had a period of 23 seconds.[17] The total energy of such waves is enormous, and would exert pressures of several tons per square foot if expended against shore installations.

Breaking Waves. When waves approach a shore they begin to drag on the bottom when the depth of water is about one half of the wave length. The wave crests continue to move forward, but the movement of water particles changes from a circular to an elliptical orbit. The wave length is shortened, but the period remains the same. Height of the waves appears to increase, but except under special conditions the actual increase is comparatively small. As the water continues to shoal, the fronts of the advancing waves become steeper and steeper until they break to form surf. Waves that pile up until the fronts become concave and topple forward are known as plunging waves; those that slide forward without toppling of the crests are called gliding waves. Little energy is lost until a wave reaches the plunge point, after which the energy loss is rapid and reaches zero at the limit of the rush up the shore.

Refraction and Related Phenomena. Ocean waves, like other forms of wave motion, may be refracted or may form various types of interference patterns. Refraction is caused by lower-

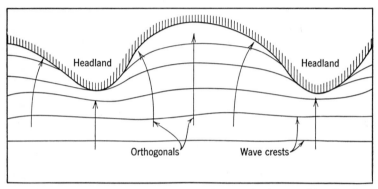

Fig. 93. Sketch illustrating refraction of waves by bays and headlands.

ing of the velocity of propagation by frictional drag along the bottom. As a result, the portion of a wave in deeper water moves faster than the portion in shallower water, causing the wave front to become curved. Owing to refraction, wave crests seldom make an angle greater than 20° to the shore at the plunge line, regardless of their angle of approach in deeper water.[7] Refraction is best visualized by drawing lines, known as orthogonals, at right angles to the advancing wave fronts. Convergence of orthogonals indicates a concentration of wave energy; divergence indicates a dissipation of energy. Figure 93 illustrates the concentration of energy on headlands bordering a broad and relatively deep bay. If more than one set of waves of the same wave length approaches a shore they may reinforce or interfere with each other. In the event that both sets are in phase (moving in the same direction in the cycle of oscillation) and have the same wave length reinforcement results; if they move in opposite directions in the cycle (out of phase) they are said to interfere. Reinforcement adds to the height of the compound wave; interference reduces it. If two sets of waves have different wave lengths, they combine to form what are known as beats.

Tsunamis. A special type of wave, also mentioned in Chapter 13, known variously as a tidal wave or *tsunami*, results from submarine earthquakes or volcanic eruptions. Tsunamis travel thousands of miles at speeds up to several hundreds of miles per hour and have been known to sweep over coasts with devastating results. The height of these waves is not very great in the open ocean, but configuration of the coast line may cause this to be multiplied many times.

OCEANIC AND LONGSHORE CURRENTS

The mass movement of sea water relative to some fixed point is known as a current. Currents are of two principal types: (1) oceanic currents which originate in the open ocean, and (2) longshore currents which result from the interaction of waves and shore lines. The longshore currents are of much the greater importance from an engineering point of view.

Oceanic Currents. Origin of great oceanic currents like the Gulf Stream, which originates in tropical latitudes and flows to the Arctic Ocean like a mighty river, has been much disputed, but prevailing winds, temperature gradients, rotation of the earth, and shore-line configuration all seem to play a part. The velocity is often surprisingly high, 3 to over 4 knots in certain portions of the Gulf Stream, and the transition zone between the current and surrounding water may be exceedingly narrow, sometimes less than the length of a ship. Some oceanic currents, the Gulf Stream for example, are warm; others like the Labrador Current are cold. Warm and cold oceanic currents have considerable influence on the climates of large continental areas. The direction and speed of oceanic currents are indicated on nautical charts under the headings of set and drift.

Longshore Currents. When waves break obliquely across a straight stretch of shore line, a longshore current is set up which flows away from the direction of wave approach. The southwest shore of Lake Michigan furnishes a good example. Here the waves generally approach the shore from the northeast and a

southward longshore current results. Sand carried by this current
eventually comes to rest at the south end of the lake where it is
reworked by the wind to form dunes. The city of Gary, Indiana, is built on these dunes, and sand derived from leveling the
site was used in embankments of elevated railways in the south
Chicago area. Longshore currents are also set up by impingement
of waves on irregular shore lines (see Fig. 94).

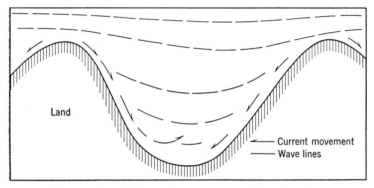

Fig. 94.   Sketch illustrating current conditions in a bay.

Rip Currents and Undertow. It is a common belief that after
waves have spent their force in uprush over a beach the returning
water sets up a current known as undertow. It is unlikely that
strong return currents are created in this fashion, for most of the
energy is dissipated too rapidly. Tide rips, or rip currents, are
very common wherever large breakers are present, but they are
surface or near-surface currents. Shepard [12] has proposed that
they be called rip currents, inasmuch as no tidal relationship exists.
The speeds of such currents are estimated at as much as 2 knots.
Rip currents are caused by the meeting of longshore currents,
which turn seaward forming a zone which is rarely more than 100
feet in width. The outward-moving current generally is dissipated as mushrooming eddies beyond the line of breakers. Research has shown that such currents occur chiefly where the wave
advance is essentially normal to the shore line. Rip currents are
irregular, and inside the breaker belt there may be very shallow
water on one side or the other, or even on both sides of the
current.

Tidal Currents. Tidal currents are caused by rise and fall of the tides. They generally recur at regular and predictable intervals, and the tidal range determined on a beach in one tidal cycle may generally be assumed to be relatively accurate for a period of weeks or months. Along straight coast lines, tidal currents

Fig. 95. Photograph illustrating rapid erosion of an exposed headland, Coos Head, Oregon. Materials eroded from the headland go to make up the beaches shown on either side. (U. S. Coast Guard.)

may not cause any net transportation of sediment toward or away from the shore, but in estuaries fed by large rivers the volume of fresh water added during rising tide causes at least portions of the outgoing current to be faster than the incoming flow. Consequently, in the improvement of some types of tidal rivers care should be exercised not to interfere with the free movement of the tide in and out of the river channel. Experience shows that poorly planned regulation works may result in accelerated deposition in those parts of the estuary from which tidal scour and fresh-

water discharge have been diverted. This deposition reduces the
tidal capacity of the estuary with consequent diminution of the
tidal flow passing the outlet. The result is a rapid extension of
deposition in a seaward direction. Jetties at the mouths of tidal
rivers should be spaced properly, and in the case of smaller
streams it may be necessary to substitute converging breakwaters.
Dredging of offshore bars which tend to obstruct entrance of the
tide may also be beneficial.

Density Currents. These phenomena, also called suspension
currents, owe their existence to suspended sedimentary materials,
and are found in reservoirs and lakes as well as in the ocean. Such
currents may flow along the bottom or at the contacts of layers
of different densities. They are commonly encountered off the
mouths of large rivers where they may extend seaward for many
miles. It is thought that density currents may be responsible
for the transportation of vast amounts of fine-grained materials,
and it has even been suggested that they have cut the vast submarine canyons encountered off many coasts.

WAVE EROSION

Wave erosion is a phenomenon too familiar to need extended
description and a good account of its place in the origin of coastal
scenery may be found in reference 5 cited at the end of this
chapter. Consequently, the present discussion is confined to
aspects of the subject that are of greatest interest to the civil
engineer.

The Profile of Equilibrium

Along an eroding shore line, the waves eventually cut a notch,
or wave-cut bench (Fig. 96). The material removed from the
wave-cut bench is thought to be deposited on the seaward side
to form the so-called wave-built terrace. The submarine platform resulting is called the profile of equilibrium, and at any time
its slope is adjusted to the velocity of currents and the type and
quantity of bottom materials. Interference with any element of
the profile results in readjustments of other elements, and dredging of sands and gravels offshore has been known to result in
accelerated erosion of the shore line. It is apparent that the longer

sea level stands at a given position the broader the platform will become. Given sufficient time, the stage must eventually be reached where the incoming waves will exhaust all their energy in traveling over the platform and can no longer accomplish erosion of the landward side. Owing to the numerous factors involved and the delicate nature of the equilibrium, however, this final stage appears to be seldom attained in nature. The narrowness of the wave-cut platforms along many coasts is apparently the result of changes in sea level incident to melting of the glaciers

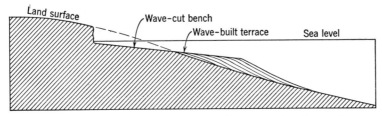

Fig. 96. Section illustrating development of wave-cut benches and wave-built terraces.

which some 30,000 or more years ago covered large portions of the continents.

Interruptions of the Cycle of Wave Erosion. As in streams, the process of wave erosion may be interrupted at almost any stage. If the coast is uplifted (or sea level drops), the former wave-cut and wave-built platforms are raised above sea level forming what is known as a marine terrace. This terrace will at first be covered by a thin veneer of marine sediments, but after uplift streams draining from the land will cover these deposits with a layer of non-marine material, which may be called the non-marine terrace cover as distinguished from the underlying marine deposits.[4] While these events are transpiring on land a new wave-cut and wave-built platform is in process of formation offshore. Thus, repeated uplifts may give rise to multiple terraces. Uplifted terraces, resembling gigantic steps, are common features along many coasts and are especially striking along the southern California coast and on the north coast of Cuba. Marine terraces tend to parallel the shore line, and they are characterized by fairly level topography. They are, therefore, much favored by civil engineers in the alignment of transportation routes.

If the coast should subside, or sea level rise, the sea will enter shallow embayments and the mouths of rivers forming numerous bays and estuaries. In regions of strong relief, a very irregular shore line results. Such coasts are usually well provided with harbors, of which San Francisco Bay is an excellent example. After subsidence is complete, waves attack the coast, the exposed headlands are cut back, and in due course all irregularities are eliminated and a relatively straight shore line is established.

Coral Reefs and Atolls

In tropical regions coral reefs are very common. Corals are animals that live only in warm, clear ocean water of moderate depth; they secrete calcium carbonate to form external supports, or "anchors." With their continued growth a calcareous deposit known as a coral reef is formed. Because of their restricted range of habitat, coral reefs are sensitive indicators of earth movements and climatic changes and are of great importance in working out the geological history of a coast line. In the South Pacific and Indian Oceans three types of coral reefs are known. They are: (1) fringing reefs, (2) barrier reefs, and (3) atolls. Fringing reefs are found near the coast and below the level of high tide. Generally, they tend to parallel the shore. Growth of corals takes place mainly on the seaward side in both upward and outward directions. Barrier reefs are similar to the fringing type, except that on the landward side they do not come into close proximity of the shore, the space between the reef and shore being occupied by a lagoon. Atolls are roughly circular-shaped coral reefs, the central portions of which are occupied by relatively shallow lagoons, but on the seaward side the ocean descends to considerable depth.

Origin. Charles Darwin observed that among the volcanic islands of the South Pacific there appear to be all gradations from fringing reefs to barrier reefs and atolls. He suggested that barrier reefs and atolls originate as fringing reefs, but as an island platform subsided the corals grew upward and outward, leaving the space between the reef and diminishing land area as a lagoon.

In time, subsidence of islands progressed to the point where they disappeared completely, leaving only atolls to mark their former presence.

Daly [3] has proposed that the colder climate and lowered sea level of the glacial epoch resulted in the destruction of the preglacial fringing reefs. The islands were thus deprived of the pro-

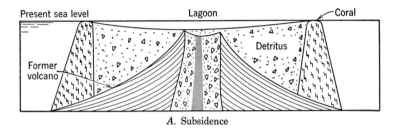

A. Subsidence

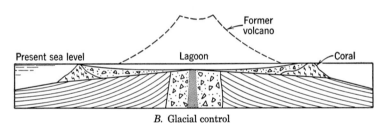

B. Glacial control

Fig. 97. Sections illustrating two theories of atoll origin. Vertical scale greatly exaggerated.

tection afforded by the reefs and subjected to wave erosion, which in some islands formed terraces on their flanks and in others cut completely across them to form submarine platforms. As the climate became warmer, the glaciers began to melt, causing sea level to rise and coral growth to become re-established on the outer margins of the wave-cut platforms and benches. Barrier reefs were formed on the peripheries of islands notched by terraces, whereas atolls were developed on submarine platforms marking the existence of former islands.

Borings at Bikini Atoll [3] give indications that the coral deposits extend to very considerable depths below sea level, much too deep to be accounted for solely by lowering of sea level. The evidence at hand seems to indicate subsidence of at least 500 feet, some or all of which may be preglacial. Origin of barrier reefs and atolls is doubtlessly complex, but in this instance at least a certain amount of subsidence seems definitely established.

DEPOSITION BY WAVES AND CURRENTS

The movements of sediments onto, away from, or along the shore greatly modify shore and harbor areas and are of particular interest and importance to the civil engineer, for they affect drastically protective works and other installations. Consequently, it is essential for him to understand these phenomena. Beaches, for example, are not the permanent features they appear to be. Some go through regular cycles of accretion and depletion but over a period of years may show no appreciable change. On the other hand, a beach may be entirely removed during a single severe storm, leaving only a rocky and boulder-strewn shore.

Detailed studies of transportation and deposition along shore lines indicate that very little, if any, material is derived from the deep waters offshore. Most of it is derived from erosion of the coast or is carried in by rivers and longshore drift. It follows, therefore, that studies of transportation and deposition of sediments cannot be confined to an arbitrarily selected expanse of shore but must be based on a natural subdivision. Mason [9] has called this natural subdivision the physiographic unit, which may be defined as "a shore area so limited that shore phenomena within the area are not affected by the physical conditions in adjacent areas." In other words, the physiographic unit corresponds roughly to what the physicist would call an isolated system. Monterey Bay, south of San Francisco, California, is a good example of a physiographic unit. Once the physiographic unit is defined, the over-all tendency toward erosion or deposition in a specified area will depend on the balance between the rate of supply and the rate of loss over a specified period of time, in other words, on what Mason has called the "material-energy" balance. Factors involved in the material-energy balance are: (1) rates of littoral drift to and from the area; (2) rates of supply and loss from and to the adjacent sea bottom; (3) rates of supply and loss from and to the adjacent land; and (4) rate of supply of material contributed by streams and other drainage features.

Beaches

Many coasts show few, if any, beaches, except at the heads of bays and near stream mouths. These are chiefly rocky and cliffed shore lines. On the other hand, beaches are rather perma-

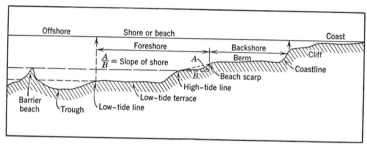

Fig. 98. Section illustrating terminology used to describe beach slopes and associated features. (After F. P. Shepard, *Submarine Geology,* copyrighted by Harper & Brothers, 1948.)

nent features along low coasts, like those of Florida and the Gulf states. Here they may be wide and continue for many miles. The origin of beaches at the heads of bays is found in the applica-

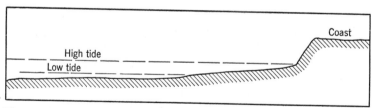

Fig. 99. Section through a narrow beach with only a foreshore. (Adapted from F. P. Shepard, *Submarine Geology,* copyrighted by Harper & Brothers, 1948.)

tion of the principles of wave refraction and the longshore currents, or littoral drift, generated by the oblique approach of waves along the sides of the headlands (see Fig. 94). Long continuous beaches owe their existence to an abundant source of sand, such as a stream mouth or rapidly eroding headlands which supply a steady stream of debris.

Profiles. There are three general patterns of beach profiles. (1) A narrow beach with a regular slope seaward with only a foreshore (Fig. 99). The beach slope flattens seaward. (2) A

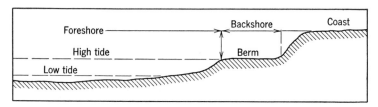

Fig. 100. Section through a beach with a backshore. (Adapted from F. P. Shepard, *Submarine Geology*, copyrighted by Harper & Brothers, 1948.)

beach having a backshore in addition to a foreshore (Fig. 100). The berm is seldom subject to submergence by waves. (3) A beach with or without berms, but having a terrace exposed at low tide (Fig. 101). Breakers may form where waves move over

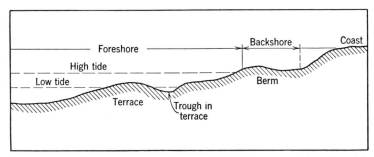

Fig. 101. Section through a beach having a terrace exposed at low tide. (Adapted from F. P. Shepard, *Submarine Geology*, copyrighted by Harper & Brothers, 1948.)

the terrace, and when a trough is formed on the terrace inside of the point where the waves break strong currents may develop. Beaches are said to be in equilibrium when the continued attack of waves over an extended period produces relatively little change in beach characteristics. It seems doubtful that absolute equilibrium is ever attained, owing to the constantly changing forces applied by winds, waves, currents, and tides.

Beach Slopes and Material Grain Size. The beach slope, both in the foreshore and offshore areas, is intimately related to the grain size of the sedimentary materials involved. The coarser debris forms steep slopes, and in general the finer the materials the gentler the slopes. Temporary irregularities may occur where storm waves steepen a berm or deepen a trough on the foreshore. Materials eroded from a beach are subjected to continuous handling not only inside the plunge point of the waves but also as they are transported to and from and along the shore. The particles are steadily reduced in size both by abrasion and solution. Because of the difference in rock composition and the shapes of individual particles, a considerable amount of sorting and lensing occurs. Pieces of uniform size and shape tend to remain together in thin layers, patches, and lentils. In the backshore areas, fine-grained sands tend to form relatively flat to gently seaward-, or even landward-, sloping berms. Cobbles and coarse debris generally form landward-sloping berms. Whereas there may be many variations and transitional relationships between particle sizes on a beach, there are often abrupt contacts between coarse and fine debris. Table VI below illustrates the relationships between grain size and beach slopes.

TABLE VI. RELATIONSHIP BETWEEN GRAIN SIZE AND BEACH SLOPES

(After F. P. Shepard, *Submarine Geology*, copyrighted by Harper & Brothers, New York, 1948.)

Constituents	Median Diameter, in Millimeters	Depositional Slopes	
		Degrees	Per Cent
Very fine sand	0.06		
Fine sand	0.12	2	3.5
Medium sand	0.25	4	7.0
Coarse sand	0.5	8	14.0
Gravel	2.0	12	21.3
Cobble	64.0	20	36.4
Boulder	256.0		

Growth and Retreat. Changes in beach characteristics may be rapid or slow, and along the coasts of North America there is a striking annual cycle.[14] This results from the occurrence of a stormy season and high waves in winter during which the beaches are eroded and a summer season of gentle seas when the beaches are restored. A less pronounced cycle is associated with the spring and neap tides. The first effects of beach retreat are seen on the berm edges where, depending on severity of wave attack, the berm regresses gradually without appreciable change in slope,

Fig. 102. A narrow and very steep beach composed of large cobbles, Antipodes Island, South Pacific Ocean. (U. S. Coast Guard.)

or a scarp is developed. The beach slope and degree of consolidation of the sands and gravels also have a marked effect on the rapidity and nature of beach retreat. The steeper beaches, composed of coarser and less compact materials, retreat much more rapidly than those having gentler slopes. During periods of storm, offshore bars and troughs retreat seaward and develop more relief. Beaches grow during periods of mild weather and small tidal fluctuations. Much of the finer suspended material is dropped as water seeps into the beach, owing to the less violent runoff. Filling of the voids between sand grains with fine materials makes a more compact beach that is less susceptible to excessive surface sculpture. Deposition in the foreshore area may produce a double berm.

Rip currents locally may move large quantities of sand toward the sea. These materials are derived from longshore currents, and

Fig. 103. Photograph illustrating effect of groins on beach growth near Tuna Canyon, California. (Spence Air Photos.)

the net movement of debris along a beach is usually much greater than either toward or away from the shore. Drift of sand is most pronounced close to the shore, and movements normal to the shore are greatest where the water is most turbulent, near the plunge point. The effectiveness of longshore currents is best appreciated in areas where engineering works have been constructed. Here deposition of materials is greatest on the up-current side of the structures (see Fig. 103).

Beach Cusps. These are roughly triangular-shaped beach projections the apices of which point in a seaward direction.

Depending on the physical characteristics of the constituent materials, they may have a relief of a few inches when the debris is fine, or several feet when it is coarse. They are more common along coarse beaches and are found where the wave approach is normal to the beach, or where sets of waves intersect in such a manner that longshore currents are not formed. Cusps develop where concentration of runoff from a beach forms cones of debris which are modified on their seaward margins and surfaces by wave action. Any change in wave direction resulting in an oblique approach is sufficient to efface the cusps. These features may grade in spacing, shape, and size from very small bodies to great cuspate features like Capes Hatteras and Canaveral on the Carolina coast.

Sand Bars, Barrier Beaches, and Keys

Much confusion exists with regard to these terms, and in many cases they have been used more or less synonymously. Shepard [13] has proposed the following revised nomenclature.

Bars and sand ridges are shoals that are covered by water at least at high tide. Barrier beaches are sand ridges which stand above the level of high tide. They trend parallel to the coastline and are separated from it by a lagoon. This definition applies to islands and spits, so far as their position fits the description. A barrier island is similar to a barrier beach, but generally has multiple ridges and dunes. Areas covered by vegetation, and often swamps, may occur on the lagoon side. A barrier spit differs from a barrier island only in that one end is attached to the mainland. Spits are often temporary features, for their connection with the mainland may be severed by changes brought about by storms. Deposition in the lagoonal area may also eventually unite a barrier spit to the mainland. A bay barrier is formed when a spit grows completely across the mouth of a bay and closes it off from open water. A bay head barrier is the term applied to a similar feature near the head of a bay. A barrier chain is the term used to designate a continuous chain of barrier islands and spits along a considerable section of coast. Longshore bars are sand ridges extending parallel to the shore line and are generally submerged at high tide. They may form single or multiple ridges and may be observed either from the air or may be inferred from the presence of lines of breakers that remain in constant position. Transverse bars trend approximately at right angles to the shore line, and like other types of bars they are usually submerged at high tide. Reticulated bars are found chiefly in bays and lagoons. They form intersecting patterns in which each set of bars is oriented diagonally to the shore line.

Sand keys, or sand cays, are small islands, not particularly elongated, trending parallel to the shore. The term "key" is also applied to elevated coral reefs and barrier islands. Lunate bars and sand keys are crescent-shaped bars and small islands, usually located off "passes" between barrier islands and harbor entrances. The surface is modified by waves and currents and is somewhat variable. Channels may be maintained at the ends or even through an aggregation of these bars.

SHORE-LINE PROTECTION

As previously mentioned, it is fairly generally realized that all types of coastal engineering must take into consideration the natural processes acting on shore lines, and few projects are undertaken without careful study. The magnitude of the project and land values involved govern, to a large extent, the scope of these preliminary studies. A study of a shore line for engineering purposes should include an investigation of its past history and the present conditions. The source and rate of movement and accumulation of shore deposits are particularly important, and it is in this phase of the study that geology is most useful.

TYPES OF SHORE-LINE STRUCTURES

Shore-line structures may be built for various purposes, the most important of which are the improvement of navigation and the prevention of property damage associated with wave attack. In some situations these purposes may be achieved without the use of structures; for example, the deepening of harbors and river mouths by dredging and the protection, or even construction, of beaches by direct placement of sand or gravel. In places where there is some assurance that the artificially placed ma-

Fig. 104. Model constructed at the Waterways Experiment Station for the study of sedimentation in Charleston Harbor, South Carolina. The technician is checking elevation of the water surface. (Corps of Engineers, U. S. Army.)

terials will not be removed by waves and currents, the "construction" of beaches is feasible and has actually been undertaken at Biloxi, Mississippi, and other places.

Seawalls and Bulkheads

Seawalls are massive structures placed parallel to a particularly valuable strip of shore for the primary purpose of preventing damage by wave attack. They may also protect the shore against slumping, but this purpose is generally a secondary one. Bulkheads are the retaining walls of the water front, for their chief purpose is to hold up the banks. They may also furnish a measure of protection against wave attack. The location of seawalls and bulkheads depends on the characteristics of the area, the intensity of wave action, foundation conditions, shoreline slope, etc. The alignment should be kept as straight as possible, and sharp angles should be avoided. The proper height depends on the elevation of the ground on the landward side; the maximum heights and frequencies of storm tides; wave heights; and the probable effects of other structures, such as groins and breakwaters. Experience shows that bulkheads should not be relied on for shore protection; and seawalls should generally not be considered as more than a final line of defense during especially severe storms. Seawalls and bulkheads generally do not aid beach accretion, and when relied on as an important means of protection against wave attack the implication is that associated structures are poorly conceived or designed.

Breakwaters

These are massive masonry structures (usually rubble or blocks) placed some distance from the shore in order to prevent wave attack or to provide anchorage for vessels. Occasionally, converging breakwaters may be placed at the mouths of tidal rivers in order to direct the maximum tidal flow possible into the river. When parallel breakwaters are used for shore protection, they tend to create a zone of objectionable stagnant water in the protected region, while no change may be brought about in the deep water outside the breakwater.

Groins

These structures are placed at an angle to the shore line and are designed to cause the creation of a beach or the widening of existing beaches. Improvement of the beach may be desired mainly for protection against wave attack, or merely for recreational purposes. When properly designed and spaced, they cause deposition of materials transported by littoral currents on the up-current side of the structure. Scour may occur on the outer end and on the lee side. The design of these structures must be governed by the shore-line profile and should provide for anticipated daily fluctuations of the water level. Groins may be either solid or permeable, the permeable type being designed to slow littoral currents to a point where deposition is encouraged.[19] The principal advantage of permeable groins is their lack of opposition to beach-building processes. They modify the processes but do not obstruct them. Beach widening is favored, and the foreshore area moves seaward until a balance is reached between constructive and destructive forces. When the groins are solid they act as complete barriers to longshore currents.

Where there is no great variation in storm conditions and the supply of debris on a relatively straight shore line is abundant, this type of structure may be satisfactory. It is inadequate, however, where the supply of debris is small and there is a strong longshore drift. If there are rapidly moving surface currents, a cataract effect is produced near the outer ends of the structures, causing removal of material from the lee side.

Length. The length of groins requires careful planning. They should be of sufficient length to form a beach wide enough to prevent storm waves from reaching the bulkhead, if one is present. Ordinarily, one should be satisfied with a beach about two thirds of the length of the groins and the formation of an area of permanently shallow water near the outer ends of the groins. The beach formed by the groins should have a gradual slope so that approaching waves are deprived of most of their energy before reaching the shore. The grain size of beach materials has a functional relationship to the slope of the beach and, consequently, there is no standard length which meets all conditions.

Angle. It is customary to build groins at about 90° to the bulkhead line, which is also approximately normal to the direction of longshore drift. When groins are built at an acute up-current angle to the shore line, the lee side suffers from severe scour and a steep beach of coarse materials is formed on the up-current side. The more acute the angle on the lee side, the more the groins lose their ability to form a beach.

Spacing. As a general rule the ratio between groin length and distance to the next groin is between 1:1 and 1:3.[2] When the length of the groin has been fixed, a line is drawn through the end of the groin parallel to the direction of storm-wave approach. The projection of this line on the bulkhead determines the space allowable between groins. For permeable groins, spacing under uniform conditions can be at least 25 per cent greater than in solid structures. Wood [19] estimates that on long straight shore lines the distance separating groins 350 to 400 feet in length may be as much as 1200 to 1500 feet. Where the available sand supply is low and the longshore drift is considerable, the spacing must be less.

Jetties

Jetties are parallel walls extending some distance out to sea; they are placed at the mouths of rivers for the purpose of improving navigation. Jetties confine the flow and thus tend to prevent deposition. Under favorable conditions, they may also promote scour and deepening of the channel. When properly planned and constructed, they may be very effective, those at the passes at the mouth of the Mississippi River being an outstanding example. It should be noted, however, that unless the submarine topography and/or offshore currents are favorable, periodic extensions will be required. In situations where there is a pronounced longshore drift, the construction of jetties may cause sand to pile up on the up-current side while the down-current side may be deprived of beach materials. Large massive groins, projecting into deep water, constructed to protect long open beaches and inlets are also often referred to as jetties.

REFERENCES

1. H. B. Bigelow and W. T. Edmonson, Wind Waves at Sea, Breakers and Surf, *U. S. Navy Dept., Hydrographic Office, Publ.* 602, 1947.
2. E. I. Brown, Beach Erosion Studies, *Trans. Am. Soc. Civil Engrs.,* Vol. 66, Paper No. 2076, 1940.
3. R. A. Daly, *The Changing World of the Ice Age,* Yale Univ. Press, New Haven, 1934.
4. W. M. Davis, Glacial Epochs of the Santa Monica Mountains, California, *Bull. Geol. Soc. Amer.,* Vol. 44, 1933.
5. Arthur Holmes, *Principles of Physical Geology,* The Ronald Press Co., New York, 1945.
6. Douglas W. Johnson, *Shore Processes and Shoreline Development,* John Wiley & Sons, New York, 1919.
7. W. C. Krumbein, Geological Aspects of Beach Engineering, in *Application of Geology to Engineering Practice* (Berkey Volume), Geol. Soc. Amer., 1950.
8. H. S. Ladd, et al., Drilling on Bikini Atoll, Marshall Islands, *Proc. Geol. Soc. Amer.,* 1947.
9. Martin A. Mason, Geology in Shore-Control Problems, in *Applied Sedimentation,* John Wiley & Sons, New York, 1950.
10. W. H. Munk, Increase in the Period of Waves Travelling over Large Distances, *Trans. Amer. Geophys. Union,* Vol. 28, 1947.
11. Francis P. Shepard, Revised Classification of Marine Shorelines, *J. Geol.,* Vol. 45, 1937.

12. Francis P. Shepard, *Submarine Geology*, Harper & Brothers, New York, 1948.
13. Francis P. Shepard, Revised Nomenclature for Depositional Coastal Features, *Bull. Am. Assoc. Petroleum Geol.*, Vol. 36, No. 10, 1952.
14. Francis P. Shepard and E. C. La Fond, Sand Movements along the Scripps Institution Pier, California, *Am. J. Sci.*, Vol. 238, 1940.
15. John Q. Stewart, *Coasts, Waves, and Weather*, Ginn & Co., Boston, 1945.
16. H. V. Sverdrup, M. W. Johnson, and R. H. Fleming, *The Oceans*, Prentice-Hall, New York, 1942.
17. U. S. Navy Dept., Hydrographic Office, Ocean Waves, *The Practical Navigator*, No. 9, 1939.
18. U. S. Army Corps of Engineers, Committee on Tidal Hydraulics, Evaluation of Present State of Knowledge of Factors Affecting Tidal Hydraulics and Related Phenomena, Report No. 1, *Waterways Experiment Station*, Vicksburg, Miss., Feb., 1950.
19. Sidney M. Wood, Erosion of Our Coastal Frontiers, *Bull. Assoc. State Eng. Socs.*, Vol. XIV, Pt. 1, No. 4, 1938.

CHAPTER 10

Glaciers and Glaciation

Although all existing glaciers are situated either in polar regions or in high mountains, glacial geology influences civil engineering in many ways. Civilization is expanding into regions of glaciers, and before long engineers may have to deal directly with glaciers and problems connected with glacier motion. Snow and ice engineering is becoming a profession. For example, in high latitudes and mountainous regions water is scarce in winter when power requirements are greatest. Consequently, Swiss engineers have come to look on melting glacial ice as a source of water. In order to minimize loss of altitude, it is necessary to place the water intakes in the bed of the glacier, and excavation of tunnels up to a kilometer in length has been completed under some of the large glaciers of the Valais region.[2]

The engineering importance of existing glaciers is small in comparison with those of the past. Large portions of the northern and southern hemispheres were covered by ice in comparatively recent geologic time, and many of the world's large cities are situated on glacial deposits. In building the Chicago subway tunnels, for example, fluctuations in the level of Lake Michigan resulting from climatic changes during retreat of the continental glaciers were found to have exercised an important influence on the types of soils encountered. So profound was the influence of past glaciation that even the engineer working in unglaciated areas such as the lower Mississippi River valley may find his problems related to changes in sea level caused by melting of the ice, and airfield construction on Pacific atolls may also involve problems originating from the indirect effects of past glaciation. So numerous are the ways in which glaciation has influenced the physical background of civilization that merely to mention them in detail would expand this chapter beyond its proper limits. It follows, therefore, that the civil engineer, and more especially the soil mechanics specialist, should know something about glaciers and glaciation. The literature of the subject is vast, and only specialists in the field can hope to become familiar with more than a small part of it. The following pages present briefly those portions of the subject that are of greatest importance and interest to the engineer. Studies dealing with specific localities are listed in bibliographies published by the United States Geological Survey and other geological organizations.

MECHANICS OF GLACIER MOTION

Although glaciers have been studied for over a century, it was only around 1940 that the mechanics of glacier movement became well enough understood to permit a classification based on this fundamental characteristic. Using mechanics of motion as a primary basis, Matthes *[18] has classified existing glaciers into two broad categories: ice streams and ice caps; the following description is an abridgement of the details of Matthes' classification. Ice streams include those glaciers which because of the steepness of their beds flow chiefly under the direct influence of gravity and, owing to secondary topographic controls, tend to assume elongate, stream-like forms. To this class belong mountain, valley, intermontane, and outlet glaciers. The first originate near the crests of high mountains and follow valleys much as streams of water follow channels. Mountain glaciers may unite to form valley glaciers which flow down the principal valleys of a mountain chain. Converging mountain and valley glaciers may join to form intermontane glaciers, which occupy large trough-like depressions between mountain ranges. By analogy with rivers, an

* Adaptations from Matthes' essay on Glaciers, from *Hydrology* by O. E. Meinzer, are printed through permission of the National Research Council and Dover Publications, Inc., New York 19, N. Y.

Fig. 105. Shrinking valley glacier in southern Alaska. Wasting of the ice has exposed large lateral and medial moraines. (R. E. Frost, Purdue University.)

intermontane glacier may be compared with the trunk stream, valley glaciers with its principal tributaries, and mountain glaciers with minor tributaries.   Outlet glaciers flow from the margins of ice caps and carry the surplus ice through deep valleys and fjords.   They include some of the largest and longest ice streams

Fig. 106.   Glacial features in the Baffin Bay region, Greenland.   Mountain glacier and cirque lake in foreground; faceted spurs and hanging valleys in background; icebergs cover the fjord in center.   (U. S. Coast Guard.)

in existence.   According to Matthes, gravitational flow in ice caps takes place only locally, and such masses usually exhibit radially divergent flow systems actuated primarily by differential pressure within the ice and only partially controlled by the underlying topography.   Therefore, they tend to spread out in all directions and are usually characterized by lobate borders.   To this class belong the ice sheets of Greenland and Antarctica.   The ice sheets of Norway, Spitsbergen, Franz Josef Land, Novaya Zemlya, and Iceland are comparatively thin, and configuration of the underlying topography is reflected in undulations on their surfaces. Matthes believes that their motion is probably at least in part gravitational, and they are therefore on the borderline between ice caps and ice streams.

Formation of Glacier Ice

The only condition necessary for formation of a glacier is that more snow accumulates throughout a number of years than wastes away by melting and evaporation. Low temperature alone is not enough, for parts of Siberia that have an intensely cold climate have no glaciers, owing to the extreme aridity of the region. Precipitation and low temperature are, therefore, the primary condition of glaciation. Snow is the raw material of glacial ice. The transformation of the former into the latter has been demonstrated by numerous observations, but details of the process are still rather incompletely understood. The first step in this transformation is the change of flaky snow to a firm, granular material midway between snow and ice known as *névé* or firn. This change is promoted by sublimation, migration of molecules, freezing and thawing, and the infiltration of melt water. Firn granules are about 1 millimeter in size, and the mass as a whole has a density of about 0.45 gram per cubic centimeter.[18]

The change of firn to glacial ice has been studied by a party of British scientists who investigated the Grosser Aletsch Glacier in Switzerland.[20] Their findings may be summarized as follows. Owing to weight of the overlying load, the density of the firn gradually increases. The granules grow but little in size, and interstitial air is expelled as the particles settle closer together. After the firn is buried to a depth of 100 feet, it attains a density of about 0.84 and the alteration to glacial ice begins. At this stage all the interstitial air has been expelled and the remaining air is contained as bubbles trapped within the crystals. This air is under considerable pressure, as is evidenced by the spontaneous explosions of icebergs that have been observed from time to time. As the firn changes to glacial ice the crystals begin to assume a definite orientation at right angles to the direction of the glacier's motion. This is probably the result of recrystallization rather than rotation of the granules. Thus, from a geological point of view, glacial ice should be considered a metamorphic rock. Crystal orientation is an important factor in aiding mobility of glacial ice, for it facilitates shearing along gliding planes within individual crystals.

The Problem of Glacier Motion

The mechanics of glacier motion have been disputed from almost the very beginning of glaciology. For a long time opinion seemed to be almost equally divided between two opposing camps: one maintained that glaciers move by viscous flow, the other contended that shear is the principal mechanism. Each cited very convincing evidence for its view. Although the problem is still not completely solved, research indicates that both types of movement actually occur. Explanation of this apparently paradoxical behavior of glacial ice is comparatively simple. The key to solution of the problem was furnished by observations that disclosed that in certain mountain glaciers movement of the upper layers is insufficient to account for disposal of all the snow furnished to the upper portions of the glacier. The conclusion seems inescapable, therefore, that under certain conditions the bottom layers move faster than the upper, forming what may be termed a bottom current. These findings appear to indicate that where the gradient is sufficient for the force of gravity to overcome the frictional resistance of the bed, but not high enough to cause the ice to form a cascade, a mountain glacier flows much like a stream of water. Under these conditions, the maximum velocity of flow is at, or near, the surface. Where the bed is basin-shaped, or the gradient is too low for gravity to overcome frictional resistance, flow results from differential pressure within the ice mass. In this event, flow is most rapid in the basal portion of the glacier. The first type of flow may be termed gravity flow, the second extrusion flow.[8] The first is invariably in a downslope direction; the second can take place on a level surface or even in an upslope direction. Thus, as is illustrated by Fig. 107, a mountain glacier may have gravity flow in one part of its course and extrusion flow in another, the type of flow depending on irregularities in the longitudinal profile. Shearing movements are caused by blocking of the normal flow in such a manner that the ice cannot pass over or around the obstruction by continuous plastic flow.

Motion of Ice Caps. Ice caps are less accessible to study than ice streams, and the mechanics of motion of these bodies are therefore somewhat conjectural. The problem has been discussed in detail by Matthes, and only a brief summary is needed here.

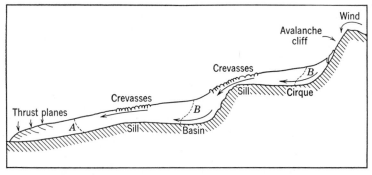

Fig. 107. Idealized profile through a mountain glacier, showing the maximum velocity of movement in various segments. The vertical velocity curves are of two contrasting types, *A* and *B*. (After Matthes, from *Hydrology*, by O. E. Meinzer, reprinted through permission by Dover Publications, New York 19, New York.)

Flowage of ice under the influence of pressure has been amply demonstrated, and under suitable conditions this type of flow may take place in masses only 100 to 150 feet thick. Therefore, it seems almost certain that vast ice caps like those of Antarctica and Greenland move primarily by extrusion flow. Such movement is typically multidirectional and proceeds from a central area of greatest height and spreads out radially toward the thinning margins.

SNOW MECHANICS

In Switzerland where snow and ice are not only economic resources and tourist attractions but a source of danger to life and property as well, the mechanics of snow and ice movement, especially when occurring in such catastrophic forms as avalanches, have been intensively studied. In the early 1930's the Swiss government appointed a Snow and Avalanche Commission which operates a well-equipped laboratory. Work done by Niggli,[1] Bucher,[3] Haefeli,[13, 14] Bader,[1] and others has served to place snow mechanics on a sound basis, and in a sense it may be regarded as a branch of the more general subject of soil mechanics. In one of his papers Haefeli [13] points out certain analogies between soil and snow mechanics and demonstrates how each may aid the other. Space does not permit a thorough treatment of this new and interesting field, and discussion will be confined to a few remarks concerning avalanches.

Avalanches. Of the many types of avalanches, according to Church,* [4] the loose-snow, wet-snow, and ice avalanches are the most important. Avalanches are similar to landslides, the chief differences being in the type of material involved and the usually more rapid movement of the former. In fact, avalanches may move so rapidly that air blasts caused by their passage may destroy trees or buildings on either side of their course. A mean velocity of 217 miles per hour has been estimated for a snow avalanche in Switzerland. The so-called wild-snow avalanche may even move through forests. If the lower slopes along which the avalanche moves are steep, it may even float in air. In addition to the factors ordinarily involved in slope stability, condition of the material (whether powdery snow, névé, or ice), temperature, wind, etc., enter into the genesis of avalanches. Snow slopes steeper than 22° are generally unsafe, but mature firn may be stable on slopes as steep as 50°. On the other hand, avalanches of wet snow may occur on slopes as low as 15°. Trees and other objects may serve to anchor snow to the slope, but unless they extend through the snow cover slip planes may be formed above them.

The Swiss Avalanche Commission has designed instruments for testing the pressure, creep, cohesion, tensile strength, and compressibility of snow masses. Profiles are kept which show the changing snow cover with time, and which in conjunction with other data are used in forecasting avalanche occurrences. Ava-

* Data from Church's essay on Snow and Snow Surveying, from *Hydrology* by O. E. Meinzer, are printed through permission of the National Research Council and Dover Publications, Inc., New York 19, N. Y.

lanches are most common in winter and during storms. Dry-snow avalanches usually take place 6 to 24 hours after a snowfall. In spring wet-snow avalanches usually occur in the afternoon when the sun has lowered the cohesion of the material. During storms, avalanches may occur at any hour.

Avalanches can at times be stopped at their source by construction of contour trenches, terraces, walls, and screens of trees. Deflection walls have also been built to detour the slide around valuable property. In other instances, incipient avalanches may be precipitated before the snow and ice reach dangerous proportions by undercutting the base or by blasting.

EROSION AND DEPOSITION BY GLACIERS

In the geologically recent past the glaciers which now cover the polar regions advanced and retreated at least four times, and at their maximum extent covered large portions of the northern and southern hemispheres. Geologists call this epoch the Pleistocene, but it is more popularly known as the Great Ice Age. Several more ancient glacial epochs are known, one of which is noteworthy for having occurred in what are now tropical regions. Causes of changing climates leading to glaciation are not well understood, but are probably chiefly astronomical in character. In order to appreciate the profound changes brought about by glaciation, an examination of glacial erosion and deposition is necessary.

Glacial Erosion

Glaciers erode by the abrasive action of rock fragments frozen in the base, by plucking blocks of jointed rock, and by the action of streams formed during temporary melting of the ice. As contrasted with other types of erosion, glacial action is not accompanied by chemical weathering and has no definite base level such as sea level, the water table, or the effective base of wave action. In discussing glacial erosion it is convenient to distinguish between the mountain and continental types.

Mountain Glaciation. In mountainous regions glaciers form in the higher altitudes and send out long tongues into the valleys below. The chief erosive effects are confined to former stream valleys occupied by ice. As contrasted with streams of water, mountain glaciers tend to fill the entire valley rather than a relatively small channel at the bottom. As a result, stream valleys with characteristically V-shaped cross sections are remodeled into U-shaped forms. The latter represent the configuration that permits the greatest volume of flow with minimum fric-

Fig. 108. A portion of the Greenland coast, deeply scoured by glacial action. (U. S. Coast Guard.)

tional resistance. As further contrasted with stream action, mountain glaciation does not tend to establish smoothly graded longitudinal profiles. In places where close jointing predominates glaciers erode by picking up blocks of rock, but, in places where joints are very widely spaced or absent, they can erode only by the much slower and less efficient process of abrasion. As a result, jointed areas are eroded much deeper than more massive zones, leading to a series of step-like discontinuities in the longitudinal profile. Under favorable conditions glacial erosion is known to have deepened pre-existing stream valleys by as much as 2000 feet.

The heads of glaciated valleys are often characterized by semi-circular-shaped widenings of the valley walls the floors of which are concave downward and usually contain small lakes. Such features are known as cirques. Frequently, a number of cirques may be present, forming an ascending series. Origin of cirques has been much disputed, but it seems fairly well agreed that alternate freezing and thawing of glacial meltwater aids in breaking up the rock. The resulting debris is then removed by the glacier. The concavity of cirque floors is perhaps attributable to rotational slipping along an arcuate thrust surface more or less parallel to the valley floor in a glacier too thin to flow plastically. In other words, movement of the ice is similar to that of certain types of landslides caused by rotation of blocks along semicircular-shaped shear surfaces.[19]

As still further contrasted with stream erosion, mountain glaciation does not tend to grade the profiles of tributary streams to the same level as the main valley. The latter carries the greatest volume of ice, and hence is eroded much deeper than the side valleys. Thus when the ice melts back the tributary valleys are left far above the main valley and form what are known as hanging valleys (Fig. 106). Some of the highest waterfalls in the world plunge from valleys of this type. Good examples are found in Yosemite Valley, California, the Alps, and the fjords of Norway. The divides between adjacent tributaries are planed off forming faceted spurs which superficially resemble a fault scarp.

Continental Glaciation. Ice sheets of continental proportions move principally by plastic flow. Consequently, their erosive action is not confined to pre-existing valleys, nor does the ice invariably move downslope. In fact, many instances are known in which the ice has actually moved uphill. Erosion produced by the Pleistocene ice sheets is known to have ranged from almost nothing to perhaps 60 feet or more. It has been calculated that the volume of glacial deposits in northern Europe is sufficient to fill up the Baltic Sea and the lakes of Scandinavia with enough left over to form a layer 80 feet thick over the entire Scandinavian Peninsula. An unknown quantity of glacial material is also buried under the North Sea. Since most of this material was derived from the Scandinavian Mountains, the depth of glacial erosion must have been very great.[10]

Differences in rock hardness and spacing of joints may have marked effects on configuration of surfaces fashioned by ice sheets. This type of topography is often characterized by a hummocky surface on which more resistant rocks stand out as hillocks which generally exhibit steeper slopes on the lee side than on the side facing the direction of ice advance. This form results from plucking action on the lee side and abrasion on the opposite, or stoss, face. Soft rocks are deeply gouged, giving rise to lake basins. The lakes of Finland, Canada, and portions of the United States are excellent examples of basins produced by glacial erosion. The Great Lakes of North America are also glacial in origin. A characteristic feature of the topography produced by glacial erosion is the almost complete absence of chemical weathering. Fine scratches, or striae, produced by the ice rasp are frequently perfectly preserved. Striae can be produced by geologic agencies other than glaciers, and when unsupported by other lines of evidence their presence does not prove that a region has been glaciated.

Deposition by Mountain Glaciers

Glacial deposition is caused by melting of the ice at a rate sufficient either to arrest its advance or to cause its retreat. Rock materials carried by the ice are dropped at the wasting margins to form elongated ridges of heterogeneous composition known as moraines. Both end and lateral moraines may be formed. The former represent materials deposited at the glacier front; the latter consist of materials carried at the lateral margins of the ice. After a valley glacier recedes, the oversteepened side walls are subject to rapid weathering and erosion, and the valley tends to fill with colluvial deposits.

Depth of alluvium in formerly glaciated valleys is a matter of some engineering importance. During construction of the Lötschberg Tunnel in the Alps the depth of loose material deposited in an old glacial valley was greatly underestimated. As a result, water-saturated debris was encountered which filled a part of the tunnel and killed several men. This material had to be bulkheaded off and a new alignment adopted before the tunnel could be completed.[9]   Borings in the St. Gotthard Tunnel in 1944 encountered sand and gravel only 130 feet above the rails, indicating that a similar catastrophe was missed by the narrowest of margins.

Recent work in Yosemite Valley furnishes another example of the uncertainties involved in estimating the depth of alluvium in glaciated valleys from purely physiographic evidence.   Reconstructed profiles and cross sections of the valley based on detailed investigations of the geologic and topographic history indicate a thickness of postglacial alluvium ranging from 100 to 400 feet.[17] However, later seismograph work indicates a thickness of from 500 to 2500 feet.[12]   Although both estimates may be in error, the discrepancy illustrates the importance of using the best available method of exploration in civil engineering work.

Deposition by Continental Glaciers

As in mountain glaciers, the wasting of a continental ice sheet gives rise to moraines. If ice recession is intermittent, a series of ridges known as recessional moraines may result. These may also be associated with wasting mountain glaciers. North America is crossed by a series of recessional moraines which extend from Long Island west along the Ohio River Valley and thence in a northwestern direction until they merge into the Rocky Mountain region of Alpine glaciation. Northern Europe is also festooned with a similar sequence of glacial deposits which extend from northwestern Germany far into Russia. Between the moraines marking positions where wasting and advance were more or less balanced for a time, there are fairly level plains known as ground moraines, or till plains, which mark zones of comparatively uniform and rapid retreat.

From an engineering point of view one of the more important features of deposition by continental glaciers is the effective masking of the upper surface of the bedrock. In the Chicago area, for example, the preglacial topography was produced by stream erosion. A relief of some hundreds of feet had been developed on the limestone bedrock. Glacial erosion planed off the tops of the ridges and filled the valleys. As the glacier retreated it left an almost level plain completely covered by varying thicknesses of glacial deposits. As a result, the foundations of a single large building may encounter bedrock only a few feet from the surface at one corner, whereas at another it may be a hundred feet or so below the surface.

The meltwater from a wasting continental glacier flows far beyond the developing moraine where it may deposit fairly well-

Fig. 109.   Glacial moraine in central Denmark.   Kettle hole in foreground is filling with organic material.

sorted and stratified materials constituting what is known as an outwash plain.   Subglacial streams may also deposit materials to form elongated ridges of stratified materials known as eskers (Fig. 110).   At points where streams issue from the ice front at high levels, cones of stratified debris may be formed.   After the ice retreats these materials may slump in the direction toward which the ice retreated and form roughly conical-shaped mounds called kames.   Glacial-fluvial deposits generally consist of sands and gravels and form abundant sources of road-building materials and high-grade concrete aggregates.

Subglacial streams may, under certain conditions, cut shallow valleys, forming what are known in Denmark as "tunnel valleys," or in Germany as *Rinnentäler*.   Following retreat of the ice, they may become more or less filled with fine-grained deposits.   Their

Fig. 110. Glacial deposits in the Copper River region, Alaska. Esker delta and esker in foreground; glacially grooved bedrock in lower right. (Bradford Washburn.)

topographic expression as long winding belts of marshy to swampy ground usually persists, however, and aids in recognition of these features on aerial photographs.

As the glaciers retreated from North Germany and Poland, the higher land to the south forced the meltwaters to drain in a direction more or less parallel to the ice front. Broad valleys, known as *Urstromtäler* (old stream valleys) subsequently filled either with later drift or water-laid sands and silts, were eroded into the drift. These features are also found in southwestern Russia and in southern Alberta. *Urstromtäler* have considerable engineering interest. Their almost level surfaces afford almost ideal canal routes, and German canals often follow *Urstromtäler*. In places where they have been filled with sands and the surface drainage is satisfactory, they furnish good sites for airfields.

In many till plains peculiar streamlined hills called drumlins are found. Their origin has been much disputed, but there seems to be little doubt that they were formed primarily by depositional activity of the ice. Drumlins are usually about a mile long and a quarter of a mile wide. The long axis parallels the direction of ice advance, and the side facing the direction of ice advance is always steeper than the lee face. They seldom occur singly, and usually form "fields" that may include a hundred or so individual mounds. Drumlins are composed of clay-rich glacial materials and are usually unstratified. In some instances they may contain rock cores. Many of the hills in the Boston area are drumlins. They also occur in New York State, Wisconsin, and other midwest localities.

GLACIAL DEPOSITS

In classifying glacial deposits a fundamental distinction may be made between stratified materials that have been reworked by meltwater derived from the ice and unstratified deposits formed directly by the ice sheet. Collectively they are spoken of as glacial drift. Till and stratified drift often intergrade to form a complex sequence of stratified and unstratified materials.

Glacial Till

The term glacial till is applied to unstratified deposits which include all ranges in grain size from enormous boulders to very fine rock flour. Usually, the particles are rather angular and more or less mixed with clay. In fact, the term "boulder clay" is more or less synonymous with till. In contrast with most sedimentary deposits, even the finest particles composing a till ordinarily show very little weathering. Unstable minerals like hornblende, mica, and plagioclase feldspar seldom exhibit the effects of chemical alteration. The thickness is, of course, highly variable, but in some areas it may reach as much as 700 feet.

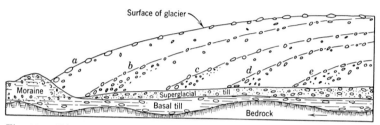

Fig. 111. Diagram illustrating origin of basal and superglacial tills. Basal till was laid down as the glacier advanced in the direction indicated by the arrow; superglacial till was formed by deposition of debris carried in and on the ice as the glacier gradually receded, occupying successively the positions indicated by letters *a* through *e*.

When exposed to prolonged weathering till is transformed into a sticky, more or less homogeneous clay known as gumbotil.

Although not stratified, many tills possess rather marked horizontal planes of discontinuity imparting to them a degree of fissility. Fissility is probably the result of flowage of the clay constituents under weight of the ice. Pressure exerted by the ice during till deposition may also compact the material to the status of hardpan. In some instances, the direction of ice movement may be indicated by the arrangement of the constituent particles, or fabric. Because of the heterogeneous composition, abundance of fine materials, and compacting action of the ice, tills are usually rather impervious.

It has long been recognized that till may be deposited in two ways: beneath the ice, forming what is known as basal till; and by the gradual decay, or ablation, of a glacier with consequent letting down of included rock fragments giving rise to what is spoken of as superglacial till. Basal till is usually rich in clay, for during its deposition there was little opportunity for washing out of fine particles by flowing water. It also tends to be dense and tough, owing to the compacting action of the overriding ice. In contrast, superglacial till has been exposed to the sorting action of flowing water and usually consists principally of coarser particles. The texture is looser and the constituents less worn. In places, a thin layer of stratified drift may separate the two types of till.

As a rule, the constituents of till reflect the bedrock composition in the immediate vicinity. Evidence is cited by Flint [10] that 50 per cent of the materials in glacial deposits of a Massachusetts locality was derived from within 10 miles of their present site. However, it is possible for resistant materials to be transported as far as 500 to 600 miles. Areas underlain by calcareous rocks tend to be covered with calcareous tills, whereas those underlain by sandstones tend to yield a sandy till. Granitic rocks normally produce a till composed of boulders. Tills derived from shales are notably clayey.

Under certain conditions glaciers may transport huge boulders many thousands of tons in weight. As the ice melts they are left behind as erratics. Such boulders do not impart information concerning character of the bedrock, and in geological field work it is necessary to discriminate carefully between erratics and outcrops. Under other conditions, boulder trains may be formed. These consist of lines of boulders which may extend for a distance of several hundred yards to many miles. They may be free on the surface or incorporated in the till. When traced to their source, they are usually found to fan out with their long dimensions paralleling the direction of ice movement.

Till occurs in end moraines, recessional moraines, ground moraines (till plains or till sheets), and in drumlins. Till deposits in end moraines when followed in the direction of ice advance frequently grade into stratified deposits. Ground moraines are usually characterized by rather undulating topography and a fairly uniform covering of till. Local depressions in till plains are generally occupied by ponds or by deposits of organic silts and clays.

Tills are good sources of material for earth dams and embankments, but they are not as a rule suitable for concrete and bituminous aggregates. Because of rapid changes in composition and permeability, they form rather treacherous foundations. Important structures of large size are seldom founded directly on till, the present practice being to carry the foundations through it into the bedrock.

Stratified Drift

As pointed out by Flint,[10] a distinction should be made between proglacial deposits formed beyond the limits of the glacier and ice-contact features built in the immediate vicinity of the margin of the wasting ice. The most important proglacial deposits are the outwash sands and gravels formed by streams draining a melting glacier. Such deposits may form fans with the apices resting against the former terminus of the glacier. More often they consist of a row of coalescent fans with the individual apices still recognizable. In deposition by mountain glaciers, the entire valley may be filled with outwash, forming what is known as a valley train. The upstream portion of such deposits may be pitted by kettles (small depressions caused by melting of ice blocks buried in outwash materials). Kettles are not confined to valley trains, but may also occur in other types of glacial deposits, till plains in particular. The grain size of outwash tends to diminish when the deposits are followed away from the glacier front. At the same time the degree of rounding increases. The presence or absence of outwash depends chiefly on the composition of the till. Very clayey tills do not give rise to much outwash simply because the particles are so fine that they are swept beyond the glaciated region, probably into the ocean. Sandy and pebbly tills tend to produce the most abundant outwash deposits.

Ice-Contact Deposits. Although stratified, ice-contact deposits differ from proglacial accumulations in their extreme range and abrupt changes in grain size and more or less intimate association with till. They are, therefore, less desirable sources of construction materials than outwash deposits. Ice-contact deposits occur in kames, eskers, and kame terraces. The first two have already been mentioned, and only the last needs to be discussed. The term kame terrace is applied to an accumulation of stratified drift laid down chiefly by streams between the margin of a glacier and an adjacent valley wall. Upon disappearance of the glacier, the deposit is left as a constructional terrace which differs from a stream terrace in that it does not represent the remnants of a mass of sediment formerly extending completely across the valley. The top is usually flat and narrow, and it is often pitted by kettles. In some areas, several terraces occur in a series marking successive positions of ice-marginal streams as the glacier thinned by wasting. It is not unusual for kame terraces to grade into valley trains in a downvalley direction.

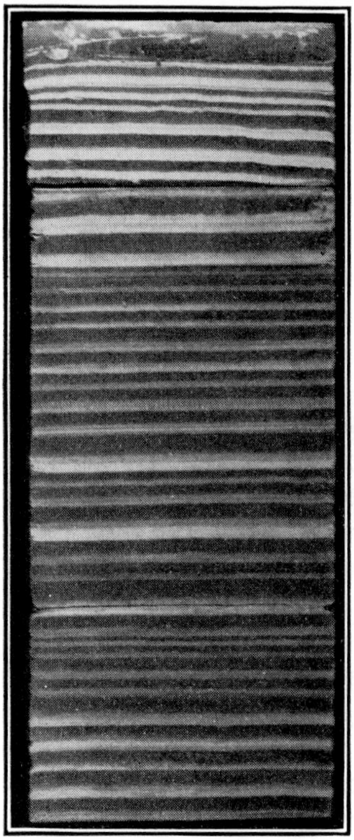

Fig. 112. Undisturbed sample of varved clays. Sample is approximately 11 inches in length. (M. Juul Hvorslev.)

Glacial Lake Deposits

The environs of a wasting glacier are peculiarly favorable to the formation of lakes, which if occurring along the margin of the ice are spoken of as proglacial lakes. Sediments deposited in proglacial lakes are fairly widespread in glaciated regions, and they possess certain unique features that are worthy of mention. The sediments consist of deltas, bottom deposits, and rafted erratics. Deltas built into proglacial lakes resemble ordinary deltas except that their apices either are at the former margin of the ice or they may grade into eskers forming what are known as esker deltas. The sediments grade rapidly from coarse to fine.

Bottom sediments are usually fine grained and frequently consist of alternate thin bands of light-colored silt and darker laminae of clay. The coarse materials grade upward into finer deposits, but they are sharply separated from the next underlying fine lamina. A pair of layers is known as a varve, and the sediments are known as varved clays. It is generally agreed that the sediment was brought into the lakes during spring and summer. The silt settled out relatively rapidly, but the clay remained in suspension and did not settle out completely until toward the end of the following winter. Each varve thus constitutes a time measure of one year, and attempts have been made to correlate glacial deposits by varve counts. Estimates of the length of post-glacial time have also been made on this basis. Lake sediments may also contain large boulders which are interpreted as having been rafted to their present position while they were frozen in icebergs. The fine-grained bottom sediments of former glacial lakes form much of the best farm land in the United States and Canada.

The shores of large glacial lakes may develop typical shore-line features such as wave-cut cliffs, beach ridges, spits, bars, etc., which have been described in the preceding chapter. Beaches of many glacial lakes have been warped and tilted indicating uplift subsequent to their formation.

INDIRECT EFFECTS OF GLACIATION

The indirect effects of glaciation may have important engineering implications, among which the influence of changes in sea level has already been mentioned. As in other phases of the subject, the problem is complex and the literature extensive. Space permits only a brief summary.

Changes in Sea Level. Almost as soon as glaciers were recognized as geological agents, Maclaren [16] pointed out that ice sheets of the size of those of the Pleistocene represent a volume of ocean water the solidification of which must have lowered sea level by some hundreds of feet. During the present century, Daly [6] calculated the maximum lowering of sea level during the height of the glacial epoch as about 400 feet. More recently, Shepard [22] invoked a still greater lowering of sea level to account for the deep submarine valleys which occur off the mouths of many rivers. However, the majority of geologists do not seem prepared to accept a figure much larger than that proposed by Daly. About 11,374,000 cubic miles of glacial ice still remain, and if it melted it would raise sea level 130 to 190 feet, causing all the world's seaports and large areas of fertile land to be submerged.

Lowering of sea level by about 400 feet not only caused many great rivers to entrench themselves into bedrock but also caused formation of terraces along many marine shore lines. The barrier reefs and atolls discussed in the preceding chapter have also been attributed to this cause. From an engineering point of view, one of the more striking influences of changing sea level is found in certain clay deposits formed in near-sea-level environments as the sea gradually rose after final retreat of the continental glaciers. Clays of this origin occur in southern England, Scandinavia, the St. Lawrence River valley, and the Boston region. Their geologic history is known in considerable detail, and these strata have been studied intensively by engineers interested in explaining the varied behavior of clays as foundation materials.[23]

Glacial Lakes. During the glacial epoch the climate of arid regions was considerably more humid than it is now. In regions lacking exterior drainage, such as the Basin and Range province of western United States, increased precipitation resulted in formation of lakes the remains of which may be seen in the sediments deposited in the basins and various shore-line features such as beaches, bars, and cliffs. Many of these lakes were very large. The present Great Salt Lake, for example, is but a remnant of a much larger body of water known as Lake Bonneville. Terraces formed by wave action at various levels of Lake Bonneville form conspicuous features of the mountain sides south of Salt Lake City. Other large glacial lakes were Lake Lahontan in Nevada; Lake Manix, California; and Lake San Agustin, New Mexico. Similar features are known in other parts of the world.

Postglacial Uplift. Geodetic work in Europe and the United States [6, 11] has demonstrated that areas formerly covered by the continental ice sheet are at present slowly rising. The north shore of Lake Superior, for example, is estimated to be rising at a rate of about 6 inches per century. The explanation of this behavior is that the weight of the ice, which reached a thickness of possibly 10,000 feet or more at the centers of accumulation, dented the earth's crust in much the same manner as a soft rubber ball may be dented by pressure of the fingers. Mechanisms by which the denting occurred are to be found in deep-seated plastic flow of material away from the depressed area, and to some extent in elastic distortion of the crust. With removal of the superincumbent load equilibrium is being restored by subcrustal flow back into the depressed areas and elastic readjustment of the crust. Although slight, uplift following glaciation has had an adverse effect on shallow harbors of some areas.

REFERENCES

1. H. Bader, E. Bucher, O. Eckel, J. Neher, P. Niggli, and C. Thams, Der Schnee und seine Metamorphose, *Geol. der Schweiz, Geoteknische Serie, Hydrology,* Lieferung 3, 1938.

2. H. Bader, Trends in Glaciology in Europe, *Bull. Geol. Soc. Amer.,* Vol. 60, 1949.

3. E. Bucher, Weissfluhjoch Research Institute, *J. Glaciol.,* Vol. 1, 1948.

4. James E. Church, Snow and Snow Surveying; Ice, *Hydrology,* Physics of the Earth, Vol. IX, McGraw-Hill Book Co., New York, 1942. Reprinted by Dover Publications, Inc., New York 19, N. Y., 1949.

5. C. A. Cotton, *Climatic Accidents in Landscape-Making,* John Wiley & Sons, New York, 1948.

6. R. A. Daly, *The Changing World of the Ice Age,* Yale Univ. Press, New Haven, 1934.

7. Max Demorest, Ice Flowage as Revealed by Glacial Striae, *J. Geol.,* Vol. 46, 1938.

8. Max Demorest, Glacier Flow and Its Bearing on the Classification of Glaciers, *Bull. Geol. Soc. Amer.,* Vol. 52, 1941.

9. *The Engineer (London),* The Lötschberg-Simplon Railway and Its Construction, Vol. 112, 1911.

10. R. F. Flint, *Glacial Geology and the Pleistocene Epoch,* John Wiley & Sons, New York, 1947.

11. Beno Gutenberg, Tilting due to Glacial Melting, *J. Geol.,* Vol. 41, 1933.

12. Beno Gutenberg and J. P. Buwalda, Geophysical Investigation of Yosemite Valley, *Proc. Geol. Soc. Amer.,* 1937.

13. R. Haefeli, Erbaumechanische Probleme im Lichte der Schneeforschung, *Versuch. für Wasserbau und Erdbau,* Nr. 7, 1944.

14. R. Haefeli, Snow and Glacier Research in Switzerland, *J. Glaciol.,* Vol. 1, 1948.

15. Chauncey D. Holmes, Glacial Erosion and Sedimentation, *Bull. Geol. Soc. Amer.,* Vol. 60, 1949.

16. Charles Maclaren, The Glacial Theory of Prof. Agassiz, *Am. J. Sci., 1st Ser.,* Vol. 42, 1842.

17. F. E. Matthes, Geologic History of Yosemite Valley, *U. S. Geol. Survey, Prof. Paper* 160, 1930.

18. F. E. Matthes, Glaciers, *Hydrology,* Physics of the Earth, Vol. IX, McGraw-Hill Book Co., New York, 1942. Reprinted by Dover Publications, Inc., New York 19, N. Y., 1949.

19. Louis L. Ray, Alpine Glaciation, *Bull. Geol. Soc. Amer.*, Vol. 60, 1949.
20. Gerald Seligman, The Structure of a Temperate Glacier, *Geog. J.*, Vol. 97, 1941.
21. R. P. Sharp, Glacier Flow: A Review, *Bull. Geol. Soc. Amer.*, Vol. 65, 1954.
22. F. P. Shepard, *Submarine Geology*, Harper & Brothers, New York, 1948.
23. A. W. Skempton, Geotechnical Properties of Post-Glacial Clays, *Géotechnique*, Vol. 1, 1948.

C H A P T E R   11

Frost Action in Soils

Frost action in soils is of great engineering importance, for the expansion that takes place during freezing of frost-susceptible materials may damage structures founded on them. Moreover, the moisture derived from the thawing of frozen soils may greatly lower their bearing capacities, leading to settlement of structures and the failure of highway and airfield pavements. Consequently, civil engineers have devoted considerable attention to the subject, but until the 1920's little was known concerning the actual processes involved, and in situations where empirical data were unavailable it was often impossible to predict the frost susceptibility of certain types of soils. Although much remains to be done, frost action is now well enough understood to permit rational methods of design based on the known or predicted behavior of the materials when subjected to frost.

Colonization and development of arctic regions, which are proceeding at a rapid pace, have brought into prominence a number of interesting and formerly little-known features associated with permanently frozen soils, or permafrost, which present the civil engineer with many difficult problems. The study of permanently frozen soils is beset with many difficulties, not the least of which is the general inaccessibility of permafrost regions, and of necessity progresses somewhat slowly. Despite these handicaps, however, results of great practical importance have been achieved.

As the main body of this chapter indicates, geology is of considerable importance in the study of frost action in soils and is basic to the understanding of permafrost phenomena. The views expressed reflect the present state of thought on the subject and cannot in all cases be considered final, for future research may lead to considerable modifications.

FROST ACTION IN TEMPERATE CLIMATES

In temperate climates, freezing extends to a depth of a few feet at most, and the ice formed in winter is completely melted in spring and early summer. Under these conditions, frost action is confined largely to the expansion of certain soils, accompanied at times by the formation of mounds and "boils," and the softening and loss of bearing capacity that take place when these materials thaw. This is the mildest form of frost action with which the engineer has to deal, but it is common in the most highly developed regions of the world and, therefore, is of considerable engineering importance.

Mechanics of Frost Heaving. All except the inhabitants of tropical regions have some degree of familiarity with the expansion of certain soils that takes place during freezing, but the mechanism responsible for this phenomenon has been much disputed. The 9 per cent increase in volume involved in the change of water into ice cannot be the sole cause, for the volume increases observed are far too large, often reaching 30 per cent or more. It has been found that all the water in soils does not freeze at the same temperature. With decreasing size of interstices the freezing point is lowered. In supercapillary openings water freezes at zero degrees centigrade, but in subcapillary openings the freezing point may be as low as $-78°$. Thus, freezing may take place in the presence of water. Taber [19] has shown that during freezing at temperatures near zero degrees water in capillary openings is apparently pulled toward the growing ice crystals, furnishing a steady supply for their continued growth. The moisture from a large volume of soil is concentrated in this manner in the upper layers in the form of ice. It is apparently the growth of these segregated ice masses, rather than the expansion which takes place when water changes to the solid form, that accounts for frost heaving in soils.

A useful distinction may be made between "massive" and "ice-stratified" masses. Massive formations are well illustrated by gravels and coarse sands. All the interstitial water freezes at a temperature of about zero degrees centigrade; and, consequently, no segregation of ice lenses and little or no heaving results. In fine-grained soils the freezing point is distributed over a wide range in temperatures, and where there is a local widening of the interstices the freezing point will be higher than in capillary openings. Ice begins to form in the larger openings, moisture is drawn from below to feed the growing ice crystals, which tend to force aside and widen the natural openings, leading eventually to an ice-stratified zone. Continuation of this process results in the formation of ice lenses which as freezing progresses are formed at successively lower levels until the frost line, or lowest depth of freezing, is reached. The material between two ice layers may often remain unfrozen.

From the above it is evident that the three most important factors in frost heaving are size of particles forming the soil mass, permeability, and availability of excess moisture. The last is governed largely by position of the water table and/or capillary fringe. Taber found that segregated ice forms readily in materials with a particle size of 0.001 millimeter or less. According to Casagrande,[6] the critical diameter of particles is about 0.02 millimeter. If the amount of these particles is less than 1 per cent, heave danger is negligible. Heave may result in non-uniform soils if the proportion of particles of this size exceeds 3 per cent. In uniform materials, the proportion must exceed 10 per cent in order for heave to occur.

Preventive Measures. Damage caused by frost action may be prevented or minimized in four ways: (1) placing the foundations of structures below the maximum depth of frost penetration; (2) preventing freezing of the soil by the use of sand or gravel blankets, or other suitable insulators; (3) removing frost-susceptible materials and backfilling with clean sand or gravel; and (4) designing on the basis of a lower bearing capacity during the thawing season. The first method is the one generally followed in the case of building foundations, but it is generally not applicable to highway and airfield pavements where either the second or third methods are usually employed. Reasonably precise methods for determining the depth of frost penetration in various climates and the thickness of insulator required to prevent freezing of highway and airfield subgrades are available.[5] It should be borne in mind, however, that frost heaving is largely dependent on the availability of excess water, which in turn is largely dependent on position of the ground water table.

"*PERMAFROST*"

Areas surrounding a continental glacier are characterized by a so-called periglacial climate which is marked by long and very cold winters and short and relatively cool summers. Under these conditions the ground freezes to great depths and only a comparatively thin layer thaws during the summer. Consequently, periglacial areas are underlain by permanently frozen ground known variously as "perennially frozen ground," "permanently frozen ground," "pergelisol," "*tjaele*," and "permafrost." In these pages the term permafrost is used.* Permafrost may be defined as the condition in which the ground temperature remains below freezing throughout a considerable number of years. Ice may or may not be present, the absence of ice being conveniently known as "dry permafrost."

Although permafrost underlies about one fifth of the land area of the world, the sparse population of high latitudes has tended to retard the study of the phenomenon and obscured its engineering and geologic importance. Muller's book on the subject [11] has been the source of much of the material in this chapter and should be read by all having direct concern with permafrost problems.

Soil structures resulting from periglacial climates of the Ice Age have long been studied in Europe; in the United States they are generally referred to as "involutions." [16] Research on permafrost and associated engineering problems is being carried out by various agencies of the United States government, and important

---

* The term "permafrost" is employed by U. S. Federal agencies but is not universally accepted elsewhere. It is used in this book because it has been adopted in most of the United States literature.

results have been obtained. In view of the rapidly growing knowledge of the subject, Bryan [4] has proposed the term "cryopedology" to designate scientific study of intensive frost action and permanently frozen ground. It includes not only the study of the processes and their effect, but the engineering devices developed to avoid or overcome the difficulties resulting from frost

Fig. 113. Periglacial involutions (soil structures caused by frost action) in silts and sands near Windham, Montana. (J. P. Schafer.)

action. As the following discussion will demonstrate, many of the phenomena associated with permanently frozen ground are positively unique, especially in their engineering implications. Indeed, it is no exaggeration to say that they constitute one of the chief obstacles to the colonization and development of northern areas.

Origin and Distribution. Permafrost appeared as a result of the general lowering of temperature that culminated in the Pleistocene glaciation, perhaps 1,000,000 years ago, and at present it is confined to areas where the mean annual temperature is a little below zero degrees centigrade. The occurrence in frozen ground of extinct animals such as the woolly elephant with the flesh perfectly preserved furnishes striking evidence of the relict, or fossil, character of permafrost. It is most widespread in the northern hemisphere, but it is also present in the Antarctic. In North America the southern boundary is thought to extend almost to the southern coast of Alaska, and a narrow tongue may follow the Rocky Mountains almost to the United States border. It follows the south end of Hudson Bay and thence swings northeastward to the southern part of Greenland. In Eurasia the boundary extends from the Kola peninsula southeast to considerably below Lake Baikal and northeast to the northern tip of the island of Sakhalin. In addition to the almost continuous regions outlined above, there are small areas of sporadic permafrost scattered throughout the world, chiefly in high altitudes above the snow line. All in all, about one fifth of the land area of the world is included, parts of which are not known to have been glaciated during the Pleistocene. The thickness is usually measured in a few feet or yards, but in some places on the shores of the Arctic Ocean it is as much as 750 to 1100 feet.

The Active Layer. Above the permafrost is a zone subject to thawing in summer and renewed freezing in winter which is known as the active layer, or mollisol in Bryan's nomenclature. The thickness of the active layer depends on numerous factors among which are temperature, composition of the ground, type of vegetation, type of surface exposure, hydrology, and snow cover. Generally, it is thin in northern areas and becomes thicker toward the south. In Siberia variations from a few inches to over 13 feet have been measured. Muller cites the following example. On a river terrace composed of water-saturated sandy clay covered by a little over 18 inches of peat and moss supporting a sparse stand of larch the thickness was about 1.5 to 2.6 feet. Under essentially the same conditions except that grasses, and trees like birch, alder, and poplar replaced larch, the thickness was 4.9 to 8.2 feet. In the same valley, but on the next higher terrace composed of well-drained sandy material covered by lichens, dry moss, and tall pines, thickness of the active layer was found to be 8.2 to 11.5 feet. Owing to the large number of variables involved, the depth of the thawed ground forms an extremely uneven surface which by analogy with hydrology may be called the frost table. With progress of the summer thaw the frost table moves progressively downward until it either disappears by reaching underlying unfrozen ground or merges with the upper surface of the permafrost. The upper boundary of the permafrost may be termed the permafrost table. Like the frost table, its configuration is determined by the type of vegetation,

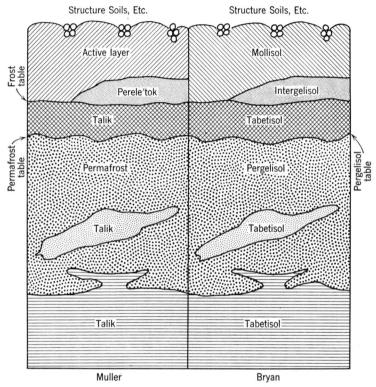

Fig. 114. Diagram illustrating Muller's and Bryan's systems of permafrost nomenclature. (Adapted from Kirk Bryan.)

heat conductivity of the ground, geographic position and character of the exposure, and hydrology. For example, in areas insulated by peat the permafrost table is higher than in those where the surface is bare.

At places, and especially near the southern boundary of the permafrost province, the permafrost lies much deeper than the depth reached by winter frosts. As a result, between the active layer and permafrost there is a zone of unfrozen ground which is commonly designated by the term *talik*, derived from the Russian and meaning "thawed ground." The term is also applied to layers or lenses of unfrozen ground that may occur within the permafrost and to the unfrozen material below the permafrost. In Bryan's nomenclature the term tabetisol is synonymous with *talik*.

Owing to an excessively cold winter or colder than average summer, the summer thaw may stop short of the usual depth, leaving a layer of frozen ground above the *talik*. This isolated frozen zone, which lasts for only a season or two, is designated by the Russian term *"perelétok,"* meaning "that which survived over the summer" (intergelisol of Bryan). The combined thickness of the ground above the permafrost (active layer, *talik*, and *perelétok* when present) is referred to as the suprapermafrost layer. In the greater part of the permafrost region the suprapermafrost coincides with the active layer, but toward the southern limits the suprapermafrost is generally thicker than the active layer. If the suprapermafrost is fairly thick (65 feet or more), in most types of construction only the active layer needs to be considered. Where it is less than 15 feet thick, however, preliminary investigations should include not only a study of the behavior of the active layer but also the effect of the permafrost on the hydrologic regime of the underlying soil as well.

Thermal Regime.   The temperature of permafrost usually shows only small variations with depth, and these are most marked in the upper portion. The range of variation diminishes gradually downward until it approaches zero at the so-called level of zero amplitude. The depth to this level is known as the depth of seasonal change. Below the level of zero amplitude the temperature gradient tends to remain stable from season to season and from year to year. However, the temperature may rise with increasing depth, or may decline still further and then rise with increasing depth. The regime in any specified locality depends on a large number of factors, and it is obvious that if any one or more of them is disturbed by natural or artificial means the thermal regime will likewise be altered. Buildings and earth fills tend to alter the regime and cause the permafrost table to change its position. Usually the new thermal regime becomes stabilized within a year or so, but in large and rapidly constructed fills the heat liberated from the fill material may cause a temporary sag of the permafrost table, which later disappears and is followed by a rise in level, resulting in the permafrost invading the lower part of the fill. Consequently, the time of year during which construction is undertaken is a very important factor in permafrost localities.

By virtue of the latent heat of fusion, layers with a high water content tend to retard lowering of the temperature gradient. In such zones the loss of heat is compensated by the heat liberated by water changing to ice so long as any liquid remains unfrozen. This condition has been termed the zero curtain. In summer a reverse effect takes place when heat is absorbed by the melting of interstitial ice. Near the base of the active layer where there is usually a high concentration of water, the zero curtain may last more than a month, and cases are known where it persisted for as many as 115 days of the year. The zero curtain is of considerable practical importance, for it permits artificial regulation of the heat balance in the ground, for example, the lengthening of the zero-curtain period, or the creation of a zero curtain, by the construction of water-tight fills around water pipes.

Degradation and Aggradation of Permafrost. In some localities, especially along the southern margin of the permafrost region, there is evidence of degradation (the gradual disappearance of permafrost by progressive thawing from year to year). Since degradation may proceed more actively from the bottom up than from the top down, it is apparent that in order to establish the occurrence of degradation the thermal regime of the entire thickness of the permafrost must be investigated. In some areas there is evidence of aggradation (the spread or first appearance of permafrost). Degradation and aggradation may result either from natural or artificial conditions. Aggradation under natural conditions has been reported in rapidly growing deltas of large rivers flowing into the Arctic Ocean and in flood plains where silting of the valley floor forms a protective blanket.

Ground Ice. The term ground ice is applied to well-defined ice masses occurring in the permafrost. Such masses may form nearly horizontal layers, or approximately vertical wedges. The latter often consist of a network surrounding irregular polygonal columns of silt, known as soil polygons, or *polygonböden*, the average diameter of which is about 20 feet.[20]   Ice wedges may reach a thickness of 8 feet or more, and are known to extend downward to a depth of over 30 feet.   In some instances ice veins have been observed at depths of as much as 300 feet below the surface, where they are enclosed in bedrock.   Ice wedges fre-

Fig. 115.   Ice wedge overlain by tundra soils, Alaska.   (R. E. Frost, Purdue University.)

quently merge with layered ice above, less commonly below. Horizontal layers or sheets may range from a fraction of an inch to well over 12 feet in thickness.   When traced laterally, they usually pinch out gradually to form lens-like bodies.   Coalescing ice layers may underlie large areas.   Ground ice occurs most commonly in silts, and is more abundant in shady locations than in areas exposed to the sun.   Origin of ground ice has been and still is much disputed, but it seems probable that the ice is younger than the enclosing materials and has formed through freezing from the surface downward.

Ground ice is obviously a treacherous foundation for any type of engineering project, airfields especially. It is to be anticipated in former river deltas, flood plains, and other sites of silt deposition. Because of their frequent association with ice wedges, presence of polygonal-shaped patterns (*polygonböden*) on the

Fig. 116. Soil polygons (*Polygonböden*), northern Alaska. (R. E. Frost, Purdue University.)

ground surface should be regarded with suspicion. If construction is undertaken in areas underlain by ground ice, care should be taken to insure that the thermal regime of the soil is not disturbed.

HYDROLOGY OF PERMAFROST REGIONS

The flow of rivers in permafrost regions is controlled by the dominant source of water. Usually the streams show a prominent rise during the late spring and early summer thaws. Rivers that depend entirely on snow fluctuate considerably, and may cease to flow after the snow is melted. Rivers that depend primarily on the rainfall reach peak stages late in summer, after the rainy season. Small streams may be fed by seepage from thawed surface soil. Streams supplied by deep springs or underground channels usually maintain a fairly steady flow through most of the year. Large rivers, and especially those that do not freeze solid in winter, have a warming effect on the underlying ground, which may remain unfrozen and usually contains percolating water. Even under streams that freeze solid the ground may remain unfrozen, and may furnish a source of water supply. Similar effects are observed under lakes.

During freezing, ice may form not only on the surface but on the bottom of the channel as well, constituting what is known as bottom or anchor ice. Bottom ice may form before the surface ice. Formation of anchor ice is most common in clear swift streams with dark rock bottoms. On cold clear nights the bottom rocks may radiate heat so rapidly that they cool below the freezing point, leading to freezing of the water immediately above them. A large spring may issue into a river channel and prevent the water from freezing locally for the whole or part of the winter. Such unfrozen spots are important in exploration for water supply, for they usually indicate the location of springs that afford dependable year-round sources.[11]

Icings. As a river freezes over the channel beneath the ice is gradually constricted so that it becomes too small for the volume of flowing water. Increased hydraulic pressure results and forces the water laterally where it permeates porous materials in the banks. The increasing hydraulic pressure may also force the water to the surface either through the ice or through the ground near, or at some distance from, the river channel. Water issuing from such breaks freezes in successive layers of ice, which may cover an area of a square mile or so and may attain a thickness of as much as 12 feet or more. Such ice incrustations are known as ice fields or icings. Icing formed in rivers is commonly known as river icing, whereas that formed on land is known as ground icing. Icings may be seasonal, or may last a number of years. Large icings may furnish dependable sources of water supply.

Ground Water. In permafrost regions the ground water may be divided into water above the permafrost (suprapermafrost water), water within the permafrost (intrapermafrost water), and water below the permafrost (subpermafrost water).

The suprapermafrost water is derived from rain and melt water, surface water, vapor condensed by contact with the cold ground surface, and seepages from within or beneath the permafrost. This water is subject to seasonal freezing and melting. If the permafrost lies at greater depth than the depth of seasonal freezing, the lower part of this water will remain liquid. In winter, the ground freezes from the top downward, and the water between the permafrost and the surface ice is confined to an ever smaller volume subjected to increasing pressure. The magnitude of the pressure is proportional to the thickness of the overlying frozen zone. Under this pressure the water may be forced to the surface to form seepage sheets (icings), or it may be wedged between frozen and unfrozen layers to form lenses of ground ice. Occasionally, the confined water may split the already frozen layers and fill the spaces between them.

Water within the permafrost is derived from infiltration of surface water, water beneath the permafrost, or both. The volume may be considerable, and may furnish a dependable water supply. On the Arctic coast the occurrence of saline waters, probably of marine origin, has been noted in the permafrost. Strongly mineralized waters may remain liquid at temperatures considerably below zero degrees centigrade. Intrapermafrost water may or may not be under greater than the normal hydrostatic pressure.

Subpermafrost water is always in the liquid state, and as a rule is under considerable hydraulic pressure. Except where too strongly mineralized, this water may be a good source of water supply.

GEOLOGICAL PHENOMENA ASSOCIATED WITH PERMAFROST

The most conspicuous features associated with perennially frozen ground may be grouped under the generic term of frost mounds. Present information is not sufficient to enable one to make a detailed account of their characteristics and origin, and the problem is further complicated by a rather confused and overlapping nomenclature. In the following discussion Muller's terminology is followed.

Pingo. The term *pingo* is of Eskimo origin and is applied to a large mound which may be as much as 230 feet high and over 3000 feet in perimeter. The top may be featured by a crater-like cavity and radiating fissures from which potable water may flow. Pingos occur in poorly consolidated deposits extending from Point Barrow, Alaska, eastward through the Mackenzie River

Fig. 117. Large *pingo*, Mackenzie River delta, Arctic Canada. (A. Erling Porsild.)

delta to the first rock outcrops east of the Horton River. Similar, though smaller features are found in Greenland [13] and in subarctic bogs in Europe and Asia. In Sweden they are known as *pals*. They are associated with poorly drained areas near river deltas or former lake basins. These mounds have attracted the attention of Arctic travelers since the early part of the nineteenth century, and the literature discussing them is fairly extensive but rather contradictory. Explorers' reports and native accounts give evidence that they grow slowly and may persist for perhaps hundreds of years. After a time, however, they tend to disappear leaving annular depressions the centers of which may contain lakes. The marshes surrounding decomposing pingos may exhibit various forms of *bodenstruktur* caused by frost action.

The pingos examined by Porsild [14] showed distinct stratification parallel to the surface contours. The silts composing the mounds contained remains of aquatic plants and shells, indicating that they once occupied the floors of lake basins. In the Kotzebue region of Alaska, the walls of a storage vault excavated in a pingo were examined by this investigator and found to consist of very fine silt traversed by a fine network of veins of pure ice. Dikes of clear ice cut across the silts and probably represent fillings of ancient fissures. When thawed, samples lost as much as 60 to 75 per cent in volume. The high proportion of ice suggests that a supply of water in excess of that contained in the interstices was available during freezing.

Porsild [14] believes that pingos are of two types which originate in entirely different ways. The first includes the mounds on the north coast of Alaska described by Leffingwell [10] and those of Greenland. These mounds always occur in pervious materials underlying a sloping ground surface. As a rule, they are not very large and exhibit a great variety of shapes. Some are oval, others are ridges which simulate moraines or eskers. This investigator concurs with Leffingwell that mounds of this sort were formed by uplift caused by hydraulic pressure of a sheet of water trapped under what is now the frozen surface soil, an origin very similar to that of frost blisters discussed in the following section. Indeed, it is entirely possible that future work may indicate that pingos of this type should either be classed as frost blisters or frost blisters should be classified as pingos. The second type includes some of the mounds on the north coast of Alaska and nearly all of those east of the Mackenzie River delta. These features are nearly always round or slightly oval in outline and may be very large in size. They invariably occur in low, lake-filled country and may have summit elevations considerably higher than the highest point for miles around. Consequently, the hydraulic pressure theory seems inadequate for their origin. Porsild believes that these features were formed by local upheaval due to expansion following the progressive downward freezing of a body or lens of water or semifluid mud or silt enclosed between bedrock and the frozen surface layer.

Frost Blister. The term frost blister is applied to an upwarp produced by local hydraulic pressure of the ground water. The pressure may exceed the elastic strength of the overlying frozen soil and the mound may rupture with a loud noise resembling a gun shot. Water may gush from a broken blister. Frost blisters usually form along sloping ground and may shift position from year to year. They are usually small and seldom exceed 25 feet in height. The ice in a blister may not melt entirely during the summer. Various names applied to such features are soil blisters, ice mounds, earth mounds, gravel mounds, etc.

Fig. 118. Frost blister, Wolf Creek region, Yukon Territory, Arctic Canada. Ice in interior of blister is indicated by white spot in middle foreground. (R. P. Sharp.)

Icing Mounds. When the active layer freezes from the top downward, the water between the permafrost and the frozen part of the active layer is confined in an ever constricting space. The resulting pressure forces the water to the surface along the path of least resistance where it may spread out to form sheets, or under certain conditions more or less conical protuberances known as icing mounds. If a building is situated so that it promotes thawing, or prevents freezing of the active surface layer, the underlying confined water may flow up into the structure and completely fill it with ice. Muller presents several photographs of this phenomenon. Icing mounds are for the most part composed entirely of ice. They are similar in origin to frost blisters, but differ in drawing their water supply from a less limited source, and usually exude water for a considerably longer period. Icing mounds, seldom more than 30 feet in height and width, are commonly surrounded by a sheet of layered ice. They are usually seasonal, but in exceptional cases do not melt completely in summer and, consequently, may last for a number of years.

Peat Mounds. These are small hummocks that are found in swamp or tundra regions. Various other names are hummocky swamp, hummocky tundra, and spotted tundra. Usually, they last only a season. In the Kola peninsula peat mounds have been

Fig. 119. Seasonal icing mound, Mackenzie River delta. (A. Erling Porsild.)

found to contain lenses of ground ice enclosed in water-saturated peat and moss. Muller is of the opinion that peat mounds are caused by localized swelling of ground resulting from the segregation of ice in soil that is unequally insulated from frost by a variable thickness of vegetation. In less protected spots the ground may freeze sooner and at a faster rate than in adjacent better-insulated areas. The first ice that forms will tend to draw moisture from adjacent areas, and when the better-insulated areas freeze the amount of water available will be considerably less. As a result, bare or less protected areas will remain in slight relief compared with adjacent better insulated areas. Redistribution of moisture during freezing may also be caused by local differences in soil texture.

FACTORS INFLUENCING FORMATION OF HEAVING, OR SWELLING, GROUND

The chief factors involved in swelling are soil texture and hydrology. Depending on the texture, some soils swell enormously whereas others hardly swell at all. In this regard three main groups may be distinguished: (1) soils that do not swell, including bedrock, and soils composed of coarse materials such as pebbles and boulders; (2) soils that swell slightly, including gravel and coarse sand with admixtures of finer particles (clay and silt), moist or saturated with water but without an available supply of additional water; (3) soils that swell markedly, which include gravel and sand mixed with clay and silt, moist or water-saturated with access to additional supplies of water; sandy clay, clayey sand, and clay; fine sand, sandy silt, silt, and wet mud; peat. The role of particle size in the mechanics of frost action has already been discussed in the first part of this chapter.

The condition in which an additional supply of water is present during freezing is known as an open system, as contrasted with a closed system in which the freezing ground does not have access to an additional supply of water. Swelling is always greater in open than in closed systems. Normally soils will swell only when the amount of water available exceeds the critical moisture content, which is defined as the maximum amount of interstitial water which when converted into ice will fill all the available pore space. Soils that contain water in excess of the critical moisture content will swell even if there is no access to an additional source of supply.

Where the active layer is comparatively thin, as along the Arctic coast, there is little danger of appreciable swelling. Farther south where the active layer is thicker, swelling is more noticeable. Presence of *taliks* tends to minimize swelling, and where *taliks* are thick it is practically non-existent. Swelling ground may often give rise to the presence of a group of irregularly inclined trees known as a "drunken forest."

Effects of Summer Thaws

During the summer thaw water is added to the active layer not only by the rainfall but also by melting of interstitial ice. Downward drainage is obstructed by the permafrost, and as a result the active layer becomes saturated with water. Fine-grained materials become more or less fluid and form muds known by the inelegant provincial English word, slud. Bryan has proposed to standardize the very confused nomenclature of phenomena associated with frost action by proposing the terms "congeliturba-

Fig. 120. Mudflow near De Salis Bay, Banks Island, Arctic Canada. (A. L. Washburn.)

tion" for the process and "congeliturbate" for the product. All forms of frost action such as heaving, differential and mass movements, solifluction, etc., are included. In Bryan's nomenclature, slud would be known as a congeliturbate transported by solifluction (a term used to indicate the slow downslope movement of a mass of water-saturated soil). In England, materials transported in this manner are known as "head."

If the thawed ground is sufficiently plastic, it flows and oozes from beneath loaded areas like taffy. Considerable damage to buildings may result. The rate of settling depends primarily on the rate of thawing. In the northern hemisphere, even light structures tend to settle more on the south than on the north side. A simple board fence running east-west may produce a noticeable difference in ground temperatures on the north and south sides of the barrier.[11]

Settling may also be caused by natural agencies such as unequal heat conduction of the surface materials. By analogy with the so-called karst topography caused by solution of underlying beds by solvent action of the ground water, the term "thermokarst" has been proposed for an irregular land surface resulting from the melting of ground ice. Thermokarst topography may result from forest fires, grass fires, deforestation, and activities of man. In inhabited areas, the dumping of kitchen waste, or a badly planned sewage disposal system may initiate a thermokarst process that may rapidly spread beyond control and cause considerable damage to buildings.

A conspicuous feature of thermokarst topography is found in the so-called cave-in-lakes, kettle lakes, or kettle-hole lakes that are common features of the north country. They result from thawing of permafrost in material that occupies less volume when thawed than in the frozen condition. Wallace [21] is of the opinion that cave-in-lakes may originate through a break in the insulating cover of vegetation, and their preservation and enlargement is the result of the storing of heat by the water occupying the lake basins. In time they may become completely covered over by growth of moss and other types of vegetation. S-shaped tree trunks and trees dipping toward a lake along with the "drunken forests" associated with swelling ground may aid in identifying permafrost conditions from the air.

Other topographic features formed by thermokarst processes are long and deep surface cracks that form in summer and as a rule last through the winter. Such cracks often precede landslides, and may open a passage for the outflow of slud. Cave-ins and funnels are small depressions that are either dry or filled with water. The shape may be conical, cylindrical, or irregular, and the sides may be smooth or uneven and traversed by numerous cracks. Sinks and saucers are shallow depressions which range from a few feet to several hundred feet in width and may be several feet deep. The side slopes are either uniformly gentle or slightly concave. The bottom may be somewhat uneven and contain small depressions and mounds. The northern slopes are usually more gentle than the southern. "Valleys," "gullies," ravines, and sag basins are usually large and may attain a length measured in miles. Their surfaces are often modified by the smaller features mentioned above.

Landslides, slump, creep, and solifluction are aided by permafrost conditions, and are very characteristic of areas underlain by perennially frozen soils. Unstable ground may be held in position by winter frosts, but when thawed becomes very mobile. Landslides and slumps are more common on south-facing slopes than on the opposite sides of hills, a feature of considerable engineering significance. Ground overlying a shoulder of the permafrost table is more susceptible to soil movement than ground overlying a more nearly level permafrost table. The moisture content of thawed material above the permafrost may be in excess of 100 per cent of the dry weight, and renders the material highly susceptible to flowage. Sliding of unfrozen soil over the upper surface of the permafrost is the commonest form of earth movement.

Preventive measures against landslides include: (1) control of the hydrologic regime by diversion of seepages and provision of drainage; (2) retention of forests; (3) planting of vegetation on unstable slopes; (4) avoidance of locations where there is a sharp drop in the permafrost table; and (5) location of quarries and pits at suitable distances from roads, railways, etc.

ENGINEERING PROBLEMS ASSOCIATED WITH PERMAFROST

In regions where winter frosts produce such features as pingos, frost blisters, and icing mounds, and summer thaws may result in widespread caving and the soil flowing like molasses, it is clear that ordinary construction methods must be considerably modified. When permafrost problems were first encountered, attempts were made to fight the forces of nature by using stronger materials, more rigid designs, and resorting to periodic and expensive repairs. The failure of these methods led to the realization that construction procedures and design must reckon with the prevailing natural conditions. Once the problems are understood and correctly evaluated, their solution is usually a matter of common sense. The frost forces may be used to work with rather than against the engineer. The understanding of natural conditions is largely a geological matter, and in few phases of engineering is a close co-operation between engineers and geologists more essential to success.

Two methods of procedure may be employed, the first and most generally applicable of which may be termed the passive method. In this procedure the existing conditions are disturbed as little as possible, or additional insulation may even be provided so that disturbance of the thermal regime caused by construction does not result in thawing of the underlying ground with consequent impairment of stability. Where the permafrost is thin and the soil has a satisfactory bearing strength when thawed, measures are taken to thaw the ground prior to construction. This procedure is known as the active method.

Building Construction

Many engineers recommend that in areas where the permafrost has a temperature of $-5°$ centigrade the passive method should be employed. In areas where the temperature of the frozen ground is between $-5°$ and $-1.5°$, either method may be employed, depending on local factors such as presence of ground ice, etc. In the southern part of the permafrost region where the temperature of the perennially frozen ground is above $-1.5°$, the active method should be employed except where elimination of frost will result in the development of plastic and unstable ground. A stable foundation can be built on bedrock by either method, but in general the passive procedure is to be preferred. The factors that must be considered in the design of foundations in permafrost areas are too numerous to be considered in detail. Those interested in this subject should consult Muller's thorough treatment of the problem.

The best way to avoid the destructive effects of swelling is to place the foundation on piles embedded in the permafrost for a depth sufficient to insure an anchorage capable of withstanding the upward thrust of the swelling soil. Continuous foundations, particularly of masonry, are not satisfactory. They do not withstand tensional strain, and because of the large contact surface and high heat conductivity tend to disturb the thermal regime. As a rule, piles driven into the permafrost for a distance equal to twice the thickness of the active layer will not be uprooted by swelling of the ground. To insure a firm grip on the pile by the permafrost, it is desirable to fill part of the hole with thick mud, which when frozen securely grips the pile. Ordinary methods of pile driving cannot be used, and it is necessary to thaw the ground by means of steam points. Piles driven into prethawed ground should be placed at the proper season so there will be no chance of uprooting by swelling ground before they have become securely frozen to the permafrost. The best period is between February and June.

In some instances where the soil consists of materials very susceptible to swelling, it may be advisable to replace the soil by materials such as clean gravel and coarse sand. Suitable precautions should be made to prevent subsequent silting of the sand and gravel fill.

Precautions against damage by settling and caving include the following: (1) Avoid construction at sites where minor relief features will be disturbed with consequent lowering of the permafrost table. (2) If present, the natural cover of grass, moss, peat, and shrubs should be maintained. (3) Structures should be painted white. (4) Foundations should rest on a raft-like platform of logs, excavations should be lined with insulating materials, and the space between the floor and ground should be ventilated. (5) The load of a foundation resting on unfrozen ground should be computed with a wide margin of safety. (6) The basal part of the foundation should be capable of withstanding considerable torsional stresses resulting from uneven thawing of underlying ground. (7) Weight of the structure should be centered with the center of gravity at the base of the foundation.

Settling may be minimized by separating the foundation from the permafrost by a layer of unfrozen sand. If the layer is of sufficient thickness the pressure of the foundation is distributed over a fairly wide area minimizing, if not entirely eliminating, the effects of uneven settling or caving. If lenses of ice and slud are encountered during excavation, they should be removed and backfilled with sand or gravel. Excessive moisture should be drained and seepages checked.

Highways

Highway construction in permafrost regions presents many difficult and unusual problems resulting from disturbance of the thermal regime by cuts and fills. It should be clear that choice of a route and details of alignment cannot be governed entirely by such factors as topography and distance. Consideration must be given to the direction of slope with reference to the highway alignment, the angle of slope, the thermal and hydraulic regimes, and nature of the soil. North slopes are less subject to landslides and are therefore preferable to south-facing slopes. Availability of construction materials must also be taken into account.

The permafrost table will rise beneath fills, and if the height of the fills is in excess of the thickness of the active layer it will extend into them. Furthermore, seasonal thawing will extend to greater depths in fills than in adjacent ground, especially on their south sides. As a result, a sag develops in the permafrost table in which moisture may accumulate and cause landslides. This may be prevented by covering the south slopes with material of low heat conductivity such as sod or peat. Particular care should be taken in constructing roads through swamps, for they are frequently underlain at shallow depths either by permafrost or by ice.

The effect of fill construction on slopes results in a rise of the permafrost table that permits drainage on the downslope side and intensifies the formerly existing swampy conditions on the upslope side. This situation is favorable to the development of landslides. Such a condition may be remedied by draining the water accumulating on the upslope side, or by the construction of berms. The materials composing the berms should be poor conductors of heat. Berms will also aid in preserving the frozen condition of the soil below fills and prevent the thawing of underground ice.

Cuts offer even greater difficulties than fills. At least a part of the active layer is removed, and often the permafrost is penetrated. A substantial lowering of the permafrost table results, and formerly firm ground is converted into an unstable mass, resulting in landslides and slumps. The only solution seems to be to replace the unstable ground by more suitable material, usually a very costly business.

A very serious obstacle to highway traffic results from icings. Seepage water from beneath the ice may freeze on or near a

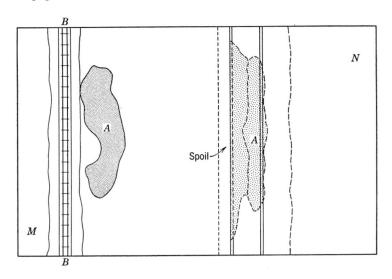

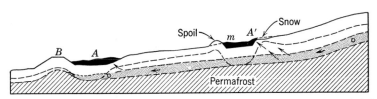

Fig. 121. Diagrams illustrating elimination of icing by the "frost belt" method. Icing *A* along road *B* is eliminated by ditch which causes icing to shift to position *A'*. (After S. W. Muller.)

road in successive sheets that may attain a thickness of many feet. In Siberia, great stretches of highway have been made entirely impassable by icings.

In instances where the water causing the icing is derived from relatively shallow sources, a corrective measure known as the "frost belt" has proved useful. As shown in Fig. 121, icing is eliminated by causing the water supplying the icing to form another icing some distance upslope and away from the road. This is accomplished by removing the vegetation and constructing a shallow ditch parallel to the highway. As a result, an icing will form either in the ditch or immediately above it. If large icing mounds form above the frost belt, they should be punctured and drained to prevent their exploding. To be effective, a frost belt should be constructed before the first snowfall, and snow should be removed from the belt until the end of January. After that time the snow may be left in place for it tends to preserve the low temperature of the ground. If the belt is placed where there is no natural insulating cover such as sod or peat, a ditch is not necessary; freezing may be accomplished merely by removing the snow. This method is not effective if the icing is fed from a deep source of water. Furthermore, after a year or so of operation the permafrost table will develop a sag under the ditch and water will percolate under the roadbed as before. This may be avoided by covering the ditch with an insulator in spring and removing it in the fall. This procedure is effective but costly.

A generally more effective method consists of cutting a narrow trench across the water-bearing layer to a depth of about 8 feet. The trench is then backfilled with clay which is well tamped. A row of sheet piling is then driven into the clay and covered with fill on both sides. The fill causes a rise of the permafrost table, and water percolating downslope will be forced to the surface by the impervious fill to form an icing on the uphill side.

Gravel-surfaced roads are often made impassable by the development of frost boils. These are moisture-saturated, semifluid pockets which develop during the spring thaw; if broken by traffic, they turn into a quagmire. Frost boils are caused by the more rapid thawing of the ground under the roadbed than on the shoulders. The snow removed from the roadbed in winter is piled on the sides where aided by the presence of sod and other thermal insulators retards the progress of thawing. Thawing of the materials under the roadbed produces an excessive amount of water which cannot escape laterally through the still frozen ground under the shoulders. Consequently, it may "boil" up along cracks and fissures, or remain undetected beneath the surface.

Frost boils may be eliminated by excavating the entire width of the roadbed and backfilling the space with at least 30 inches of gravel. Two lines of drains are laid along the edges of the fill.

Bridge Piers. It will be recalled that the permafrost table usually exhibits a pronounced depression under rivers and other water bodies. Consequently, bridge construction must be based on procedures that apply to the active layer rather than to permanently frozen ground. Proper precautionary measures must be taken to safeguard against possible damage by river icing. In places where the river channel is deep bridge piers are less likely to be damaged than where it is shallow. Shallow reaches are usually characterized by deeper penetration of frost and usually freeze more quickly than deeper spots. Freezing may even extend into the underlying river bed. As a result, considerable hydrostatic pressure may develop resulting in river icing and damage to piers. A pyramidal design has been found satisfactory, for the lateral thrust of the ice tends to push the piers downward and thus increases their stability. Where practicable, suspension bridges should be used.

Airfields

The chief problem connected with airfields consists in finding a fairly large area of level ground underlain by fairly homogeneous materials. This is usually difficult in the Arctic and subarctic regions where inhomogeneous glacial drift is widespread. Runways may be installed in tundra country if suitable drainage is provided. A surface of sand or fine gravel is best, for minor frost damage can be readily repaired. Warping of the pavement surface can be minimized by stripping off fine-grained materials and replacing them by sand or gravel. Runways may be constructed on the undisturbed ground surface by employing materials of low heat conductivity such as pumice, slag, and porous lava. Wooden runways may also be built on a gravel blanket of sufficient thickness to prevent thawing of the underlying permafrost. Runways may also be constructed on ice, and in summer the numerous lakes may be employed for landings.

Water Supply

Despite the abundance of water in summer, securing an adequate year-round water supply is often very difficult. The most dependable source of a perennial water supply is from wells that penetrate the permafrost, but this may require deep and difficult drilling. In some inhabited areas the melting of ice and snow forms the only means of water supply. Even if a satisfactory source is provided, distribution problems are so numerous and difficult that they should be considered first in the planning of a major project. If a suitable supply cannot be developed at, or near, the site, the location should be moved or the project abandoned.

The following sources of water may be utilized: (1) meteoric water and snow, (2) ground ice and surface ice, (3) surface waters, lakes, rivers, and (4) ground water. The first may be utilized directly from the runoff, by gathering the rain and snow from roofs and specially prepared surfaces, by condensation of moisture in the air in condensers constructed for that purpose, and by reservoirs built to conserve runoff. Of these only the last is practicable on any considerable scale. River, lake, and glacier ice has long been used as a water supply. In winter blocks of ice are stacked in cellars or near a dwelling and then used as needed. Rivers that can be used as a source of water supply are not very numerous, and in winter they may freeze over completely. However, even in deep streams that do not freeze over entirely, the water may become stagnant and unfit for use. Only the larger rivers and lakes furnish water throughout the year.

Generally, the ground water furnishes the best source of supply, especially in the southern portions of the permafrost province. Farther north the great thickness of the permafrost may form a formidable obstacle to development of ground water sources. Most springs go dry in winter, but some fed from *taliks* may flow throughout the year. Search for such springs is best carried out in winter when the ground is frozen to a maximum depth. Common evidences of such springs are seepages, icings, and other types of frost mounds.

The zone above the permafrost will furnish a year-round source only if the active layer does not freeze to the top of the permafrost during winter. Such conditions may occur along the banks of large rivers and lakes with a constant inflow and outflow. Unfrozen water above the permafrost may also occur at the mouths of valleys and at the heads of alluvial fans. Shallow waters are subject to contamination and require careful investigation and protection.

Water within the permafrost occurs mainly in thick alluvial deposits near rivers or old river channels. It may also be found in rocks that are strongly fissured and jointed. Water within the permafrost percolates through permeable layers or may flow freely through pipe-like openings. In developing this source of supply, care must be taken not to retard the rate of circulation or freezing will result. Excessive or accelerated pumping may also cause freezing of the underground water channel.

Water from below the permafrost furnishes the most dependable source. The yield is usually large and of good quality. Prospecting for subpermafrost water may be very difficult because of scarcity of surface exposures and indications. Those interested in this phase of the subject should consult Muller's book on permafrost.

The laying of water pipes must be preceded by a thorough study of the thermal regime. A *talik* offers an ideal medium for water pipes, but where not available ground with a prolonged zero curtain is to be preferred. A zero curtain can be produced artificially by imbedding the pipes in moist ground of low permeability surrounded with a layer of material of low heat conducting capacity such as peat and moss. The heat balance around the pipe may be further regulated by placing the pipe in a trench backfilled with sand. The sand should be wetted in summer; in this condition it aids in conducting heat into the ground. The sand is allowed to dry out before the onset of frosts and in this condition acts as a blanket which aids in storing heat. Preheating of water in the pump house is also a satisfactory method of preventing freezing. Buildings that consume the largest amounts of water should draw their supply from the end of the water main.

In this way, rapid circulation is maintained and the danger of freezing is lessened.

Geological Investigation of Frozen Ground

From the foregoing, the importance of a thorough investigation of permafrost areas prior to planning and construction should be clear. It is certainly better to spend a few hundred thousand dollars on preliminary work than to pour millions into temporary and often futile repairs. Preliminary studies should be broad in scope and should include: (1) topography, (2) earth materials, (3) hydrology, (4) climate, (5) botany and ecology, and (6) laboratory tests and experiments. Space does not permit discussion, and for further information the reader should consult Muller's [11] and Nees' [12] treatment of the subject.

REFERENCES

1. Gunnar Beskow, *Tjälbildningen Och Tjällyftningen,* Norstedt & Soner, Stockholm, 1935.
2. Robert F. Black, Permafrost, in *Applied Sedimentation,* John Wiley & Sons, New York, 1950.
3. G. J. Bouyoucos, Movement of Soil Moisture from Small to Large Capillaries of the Soil upon Freezing, *J. Agr. Research,* Vol. 24, 1923.
4. Kirk Bryan, Cryopedology—The Study of Frozen Ground and Intensive Frost-Action with Suggestions on Nomenclature, *Am. J. Sci.,* Vol. 244, 1946.
5. H. Carlson and M. S. Kersten, Calculation of Depths of Freezing and Thawing under Pavements, Presented at Highway Research Board Meeting, January, 1953, Summary in *Highway Research Abstracts,* December, 1952.
6. Arthur Casagrande, Discussion of a Paper by A. C. Benkelman and F. R. Olmstead, *Proc. 11th Ann. Meeting Highway Research Board,* pp. 168–173, 1932.
7. Bertil Högbom, Über die Geologische Bedeutung des Frostes, *Bull. Geol. Inst. Univ. Upsala,* Vol. 12, 1914.
8. A. W. Johnson, Frost Action in Roads and Airfields—A Review of the Literature, *Spec. Rept. No. 1, Highway Research Board, National Res. Council,* 1952.
9. *J. Geol.,* Vol. 57, No. 2, March, 1949.
10. Ernest de K. Leffingwell, The Canning River Region, Northern Alaska, *U. S. Geol. Survey, Prof. Paper* 109, 1919.
11. Siemon William Muller, *Permafrost or Permanently Frozen Ground and Related Engineering Problems,* J. W. Edwards, Ann Arbor, 1947.

12. Louis A. Nees and A. Morgan Johnson, Preliminary Exploration in Arctic Regions, *Am. Soc. Testing Materials, Spec. Tech. Publ.* 122, 1952.

13. A. Erling Porsild, Iagttagelser over den Grønlandske Kildeis (Grl: Sêrsineq) og dens Virkninger paa Vegetationen og Jordoverfladen, *Saertryk af Geografisk Tidsskrift*, Bind 28, Hefte 3, 1925.

14. A. Erling Porsild, Earth Mounds in Unglaciated Arctic Northwestern America, *Geog. Review*, Vol. 28, No. 1, 1938.

15. R. F. X. Ruckli, *Der Frost im Baugrund*, Julius Springer, Vienna, 1950.

16. R. P. Sharp, Periglacial Involutions in Northeastern Illinois, *J. Geol.*, Vol. 50, 1942.

17. R. P. Sharp, Soil Structures in the St. Elias Range, Yukon Territory, *J. Geomorphology*, Vol. 5, No. 4, 1942.

18. R. P. Sharp, Ground-Ice Mounds in Tundra, *Geog. Review*, Vol. 32, 1942.

19. Stephen Taber, The Mechanics of Frost Heaving, *J. Geol.*, Vol. 38, 1930.

20. Stephen Taber, Perennially Frozen Ground in Alaska: Its Origin and History, *Bull. Geol. Soc. Amer.*, Vol. 54, 1943.

21. Robert E. Wallace, Cave-In Lakes in the Nabesna, Chisana, and Tanana River Valleys, Eastern Alaska, *J. Geol.*, Vol. 56, 1948.

22. A. L. Washburn, Reconnaissance Geology of Portions of Victoria Island and Adjacent Regions in Arctic Canada, *Geol. Soc. Amer., Mem.* 22, 1947.

CHAPTER 12

Landslides and Related Phenomena

In preceding chapters it has been shown that surface materials may be transported by water, wind, and ice, the over-all process being known as erosion. In addition to these types of transportation, mass movements of materials activated directly by the force of gravity are very common. This process is sometimes referred to as mass wasting, and the movements involved may occur either in the dry state or grade into fluvial or glacial types of transportation. To the geologist, landslides and related phenomena are phases of the erosional process, and geological studies emphasize their origin and systematic classification. As a rule, the geologist confines his studies to mass movements that have already taken place, and analysis of the stability of existing slopes usually does not enter into his considerations. The civil engineer is less interested in past slides than in future possibilities, and analysis of slope stability forms an important part of soil mechanics. Since the ideal engineering classification of landslides, based on the mechanisms and causes involved, has not yet been formulated, it seems best to treat the subject from a basically geological point of view. The most recent and thorough geological study of the subject is that by Sharpe,[12] and much of the material in the following pages has been taken from this source.

Landslides and related phenomena are very important to civil engineering. Although a large proportion of destructive slides of one type or other takes place in sparsely inhabited regions and therefore goes unnoted, the annual damage to highways, railroads, buildings, power lines, etc., in Pennsylvania, West Virginia, and Ohio alone runs into millions of dollars. Furthermore, slides encountered during construction are often a very considerable item of the total cost. It is estimated that up to January 1, 1917, approximately 52,000,000 of the more than 168,000,000 cubic yards excavated in the Gaillard Cut of the Panama Canal were necessitated by landslides. In addition, a considerable number of lives have been lost from this cause.

Many destructive earth slides might have been avoided had their significance been appreciated. In many locations potential or actual slide conditions were not recognized, and in others the cost of changing the alignment of a highway or railway to areas not menaced by slides appears to have been thought too great. However, it is probable that the cost of clean-up and corrective work ran considerably higher than the difference in estimated cost of an initial change of line.

In construction projects such as dam sites, reservoirs, highways, railways, etc., the geologist should be alert to detect indications of slides, or conditions favorable to their development. He should inform the engineer of any possibilities of destructive earth movements and should be prepared to suggest corrective measures. In some instances, slides caused by liquefaction of loose sands, for example, it is impossible to estimate the danger without accurate and detailed laboratory studies. Here geological field studies must go hand in hand with soil mechanics investigations. The function of the former is to present a complete and accurate description of natural conditions in the area of interest; the role of the latter is to furnish estimates of slope stability.

TYPES OF MASS MOVEMENTS AND BASIS OF A GENERAL CLASSIFICATION

The classification of mass earth movements developed by Sharpe [12] rests on a wide variety of factors which include: (1) type, size, cause, and rate of movement; (2) water content; (3) type of material involved; (4) characteristics of internal friction and organization of material within the moving mass; and (5) relationship of the moving mass to surface materials and the substrata. The classification includes 10 major and 8 subtypes, but it is probable that if the basic causes were better known this number could be very considerably reduced. Earth movements of a compound nature are common, and, rather than multiply categories, it has been suggested that they be described as being of a transitional or compound character; for example, "combined rockslide and rockfall." The classification includes earth and rock movements on land only, but submarine slips are not uncommon. Submarine cables have been known to be damaged by them.

The distinction between the two principal types of mass movements, slides and flows, depends on the presence or absence of a slip plane separating the moving mass from stable ground. In flowage of any kind, either viscous or plastic, no slip plane is present and movement takes place by continuous deformation. True slides, on the contrary, move on a slip surface. Deformation is not continuous but involves finite shear. As in the case of most natural phenomena, no absolutely sharp boundaries can be drawn between groups. In some types treated as flows the initial movement may start with slippage; in some treated as slides the initial sliding or falling movement may develop into flow. The major types are, therefore, divided according to their predominant characteristic. Typical forms may be readily recognized in nature, and except for a few borderline cases a distinction is fairly easily made.

On the basis of type of movement, mass movements may be divided into the following major groups: (1) slow flowage, (2) rapid flowage, (3) slides, and (4) subsidence.

SLOW FLOWAGE

Slow flowage, or creep, includes solifluction and other forms of slow creep, the motion of which as a whole is by continuous deformation. Movement is generally so slow as to be imperceptible, except to observations of long duration. It affects all sizes of material from joint blocks tens of feet across to fine sands, silts, and clays. The water content may range from zero to the saturation point and may be in the form of liquid or interstitial ice. Generally, only a thin surface is affected, but the vegetation and man-made structures are carried along with the moving mass. As shown by Table VII, five classes of creep movements may be distinguished.

TABLE VII. CLASSIFICATION OF CREEP MOVEMENTS

(After C. F. Stewart Sharpe, *Landslides and Related Phenomena*, copyrighted by Columbia University Press, New York, 1938.)

Movement		Type of Material and Class of Movement		
Kind	Rate			
		Earth or rock plus ice	Earth or rock, dry or with minor amounts of ice or water	Earth or rock plus water
Flow	Usually imperceptible	Rock glacier creep, solifluction	Rock creep, talus creep, soil creep	Solifluction

Soil Creep

Evidence of creep of the soil mantle is found on almost every soil-covered slope. Curved tree trunks; tilted fence posts, telegraph poles, and monuments; lines of stone accumulation in the soil; broken and displaced retaining walls and foundations; and displacement of highway and railway alignments are only a few of the signs of its presence. A common evidence of creep is detached fragments of bedrock forming imperfect bands or drawn out into an irregular downhill pattern. Where developed, they appear as a broken line of stones, suggesting the term "stone line" applied by Sharpe.

The rate of creep on a hillside depends not only on climatic conditions and angle of slope but also on the type of soil, parent material, and numerous other factors. In materials like true, or unmodified, loess which has strong tendency to stand in nearly vertical walls, creep is at a minimum, whereas in loosely consolidated materials or soils containing a high proportion of rounded particles the rate of movement is more pronounced. Creep may be very effective without giving much evidence of its presence.

Creep is caused by a great number of agencies among which are frost heaving, alternate heating and cooling during day and

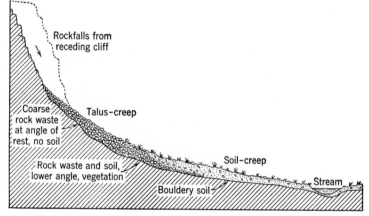

Fig. 122. Section through a cliff, illustrating relationships of soil and rock creep. (After C. F. Stewart Sharpe, *Landslides and Related Phenomena*, copyrighted by Columbia University Press, 1938.)

night, filling of desiccation cracks, moisture changes, and action of plants and animals. Movement is more rapid in partly saturated soils, but takes place even under the most arid conditions. Complete water saturation by heavy rains may greatly increase the rate of movement, and may lead to a sudden and even catastrophic displacement of the soil mantle to which the name "debris avalanche" is applied.

Talus Creep

A talus is an accumulation of rock waste at the foot of a cliff. Ordinarily, talus materials consist of moderately coarse to very large irregular blocks. According to Sharpe, talus creep is most rapid in cold regions where the major cause is the expansive force of alternate freeze and thaw of ice in the interstices. This type of movement grades into a very special type known as rock glacier creep. In warmer climates creep proceeds in more leisurely fashion and is largely due to simple diurnal temperature changes. When the talus contains appreciable amounts of shale, schist, or chlorite, mica, and talc fragments creep can take place on slopes as low as 10°. Terzaghi and Peck [15] state that the best method of stabilizing talus consists of providing adequate drainage. Dry or well-drained talus is generally stable and standard slopes can be maintained without undue difficulty.

Rock Creep

Although masses displaced by the process of rock creep may have moved many feet, their original relationship to the ledge from which they came can generally be determined. The movement consists mainly of sliding rather than flowage or floating,

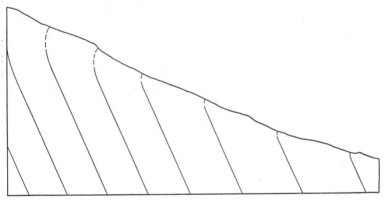

Fig. 123. Outcrop curvature resulting from downhill creep.

but the slow downhill slippage of jointed blocks is also included in this category because of the close relationship to soil and talus creep. Blocks of hard rock often creep downslope gradually widening the gap between them and the parent ledge. Frost action and prying by roots are major causes of such movement, although presence of a thin layer of argillaceous material beneath the block may cause movement even in their absence. Laminated rocks, such as shales, and thinly bedded sandstones and limestones exhibit a different type of creep. In hillside outcrops these strata often sag or may even show a reversal of true dip (Fig. 123). This phenomenon has been variously termed settling, terminal creep, and outcrop curvature. It has been known to be the cause of erroneous structural mapping.

Solifluction

Solifluction consists of the slow flowing from higher to lower ground of masses of waste saturated with water. It has been already mentioned in connection with permafrost, and though

Fig. 124. Solifluction lobe, Victoria Island, Arctic Canada. (A. L. Washburn.)

not confined to cold climates it is almost universally present on slopes in high latitudes. The material involved in solifluction is similar to that subject to soil creep in warmer climates. The bulk of the mass consists of fine debris, but blocks of considerable size may be carried along in suspension. Solifluction bodies creep either as a sheet of debris or as glacier-like masses concentrated in a system of main and tributary channels analogous to those of streams of water. In the latter instance, the rate of movement is more rapid. Movement may occur on slopes as flat as 2° or 3°. Solifluction differs from mudflow in the slower and more continuous movement. Furthermore, it is not necessarily confined to a channel, but may cover an entire slope.

Features of cold climates that favor solifluction are: (1) continual supply of water from melting snow and ice, (2) a soil relatively free of vegetation, and (3) deeply frozen ground which thaws from the surface downward, the ground water drainage being limited to the upper unfrozen layer only a few feet in thickness.[12] The striped ground commonly found in subarctic

Fig. 125. Striped ground developed in basaltic rock, Wolf Creek region, Yukon Territory, Arctic Canada. (R. P. Sharp.)

regions (Fig. 125) and at high elevations is probably related to solifluction.

Rock Glacier Creep

Rock glaciers are glacier-like tongues of angular rock waste which usually head in cirques or other steep-walled basins. In many cases they grade into true glaciers. They are most common in mountainous regions that have been moderately glaciated. It appears that a rock glacier may originally obtain its material from a wasting glacier of ice, from one or more landslides, or even from the slow continuous dropping of blocks that would normally form a talus, but which under special conditions of climate and topography form a rock glacier. The surface features of rock glaciers seem to result from flowage (see Fig. 126).

Fig. 126. Mud-rock glacier near McCarthy, Alaska. Slope in lower

right-hand corner is scarred by landslides. (Bradford Washburn.)

CAUSES OF CREEP

Creep may be caused by a great variety of conditions the most important of which are the type of material, its physical condition, angle of slope, climate, and vegetation. Unconsolidated materials and thinly bedded formations are affected most. Other factors being equal, the rate of creep tends to vary directly with slope and the amount of water present. Creep is also promoted by frequent freezing and thawing, and the presence of a vegetal cover sufficient to absorb precipitation but not sufficient to anchor the soil. The provision of adequate surface and internal drainage is the most generally applicable and effective method of controlling or minimizing creep.

RAPID FLOWAGE

In slow flowage the water content is relatively low. With increased water content more rapid movement is facilitated. The extreme is reached in fluvial transportation where the water is greatly in excess of the amount needed to counteract internal friction. Rapid flowage, in which the amount of water present is little more than that required to overcome internal resistance of the mass, is thus intermediate between fluvial transportation and landslides.

Earthflow

Earthflow is the least rapid of the three types of rapid flowage. In most instances, earthflow occurs either as a low-angle flow or as a somewhat steeper hillside flow. Flowage movements on gentle slopes are typical of humid areas.

Earthflows are common in tropical regions, where weathering may result in the formation of a somewhat coarser and more porous layer near the surface, underlain by a zone of relatively impervious clay. Owing to the abundant rainfall, the porous surface layer acts as a reservoir that furnishes a nearly continuous supply of water to the underlying clay. The earthflows that result may take place on slopes as low as $3°$.

The earthflows studied in the field are often found to be characterized by slumping at the head of the flow and bulging, which causes the development of a strongly crevassed dome, at the toe. In general, a series of down-slipped blocks, the surfaces of which are tilted back into the slope, occur at the head of the flow; a buckled and disrupted zone with wide fissures transverse to the direction of movement is situated in the central area; and in the lower portion there is an anticlinal ridge, or a series of ridges,

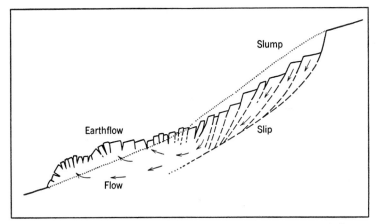

Fig. 127. Section through an earthflow, illustrating association of flowage and slumping. (After C. F. Stewart Sharpe, *Landslides and Related Phenomena,* copyrighted by Columbia University Press, 1938.)

the foremost of which may be overthrust. The upper portion is bounded by a scarp. The toe is characterized by a raised rim of debris which may resemble a natural levee. Many of the slips reported by highway and railway engineers as "landslides" are actually earthflows. As the rate of movement increases, earthflows grade into debris avalanches.

Flow Slides Caused by Liquefaction

It is known that if a mass of dense sand is subjected to a shearing stress the volume increases. Conversely, if a mass of loose sand is sheared or vibrated the volume decreases. If the material is water-saturated and poorly drained, the decrease in volume results in at least a temporary increase in the pore-water

pressure. As a result, the internal support of the particles is transferred from the contacts between individual grains to the pore-
water, causing the sand to liquefy and flow more or less like a
fluid. Flow slides caused by liquefaction of loose sands are characterized by gliding surfaces inclined on a slope as low as 3°.
They tend to be pear-shaped in ground plan, with most of the
material flowing out through the relatively narrow stem at the
bottom. If the sands are overlain by more cohesive materials,
failure of the upper layers may take the form of a series of shear
displacements which obscure the basic mechanism involved.
Many failures ascribed to slump or shear are possibly the result
of liquefaction of underlying sands. Liquefaction slides can apparently be set off by relatively insignificant forces such as surficial sloughing or the impact of waves. Peck and Kaun [11] have
described a flow slide which was started by the pulling of two
sheet piles. In less than 30 minutes about 800 cubic yards of
sand flowed from an opening only 2.5 feet wide. Flow slides
are common occurrences along the coasts of the Dutch Province
of Zeeland,[7] where no fewer than 229 were recorded between
1881 and 1946, and on the shores of the Lake of Zug, in Switzerland.[9] They are also known to occur along the banks of the
lower Mississippi River where levees and revetments have been
damaged by them. Certain types of clays may act in a manner
similar to loose sands and silts, and slides in these so-called "quick
clays" have been especially common in Sweden and Norway.[6]

Mudflows

A typical mudflow differs from an earthflow in several important respects. Because of the higher water content and steeper
slope, mudflows usually move more rapidly. They generally
follow former stream courses; and, unlike earthflows, commonly
recur in the same channels. Mudflows are produced when water
is suddenly applied to an area where a suitable and abundant load
is already available. They are strictly a flow phenomenon, not
necessarily accompanied by slides, although small slumps and
slides may occur along their margins. The load consists of a
tumbled mass of heterogeneous debris, whereas an earthflow ordinarily includes larger segments of the land surface, and, when
movement is sufficiently slow, may preserve the continuity of the

mass with only minor fragmentation, tilting, and deformation. The conditions most favorable to the development of mudflows appear to be: (1) abundant but intermittent water supply, (2) absence of a substantial vegetal cover, (3) unconsolidated or deeply weathered materials containing enough clay or silt to aid in slipping of the mass, and (4) moderately steep slopes.[12] These conditions are found very frequently in desert regions, and the mudflows associated with alluvial fan building have already been described in Chapter 8. Three rather well-defined types of mudflow can be recognized: arid or semiarid, Alpine, and volcanic. A fourth type, but one of less importance, may result from the outflow, or "bursting," of bogs.

Debris Avalanches

Debris avalanches are most abundant in humid regions where there is a well-developed cover of vegetation. Under these conditions the torrential mudflow of the arid or semiarid type is replaced by a species of flowing slide termed a debris avalanche. If the water content were lower, the same material would move as a debris slide.[12]

A typical debris avalanche has a long and relatively narrow track and occurs on steep mountain slopes or hillsides. It is usually preceded by heavy rains which increase the weight of the material and aid in its movement. In exceptional cases, the water may be "wrung" out of the earth by earthquake shocks.

Initial movement and displacement on the steeper upper part of the course is caused by slippage, at times on a smooth underlying rock surface and at times within the loosened debris. Recorded slopes range from about 20° to 40° near the top but flatten to 15° or less toward the base where the accumulated momentum and high water content cause the mass to flow in a manner similar to a mudflow.

CAUSES OF RAPID FLOWAGE

Conditions favoring rapid flowage are generally similar to those promoting creep. The necessary condition for rapid flowage is the reduction of internal friction. This usually results from the presence of excessive amounts of water from rain, snow, hail, streams, springs, etc., or, in volcanic mudflows, condensed steam or water from a crater lake. In liquefaction slides, the water is furnished from the sliding mass itself. Many rapid flows

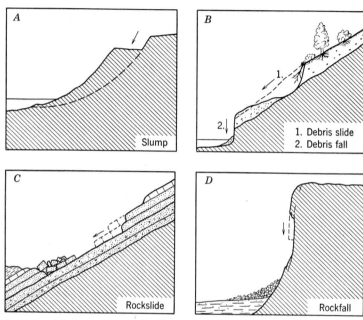

Fig. 128. Diagrams illustrating five classes of landslides. *A.* In slump the mass rotates backward so that the slope of the upper surface of the block is diminished or reversed. *B.* A debris slide may be of any size, may or may not expose bedrock, and often produces a surface resembling morainal topography. Debris fall occurs when unconsolidated material drops from a vertical cliff. *C.* Rock slides may occur along bedding, joint, or foliation surfaces, or other planes of weakness. *D.* Rockfalls occur on cliffs oversteepened by rivers, glaciers, or waves. (After C. F. Stewart Sharpe, *Landslides and Related Phenomena*, copyrighted by Columbia University Press, 1938.)

start with a sliding motion, and their initiating causes are in general similar to those of landslides. Rapid flowage is difficult and often impossible to control. Good surface and subsurface drainage are important, but in general they cannot be relied upon exclusively. It is usually better to recognize areas susceptible to rapid flowage and avoid them, if possible.

LANDSLIDES

A landslide is a "perceptible downward sliding or falling of a relatively dry mass of earth, rock, or a mixture of the two." [12] The slope is usually steep and the movement rapid to moderately rapid, but exceptions may occur. Masses involved in landslides usually have fairly high water contents, but if much excess liquid is present the mass may assume the characteristics of a debris avalanche, mudflow, or earthflow. Landslides may be divided into several classes according to the rate and character of movement and the type of material involved. The material consists chiefly of earth and loose debris, rock, or newly detached fragments of bedrock. A scheme of classification is presented in Table VIII.

TABLE VIII. CLASSIFICATION OF LANDSLIDES

(Modified after C. F. Stewart Sharpe, *Landslides and Related Phenomena*, copyrighted by Columbia University Press, New York, 1938.)

Movement		Type of Material and Class of Movement		
Kind	Rate			
		Earth or rock plus ice	Earth or rock, dry or with minor amounts of ice and water	Earth or rock plus water
Slip	Slow to rapid		Slump (or deep deformational or shear slide)	
	Very rapid		Debris slide Debris fall Rockslide Rockfall	

Shear Slides, or Slump

The terms shear slides or slump are applied to slides in rock or unconsolidated materials in which movement takes place along a definite shearing surface. This surface is usually deep-reaching and spoon-shaped. Because of the extensive investigations of this

Fig. 129. Slump in deep residual clays, Ozark Mountains, Missouri.

type of slide carried out by Swedish engineers, the gliding surface is often referred to as the "Swedish break" or "Swedish circular arc." Generally, two or more roughly parallel slip surfaces are present, and the displaced materials form a series of blocks the surfaces of which are tilted back into the slope from which the slide descended.

Shear failures on the sides of cuts and artificial fills are common. Extensive slides of this type occurred during construction of the

Panama Canal and were termed "deep deformational slides" by the investigating committee.[10]   Slides of the deep deformation, or shear, type occur on steep slopes when the height and weight of the banks exceeds the shearing strength of the materials involved.   There is thus a limiting height for a vertical cliff in any type of rock.   This height is proportional to the shear strength and ranges from as little as 40 feet or less for soft clays to an estimated 4000 feet or more for granite and other strong rocks. In theory, if a homogeneous material is assumed, origin of such slides is dependent solely on height and slope of the banks and the shear strength of the material, but in practice geologic structure is also important.   Shearing weakens the material and promotes further movements, and shear slides are among the most troublesome with which the engineer must deal.   If at all possible, areas subject to shear slides should be avoided, but often this danger can be recognized only by very careful and detailed investigations involving both geologic field work and soil mechanics studies.

It should be noted that shear slides may occur long after a cut has been opened.   Ward [18] mentions a slide in England which occurred 70 years after completion of a railroad cut.   Reasons for the time lag are to be found in the progressive deterioration of the rock by exposure to the atmosphere and surface water.   These effects are most pronounced in fissured clays, and, if subsequent slides are to be avoided, suitable allowances for aging should be made when the excavation slopes are being determined.

Terracettes.   Terracettes, also known as "cat steps" and "sheep tracks," is the term applied to small step-like discontinuities on the surface of slopes.   They may be the expression of underlying rock structure, but they are also common in unconsolidated materials.   Field observations show all gradations from true paths made by animals to miniature slip blocks in the origin of which animals had no part.   The formation of true terracettes requires loose soil, a suitable and rather steep surface slope, a certain amount of moisture, and a connected cover of vegetation. The most favorable conditions for the development of terracettes are found on slopes in unconsolidated materials whose bases are free to flow or slide following undercutting.   In some localities the steps appear to have resulted from movement of a series of small fault blocks which may be considered as offshoots of a larger slip zone of the low-angle normal fault type.

Debris Fall

The class of slides known as debris fall is of minor importance and grades on one hand into debris slides and on the other into rockfalls. A debris fall may be defined as "the relatively free falling of predominantly unconsolidated earth or debris from a vertical or overhanging cliff, cave, or arch." [12] They are usually of small magnitude, because materials too poorly consolidated to be classed as rock seldom stand on vertical cliffs of any considerable height. Loess is an important exception, and in the lower Mississippi Valley loess bluffs 40 to 60 feet high are sometimes found. Vertical slabs formed in the loess become loosened by percolation of water along cracks and fissures and topple forward.

Rockslides

By definition, a rockslide is the "downward and usually rapid movement of newly detached segments of the bedrock sliding on bedding, joint, or fault surfaces or any other plane of separation." [12] Rockslides resulting from natural causes are frequent in mountainous areas. Smaller rockslides resulting from human activity, such as undercutting by road or railway excavations, are also common occurrences. They are less common, however, than debris slides and shear slides. The characteristics of rockslides can be made clearer by descriptions of the following outstanding examples.

The Gohna Rockslide. One of the greatest of all known rockslides occurred in the valley of the Birahi Ganga near Gohna, India, in 1893.[5] A small slip took place in 1891, and the cliff then appears to have had a surface slope of about 54°. The rocks exposed in the cliff were crumpled and faulted in a complicated manner and dipped on the average about 45° to 50° toward the valley. The slide of 1893 continued for three days with deafening noise and clouds of dust that fell for miles around. Further slips occurred at intervals following heavy rains. Blocks of several tons broke from the ledge and traveled more than 3000 feet. The first part of the slide pitched forward and came to rest against a cliff on the opposite side of the valley nearly a mile away. The second did not travel as far as the first and left a saddle between the two masses of debris. The debris reached for two miles along the valley to a depth of at least 850 feet at the saddle previously mentioned.

The reservoir formed by the debris filled with water which overflowed the dam on August 25, 1894. The resulting flood released 400,000,000 cubic yards of water in about 4 hours and rose to a height of 140 feet in the valley downstream. Had not the danger been anticipated and the population warned, many thousands of lives would have been lost, but owing to intelligent control of the situation only one person lost his life. A permanent lake nearly 3 miles long, about 1 mile wide, and over 300 feet deep remains.

The chief cause of the catastrophe was the very steep dip of the limestone and shales forming the cliff aided by undercutting by the stream.

Turtle Mountain Rockslide. Another well-known rockslide took place at Frank, Alberta, in 1903.[4] The slide was of the composite type and involved both sliding and falling motions. This slide centered near the highest, or southern peak, of Turtle Mountain, the top of which stood 3100 feet above the valley. This peak is composed of massive limestones overthrust on much younger and softer sandstones and shales. Many of the limestone beds were strongly folded, and considerable material had been removed by mining of coal in the sandstone series. It is probable that percolating water and frost action also contributed to instability of the peak. Earthquake shocks some 2 years prior to the slide also may have aided in preparing the way for the catastrophe. The movement started on a well-developed system of joints which dipped toward the valley at an angle of about 45°. Where these planes were truncated by erosion the movement changed to rockfall which involved both sliding and rolling on a steep talus. The volume of the slide is estimated at 35,000,000 to 40,000,000 cubic yards. It spread across the valley and reached 400 feet up the opposite wall burying a railroad line and the entrance of a coal mine. Seventy persons lost their lives.

Gros Ventre Rockslide. An important rockslide involving 50,000,000 cubic yards of material which occurred in the Gros

Ventre River Valley of Wyoming in 1925 is illustrated by Fig. 130. Heavy rains and melting snow had saturated clay layers in the bedrock which dip toward the valley at angles of 18° to 21°. According to Alden,[1] the chief factors favoring the slide

Fig. 130. The Gros Ventre rockslide, Wyoming. (R. P. Sharp.)

were oversteepening of the lower slopes of the valley by glaciation and the presence of tilted shale beds.

Rockfalls

On vertical or overhanging cliffs and very steep rock slopes, blocks or segments loosened by weathering drop under the influence of gravity. Unless the cliff is undergoing active undercutting, evidence of the amount of rock that has fallen will be furnished by a talus. Movement may consist of a single vertical drop, or a series of bounds down a steep slope. The relatively free fall of the mass involved distinguishes rockfalls from rockslides. If the block is large or the distance of fall great, the mass may be thoroughly fractured by impact and may move outward from the base of the slope by flow.

Earth Slips and Subsidence Caused by Underground Erosion

Terzaghi [14] has called attention to these types of movement which seem to have escaped the attention of geologists. If granular deposits are present underneath a hillside, circulating subsurface waters may remove a portion of the materials leading to the formation of underground cavities, or pipes. These cavities may be held open for a time by arch action of the roof, but as the process continues and the cavities enlarge the roof eventually collapses. Surface subsidences and/or slides result. The extent to which such movements take place in nature is not known precisely, but Terzaghi attributed to this cause slides and subsidences along the steep loess bluffs near Memphis, Tennessee. It is becoming known that pipes and underground openings may exist in thick deposits of sand and similar materials, and the surface depressions often associated with such deposits may be at least in part the result of collapse of these cavities. The Carolina "Bays" have probably originated in this manner.

CONDITIONS FAVORING LANDSLIDES

Conditions favoring landslides may exist for a long time without any movement taking place, but given the proper impetus slides may be touched off either instantaneously or within a short space of time. Conditions favorable to sliding are too numerous and complex to be treated in detail, and those interested in the subject should consult Sharpe's [12] and Terzaghi's [16] thorough treatments of the subject.

Terzaghi [16] has advanced the view that landslides are actuated mainly by changes in the position of the ground water level. Although landslides may be actuated by causes other than rapid draw-down, an examination of rapid draw-down will serve to illustrate the basic process. A fundamental law of hydrostatics states that a body submerged in water is lighter by an amount equal to the water displaced. Thus, if the ground water level drops, the soil above becomes correspondingly heavier. Moreover, if the material is too fine-grained to permit free draining, its weight is further increased by an amount equal to the weight of the water trapped in the voids. This increase in the effective weight exerts pressure on the underlying submerged materials and tends to squeeze out water contained in the voids. If these materials are fairly impervious, the decrease in water content lags behind the increase in load. Excess hydrostatic pressures are created which lower the shearing resistance of the material by an amount proportional to the unit weight of the water multiplied by the hydraulic head. The excess hydrostatic pressure depends on the rate at which the water level subsides, and is less in instances where the rate of draw-down is slow than in cases where the draw-down is rapid. This explains why earth slips and subsidences along river banks are more frequent after a rapid recession of flood waters than at other times. In all cases, it is the loss of shear strength associated with the excess hydrostatic pressure, and not "lubrication," that leads to sliding.

PREVENTION AND CHECKING OF LANDSLIDES

This subject has been discussed in some detail by Terzaghi,[16] Sharpe,[12] Ladd,[8] and Baker,[2] and only a brief summary is needed here. In pervious materials, the simplest and usually the most effective method of preventing landslides is to increase the internal friction of the mass by lowering the water content. Water already present may be removed by ditches, tile drains, or other conduits, and by drainage through inclined tunnels or by pumping from shafts or wells. The entrance of additional water may be prevented by diverting the surface flow from unstable ground and causing runoff from precipitation to be as complete and rapid as possible. Artificial freezing of the material has been employed with some success, but generally this method is restricted to locations where the value of a project under construction warrants the expense. Various forms of artificial soil stabilization also offer prospects of aiding in the control of slides, but they do not seem to have been employed to any marked extent.

Slides in impervious materials are not very susceptible to treatment by draining, although every effort should be made to prevent additional water from obtaining access to the material. The best method is to use flat slopes when excavating through such materials and anticipate occasional clean-up and corrective work.

Once started, shear, or deep deformational slides cannot be

controlled and must be allowed to run their course. They can be prevented in only one way, that is, by proper design of the excavation slopes, which should be designed so as to be stable for the type of material and height of bank involved. This procedure usually involves a very close co-ordination between geologic and soil mechanics studies, the details of which are beyond the scope of this book. It should be noted that, if slides of this type once take place, the stable angle slope is considerably flatter than before deformation and weakening of the material occurred. Consequently, proper initial design of the slopes not only promotes uninterrupted progress of the work, but also leads to considerable savings in cost.

Devices for detecting impending or incipient slides have been developed.[16] They constitute a wise precaution in situations where slides may lead to a loss of life. Protective walls and fences for preventing loose rock from rolling onto highways and railways are inexpensive and are usually very effective. In one ingenious type, the movement of material against a woven-wire fence breaks an electrical circuit and closes the block to trains.

SUBSIDENCE

Subsidence may be defined as "movement in which there is no free side, and surface material is displaced vertically with little or no horizontal component."[12] It is caused by the removal or consolidation of underlying material, or from a downwarping of the earth's crust. Subsidence usually occurs as a slow settling, but rapid movements are not unknown. Kettleholes in glacial drift have been formed by subsidence following the melting of underlying ice masses. Similar shallow, saucer-like depressions 100 feet to several miles wide are common on the High Plains of western United States and have been attributed to consolidation of sediments and the solution of salt and gypsum.

Spectacular instances of subsidence from natural causes include the sinking which formed Reelfoot Lake, Tennessee, accompanying the New Madrid earthquake of 1811; the lowering and partial submergence of the city of Port Royal, Jamaica, during the earthquake of 1692; the overnight disappearance of several buildings and about an acre of land at Rosel, Kansas, in 1898, leaving a pond 70 feet deep; and the caving at Staunton, Virginia, caused by the softening of the clay cover of a large limestone solution channel.[12]

Subsidence caused by human activity includes a large number of instances where sinking has affected rather large areas. Fenlands of England and Holland have undergone consolidation following diking and draining for cultivation of the soil. The resulting oxidation and consolidation of underlying peat beds has caused parts of the drained English fens to sink more than 13 feet in the last 100 years. Subsidence has long been a problem in San Francisco, California, where consolidation, drainage, and flowage of underlying materials are responsible for the movement. Subsidence also results from the removal of oil from underground. The Goose Creek Oil Field in Texas and the Wilmington Field in California are cases in point. In the latter instance, subsidences in parts of the neighboring city of Long Beach have been as much as 12 feet and additional subsidences of almost the same amount are anticipated. Subsidence of the Mexico City region, mentioned in Chapter 5, is probably the result of lowering of artesian pressures following excessive pumping of the ground water.

Damage to farms, roads, railroads, pipe lines, and buildings often results from subsidence. As a rule, the settling is gradual and uniform over a considerable area, and under these circumstances there may be no very serious consequences. Although common over metal mines, because of their greater areal extent subsidence is more frequent and widespread over coal mines. In the Illinois-Indiana fields, the area around Pittsburgh, and the anthracite fields of eastern Pennsylvania important damage has been caused by subsidence of the ground over worked-out coal seams. Subsidence caused by mining is reflected on the surface by: (1) surface cracks and displacements, (2) pit holes or caves, and breaks caused by the collapse of the roofs of mine workings, and (3) sags or basin areas formed by the downwarping of the surface, usually over larger excavations or where the surface rocks are stronger than in the second instance. Subsidence of the ground over coal mines has been extensively studied, and there are thought to be several mechanisms of subsidence which vary according to the physical characteristics of the rock and the method of mining employed.

REFERENCES

1. W. C. Alden, Landslide and Flood at Gros Ventre, Wyoming, *Trans. Amer. Inst. Mining Met. Engrs.*, Vol. 76, 1928.

2. R. F. Baker, Analysis of Corrective Actions for Highway Landslides, *Proc. Am. Soc. Civil Engrs.*, May, 1953.

3. C. H. Behre, Jr., Talus Behavior above Timber Line in the Rocky Mountains, *J. Geol.*, Vol. 41, 1933.

4. *Canadian Geological Survey*, Report of Commission Appointed to Investigate Turtle Mountain, Frank, Alberta, *Mem.* 27, 1912.

5. *Geological Survey of India*, The Gohna Landslip, *Records*, Vol. 27, 1894.

6. Per Holmsen, Landslips in Norwegian Quick-Clays, *Géotechnique*, Vol. III, No. 5, 1953.

7. A. W. Koppejan, B. M. van Wamelen, and L. J. H. Weinberg, Coastal Flow Slides in the Dutch province of Zeeland, *Proc. 2nd Intern. Conf. Soil Mech. and Foundation Eng.*, Vol. 5, Rotterdam, 1948.

8. G. E. Ladd, Landslides, Subsidences and Rock-Falls, *Proc. Am. Ry. Eng. Assoc.*, Vol. 36, 1935.

9. A. von Moos, Settlement Measurements in the Region of the Landslide of Zug, *Proc. 2nd Intern. Conf. Soil Mech. and Foundation Eng.*, Vol. 4, Rotterdam, 1948.

10. *National Academy of Sciences*, Report of the Committee of the National Academy of Sciences on Panama Canal Slides, *Mem.* 18, 1924.

11. R. B. Peck and W. V. Kaun, Description of a Flow Slide in Loose Sand, *Proc. 2nd Intern. Conf. Soil Mech. and Foundation Eng.*, Vol. 2, Rotterdam, 1948.

12. C. F. Stewart Sharpe, *Landslides and Related Phenomena*, Columbia Univ. Press, New York, 1938.

13. Karl Terzaghi, The Mechanics of Shear Failures on Clay Slopes and the Creep of Retaining Walls, *Public Roads*, Vol. 10, 1929.

14. Karl Terzaghi, Earth Slips and Subsidences from Underground Erosion, *Eng. News-Rec.*, July 16, 1931.

15. Karl Terzaghi and R. B. Peck, *Soil Mechanics in Engineering Practice*, John Wiley & Sons, New York, 1948.

16. Karl Terzaghi, Mechanics of Landslides, in *Application of Geology to Engineering Practice* (Berkey Volume), Geol. Soc. Amer., 1950.

17. D. J. Varnes, Relation of Landslides to Sedimentary Features, in *Applied Sedimentation*, John Wiley & Sons, New York, 1950.

18. William H. Ward, The Stability of Natural Slopes, *Geog. J.*, Vol. 105, May–June, 1945.

C H A P T E R 1 3

Volcanoes and Earthquakes

Volcanoes and earthquakes are the most spectacular natural phenomena, and for this reason have perhaps an unduly profound influence on the popular mind. In ancient times they were regarded as omens, and the outcomes of important events were often decided by a volcanic eruption or a shaking of the earth. In our own day, the decision to buy the French concession in Panama, rather than adopt the Nicaragua canal route, was considerably influenced by the occurrence of volcanoes in the latter country. Although fear of earthquakes and volcanic eruptions has in some instances been excessive, they have been, and probably will continue to be, the cause of great and widespread damage, often accompanied by disastrous loss of life. The consequences of the eruption of Vesuvius in A.D. 79 which buried the city of Pompeii are known to everyone, and the great Japanese earthquakes of 1924 destroyed life and property on a scale almost unknown until the wars of the present century. In regions where they occur, volcanoes and earthquakes probably represent the greatest of all natural hazards to engineering works, and for this reason the engineer should know something about them.

It is exceedingly improbable that volcanoes and earthquakes will ever be controlled by man, but scientific study of these phenomena has dispelled the almost superstitious dread with which they are still regarded by many and has laid the groundwork for an intelligent approach to the engineering problems they present. Remarkable advances have been made in earthquake-resistant designs, and it is perhaps not unduly optimistic to assume that life and property losses resulting from earthquakes will eventually be reduced to a small fraction of present amounts. If this result is to be achieved, engineering advances must go hand in hand with continued geological and seismological investigations. Consequently, the chief purposes of this chapter are to summarize the existing state of knowledge concerning volcanoes and earthquakes and to outline briefly the present methods of investigating these phenomena.

VOLCANOES

A volcano is a vent, or orifice, in the earth's crust connecting the surface with a body of molten rock below. Masses of igneous and pyroclastic materials which accumulate around the vents are known as volcanic cones, cinder cones, lava domes, and plug domes. Crater-like basins formed by subsidence of volcanic mountains are known as calderas. It was once fairly generally believed that volcanoes draw their lava from a zone of molten rock not far below a comparatively thin hardened crust, but modern geophysical work has demonstrated that no single central lava source exists, a fact that makes the origin of the molten rock even more difficult to explain. The older view held that the heat contained in bodies of molten rock is residual, i.e., inherited from the remote period of geologic time when the earth was completely molten like the sun. The discovery of radioactivity has disclosed a hitherto unsuspected source of energy, and Joly [12] and Holmes [10] have elaborated in considerable detail hypotheses of the origin of volcanoes based on radioactivity. It is possible that both views are correct, and that the heat of molten rocks is partially residual and partially radioactive in origin.

Gases in Molten Rocks

Molten rock below the crust is known as magma; that reaching the surface is called lava. Volcanologists distinguish between two chief types of magma: basaltic and rhyolitic. The former may be derived from somewhat deeper zones in the earth than the rhyolitic type and is characterized by lower gas content, higher temperature of fusion, and greater mobility, even at relatively low temperatures. Because the gas held in basaltic magmas ordinarily can escape from the mass, explosive eruptions are comparatively rare. Rhyolitic lavas, on the other hand, have a lower temperature of fusion, but are more viscous and less mobile, even at relatively high temperatures. Consequently, gases can escape only with difficulty and explosive eruptions are comparatively com-

Fig. 131.   Lava from Parícutin, entering the town of San Juan Parangaricutiro, Mexico, July, 1944.   (A. Brehme.)

mon. It is thought that rhyolitic magmas originate at somewhat shallower depths than basaltic types. Volcanoes that erupt basaltic lava are most abundant in the Pacific Ocean basin; rhyolitic types are largely limited to continental areas. Early investigators were disposed to regard the gases as derived from sea water vaporized by contact with molten rock. It is now known that they are in large part original constituents of the magma, and thus must be derived from deep within the earth.

Gases and Eruptive Activity. It was once fairly generally believed that volcanic eruptions result from the contact of molten rock with subsurface water. Although the violent eruption of Kilauea in 1924 has been traced to the penetration of ground water into the conduit during a period of lava subsidence, most modern workers are inclined to attribute little importance to subsurface water as the cause of eruptions. Nearly all volcanologists are agreed that eruptions are caused by escape of gas contained in the lava. The chief difference of opinion centers around questions as to how much of the gases is original to the magma, and what proportion of water vapor results from oxidation of magmatic hydrogen. In order for really violent eruptions to occur, it is necessary for the gas to be confined under great pressure. This explains why violent eruptions are more frequently associated with viscous rhyolitic lavas than with the more mobile basaltic types.

Chemical Reactions and Heat of Lavas

The residual and radioactive sources of heat have already been mentioned, but a third deserves a brief consideration. Observations of existing volcanoes have demonstrated that near the surface temperature of the lava increases, rather than decreases as might normally be expected. This rise in temperature is attributable to reactions between escaping gases, burning of combustibles, and reactions between non-volatile constituents of the magma. In any event, shifts in chemical equilibria close to the surface add greatly to the original supply of heat.

Distribution of Volcanoes

By far the greater number of existing volcanoes are concentrated in two zones: one outlining the Pacific Ocean basin, the other extending from the East Indies to the region of Mexico City through the Caribbean and eastward into the Mediterranean region. Comparison of this distribution and a geologic map discloses that the volcanic belts coincide with zones of intense crustal deformation, and that the trend of volcanic belts is usually roughly parallel to the known axes of intense folding and faulting. These features suggest a causal connection between volcanism and deformation, but authorities are not agreed as to whether volcanic activity is a cause or an effect of earth movements. There seems to be a growing trend among volcanologists to consider magmas as active rather than passive agents in crustal deformation. In any event, there seems to be little doubt that chains of volcanoes mark the locations of profound rifts, or fissures, in the earth's crust which tap pockets of deep-seated molten rock. In this connection, it is interesting to note that regions of volcanic activity are also marked by high seismicity.

TYPES OF VOLCANIC ACTIVITY

Volcanologists are accustomed to classify volcanoes according to the type of eruptive activity displayed, and to label each well-defined type of activity according to the volcano or geographic region where it was first recognized or is known to occur. On this basis there are six types of activity which may occur practically simultaneously.

Hawaiian Type. This is the mildest form of activity, and is characterized by quiet flows of thinly liquid lava without explosive escape of gases and the ejection of fragmental material. Although gas is given off, it is discharged without great violence. The lava is very hot and highly fluid.

Strombolian Type. Stromboli, a volcano north of Sicily, has been continuously active since ancient times. The activity is marked by mild explosions occurring every 10 to 15 minutes. The lava does not congeal between explosions, and incandescent clouds are ejected. Consequently, a volcano is said to be of the Strombolian type when the lava does not crust over between successive eruptions and clouds of incandescent material are emitted.

Vulcanian Type. Vulcano is an active volcano not far from Stromboli. Its eruptions are characterized by the solidification of the lava between explosions, and as a result the clouds emitted are not incandescent. When very old and cold material is ejected, the eruptions are often termed ultra-Vulcanian.

Vesuvian Type. This represents a more violent extension of the Vulcanian and Strombolian types. The significant feature is that the lava is highly charged with gas during a long interval of quiescence or mild activity. In consequence of the preliminary eruption of a part of the contents of the conduit, the pressure is relieved. The magma then bursts into a froth and is expelled as large luminous clouds of "cauliflower" form.

Plinian Type. This type of activity was first described by the Roman, Pliny, in A.D. 79. It is characterized by the most extremely violent activity of the Vesuvian type. The gases are expelled with great violence to a height of several miles, and there spread out into an expanding cloud formed of globular masses of gas and vapor.

Peléan Type. Mont Pelée is on the island of Martinique in the West Indies. In 1902, after preliminary eruptive activity, an enormous cloud of superheated steam and highly heated rock particles was ejected from the flank of the mountain and swept over the town of Saint Pierre. Of its 26,000 inhabitants, only two escaped with their lives. French investigators have termed such clouds of highly heated vapor and rock material *nuées ardentes*. These clouds are ejected horizontally from beneath a plug of hardened lava filling the top of the volcanic orifice.

Causes of Various Types of Eruptions

The causes underlying differences in types of eruptions are to be sought in numerous factors, among which the more important probably are: (1) amount of gas absorbed by the magma during its ascent, (2) chemical composition of the magma, (3) degree of crystallization prior to eruption, and (4) depth of the feeding chamber. The first two have already been touched upon, but the others deserve brief mention. It has been suggested that crystal differentiation and retrograde boiling are fundamental in determining the character of the eruption and the type of lava erupted.[19] For example, if a basaltic magma crystallizes quietly for a long enough time, a residual rhyolitic melt will be formed. The vapor pressure of the gases held in the rhyolitic melt may become very high, and violent explosions may result. On the other hand, if cooling basaltic lava escapes at the surface (as it does in the Hawaiian type of activity), differentiation into a rhyolitic fraction does not take place, and the activity continues to be quiet rather than explosive. Others have advanced the view that the dominant control lies in depth of the feeding conduit.[20] The deeper the conduit, the more gas the magma can hold in solution, and hence the greater the probability of explosive activity. According to this view, the Hawaiian and Strombolian types of eruptions should be related to shallow magma reservoirs, whereas more violent types should be related to increasingly deeper sources of magma.

Cryptovolcanic Eruptions

Cryptovolcanoes are circular structures of volcanic origin which occur in otherwise undisturbed plateau regions. There are no volcanic cones, or accumulations of pyroclastic materials, and it is believed that the structures originated through muffled explosions at depth, or through the subsidence of cylindrical segments of the crust over cooling and contracting masses of magma. Cryptovolcanic structures have been described from southern Germany, the Gold Coast of Africa, and central United States.[1] The last occurrence is especially interesting because of the otherwise almost complete lack of igneous phenomena in the region.

TYPES OF VOLCANIC EDIFICES

Accumulations of loose ejected material, chiefly cinders, are known as cinder cones. They are roughly circular in plan, and usually have the form of a truncated cone with a crater at the top. Most cinder cones appear to have been formed by a single eruption, and this may account for their small size, for none of them exceeds 1000 feet in height. When the mass is built of ejected materials of many different sizes, the term "pyroclastic cone" is used.

Shields. Shields consist of the accumulation of successive lava flows around a central vent. Such structures are formed by the outpouring of highly fluid lavas, and as a result the height is low in comparison with the area covered. Numerous examples exist in the Hawaiian Islands.

Stratocones. Stratocones are intermediate in form between the steep cinder cones and flat shield accumulations. They consist of fragmental material and lava sheets, which are interstratified. The lava rarely flows from the central crater, and usually issues from the sides. Radial dikes are often associated with this type of accumulation. The profile is gently concave upwards ending in a central crater.

Plug-Domes. Very viscous lavas may pile up near the vent to form dome-like structures called plug-domes. An accumulation of this kind grows by addition of new material under the hardened surface. During growth the outer portion of the mass breaks into blocks and talus accumulations are formed around the dome. This type of edifice is confined chiefly to rhyolitic lavas.

Calderas. Calderas are crater-like basins which are characterized by their great size as compared with the depth. Some may have been formed by great explosions that blew off the tops of volcanic mountains, but most of these features probably have resulted from collapse of the surface following rapid emptying of the magma reservoir. Calderas are common in many parts of the world. The best known example in the United States is Crater Lake, Oregon.

PREDICTION OF ERUPTIONS

Considerable progress toward prediction of volcanic eruptions has been made, but much still remains to be done. It is not improbable that continued work by volcanic observatories will result in fairly precise methods of prediction. According to Williams,[23] the best criteria of impending activity are: (1) increase in the frequency of local earthquakes, (2) swelling of the volcanic edifice indicated by accumulating earth tilt, (3) unusual changes in the behavior of hot springs and in the emanations of fumaroles (far from infallible), (4) comparison with former cycles of activity, (5) strong magnetic disturbances, (6) low buzzing sounds detectable by the use of earth-contact microphones, (7) sympathetic eruptions of neighboring volcanoes, (8) when the lava stands high in the conduits and the conditions of eruption are at hand, the precise time of eruption may be influenced by lunar-solar controls,* (9) when the lava column is high the danger of flank eruptions is at a maximum, and (10) experiments with thermoelectric couples on the flanks of Merapi in Java indicate that these devices may give warning of the approach of glowing clouds and mud flows.

Engineering Significance. Even should prediction of volcanic eruptions become a precise science, the chief gain would probably be in reduction of loss of life rather than in increased safety of

---

* Attempts have been made to correlate sun spot maxima with frequency of volcanic eruptions and earthquake frequencies. No causal connection has been demonstrated, however.

man-built structures. If the paths of mud and lava flows are known, it may be possible to deflect them by diversion dams, but when occurring on a large scale lava flows in particular develop very considerable force. Diversion of lava flows by bombing has been tried with apparently successful results, but it is possible that the flows would have stopped shortly of their own accord. However, it is possible that when the topography is favorable and the lava rather fluid bombing may have good results. Where the ground surface under the lava flow slopes steeply toward a structure it seems doubtful that bombing can accomplish useful results. Even more difficult than diversion of lava and mud flows is protection against burial under thick deposits of pyroclastic materials, which in great eruptions may reach 100 feet or so in thickness. Construction should be avoided in regions of volcanic activity, but if such construction is unavoidable, reasonable protective measures should be taken. Although "new" volcanoes appear from time to time (for example, Parícutin in Mexico, which originated on February 20, 1943 in the midst of a cultivated field),[17] they are known to occur only in volcanically active regions. Consequently, regions in which volcanoes are unknown may be regarded as safe from future activity.

POWER FROM VOLCANIC SOURCES

In Italy, a country of numerous volcanoes and very little fuel, successful attempts have been made to utilize volcanic heat as a source of power. In 1941, the plant at Larderello produced 100,-000 kilowatts of electric power and over 10,000,000 kilograms of boric acid, borax, ammonium carbonate, carbon dioxide, and other chemicals.[18] The installations were damaged during the war, but have been repaired, and it is hoped that future production will exceed that in the past. During the last war, volcanic heat was piped to an airfield at Reykjavik, Iceland, and the city of Reykjavik is heated mainly by hot water and steam from volcanic sources. New Zealand, California, and other areas also possess potential supplies of energy from volcanic sources.

EARTHQUAKES

As the name suggests, an earthquake is a shaking of the earth which may range from an imperceptible tremor to a catastrophically violent shock. Seismology, as the scientific study of earthquakes is called, is one of the youngest of the natural sciences, but despite its brief life span, great progress has been made in explaining the origin and mechanics of earthquakes. Although precise prediction of disturbances may never be realized, the knowledge now available is sufficient to enable the civil engineer to estimate the danger in any given locality, and, if necessary, precautionary measures can be taken.

SEISMOLOGY

A periodic displacement that repeats itself at approximately equal intervals of time is known as a simple harmonic motion. Although the vibrations set up by earthquake waves are more complicated than simple harmonics, an understanding of this type of motion is basic to analysis of the more complicated wave types. Simple harmonic motion may be illustrated graphically by a sine curve like that shown in Fig. 132. The distance *a–b* is

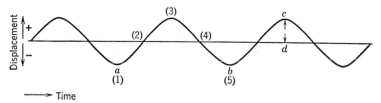

Fig. 132. Diagram illustrating simple harmonic motion.

known as the wave length; *c–d* is the amplitude. The time required to pass through one complete oscillation from *a* to *b* is known as the period, the reciprocal of the period being known as the frequency. The number of oscillations taking place per unit time (frequency) multiplied by the wave length must obviously equal the speed of propagation. Thus, if speed is constant an increase in frequency entails a decrease in wave length. Three other quantities must also be considered: (*a*) displacement, (*b*) velocity, and (*c*) acceleration. With reference to Fig. 132, it will be seen that displacement reaches a positive maximum at 3 and a negative maximum at 5. At 2 and 4 it is equal to zero. Vertical velocity reaches a positive maximum at 2 and a negative maximum at 4. Horizontal velocity reaches a maximum at 1 and at 3. Acceleration reaches a maximum when velocity is at a minimum.

Elastic Waves

Wave motions that obey Hooke's law (stress proportional to strain) are known as elastic waves. The deduction of the properties of earthquake waves from elastic theory is one of the triumphs of modern seismology. Unfortunately, the derivations are long and difficult, and must be replaced here by a few general statements. In an unbounded isotropic solid two types of waves may be transmitted: (1) those in which the particles move back and forth in the direction of wave propagation, known as longitudinal or compressional waves, and (2) those in which the particles are displaced at right angles to the line of propagation, known as transverse or shear waves. In earthquakes the longitudinal wave always arrives at a recording station sooner than the transverse wave. In seismology it is customary to refer to the longitudinal wave as the primary, or *P*-wave, and the latter as the secondary, or *S*-wave. Because both waves are transmitted through the body of the medium, they are often described as the body waves. In a finite isotropic elastic medium, a third type known as the Rayleigh wave may be transmitted along the surface. This wave causes a particle in its path to oscillate around an elliptical orbit. The major axis of the ellipse is vertical and the minor axis longitudinal. The direction of oscillation is toward the source, a type of motion known as retrograde. In the anisotropic earth, other types of surface waves are also known to occur.

Propagation of Elastic Waves. In an isotropic elastic medium, wave disturbances originating at a point are propagated outward in all directions. Because the velocity is the same in all directions, the wave fronts form concentric spheres. The circumference of the spherical wave fronts increases with distance from the source, and when the distance becomes sufficiently great any small segment of the wave front may be considered to be a plane. A ray is a line drawn perpendicular to the wave front and passing through the source. If two waves of different speeds of propaga-

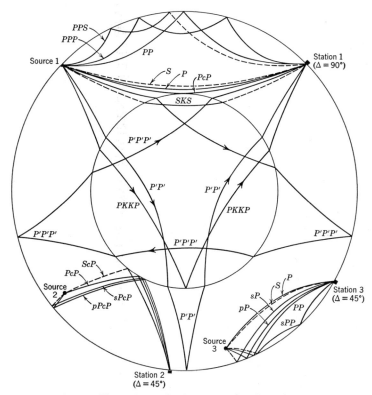

Fig. 133. Diagram illustrating reflection and refraction of seismic waves in the earth. Wave *PKKP*, for example, is refracted in the dense core, reflected at the boundary of the core, and again refracted. (After Beno Gutenberg in *Engineering and Science*, February, 1948.)

tion are generated simultaneously from the same source, as the distance from the source increases the interval between the two wave fronts becomes progressively greater and is proportional to the distance from the point of origin. In earthquakes the P-wave is always faster than the S-wave, and the interval P–S furnishes a means of estimating the distance of the shock from the record-

ing station. When waves strike a medium of different elastic
properties, they are both refracted and reflected (Fig. 133).

INSTRUMENTAL RECORDING OF EARTHQUAKES

A seismometer for measuring earthquake vibrations is a vibrating system with a natural period that results from its dimensions
and design characteristics. The ground on which the instrument
rests acts as another vibrating system, which has a period generally
different from that of the seismometer. Therefore, when the
ground is disturbed by the passage of elastic waves, there is a
tendency for the seismometer to oscillate freely with its natural
period. This tendency toward free oscillation must be reduced,
and may be accomplished by damping. A damping force resists
motion by an amount that is proportional to the velocity; it may
be either mechanical or electrical in operation. The amount of
damping employed has a marked effect on the response of the
instrument. Not only the magnification ratio, but the relationship
between the seismometer's writing point and actual ground motion is affected. Thus, in general, the response of the measuring
instrument and the ground motion will not be identical.

Types of Seismographs

A seismometer equipped with a recording device is known as
a seismograph. As previously mentioned, a seismograph consists
essentially of a vibrating system which uses the inertia of a suspended mass to detect vibrations in the ground. These vibrations
are very slight, and some sort of amplification system is necessary.
Amplification may be accomplished either mechanically or elec-
trically. Electrical seismographs may operate on either the electromagnetic induction or magnetic reluctance principles and may
be designed to record either horizontal or vertical components of
ground motion. These components do not ordinarily correspond
to the actual movement of the ground, but if the response characteristics of the instrument are known the true ground motion
may be calculated.[15] For a record to look most like the true
ground motion a displacement seismograph would be required,
but for the sharpest indication of the arrival of the shock an
accelerograph is necessary.

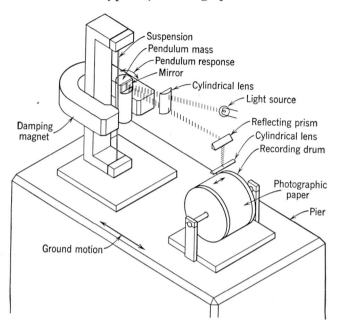

Fig. 134. Wood-Anderson torsion seismograph. (Seismological Laboratory, California Institute of Technology.)

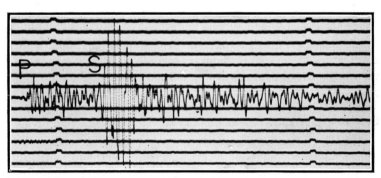

Fig. 135. Seismogram of a Kern County, California, aftershock, magnitude 4.5, recorded at the Seismological Laboratory of the California Institute of Technology, August 11, 1952. Distance to epicenter approximately 125 kilometers. (Seismological Laboratory, California Institute of Technology.)

Recording Mechanisms. Regardless of the type of instrument, recording is accomplished by a light beam or stylus. When a light beam is used the recording paper consists of a photographic film which moves past the beam at a constant speed, thereby furnishing a time axis which is transverse to the long dimension of the recording paper. Figure 135 shows a typical record of an earthquake. The difference in times of arrival of the P- and S-waves is, of course, proportional to the distance from the recording station.

Location of Epicenters

The point on the earth's surface immediately below which the movement generating an earthquake shock took place is known as the epicenter. It is obvious that if the direction of the shock relative to the recording station is known, the location of the epicenter may be determined from the difference in times of arrival of the P- and S-waves. Because of difficulties in determining the direction with sufficient accuracy, this method is not reliable. The most accurate and generally used procedure is the three-point intersection method. If an earthquake is recorded at three properly located stations, a circle is drawn from each station with a radius proportional to the distance between the shock and station concerned. The intersection of the three circles is the epicenter. For some years past recording stations have been distributed widely enough over the world to permit reasonably accurate location of all earthquakes recorded. As a result, a very valuable body of data relative to the frequency and location of earthquakes is being compiled.

Complexity of Earthquake Vibrations

Because of the complex nature of earthquake waves and the heterogeneous character of the earth, the movement of a particle during an earthquake is extremely complex. Figure 136 shows the well-known model by the Japanese seismologist, Sekiya, illustrating the motion of a particle during the first 20 seconds of the Japanese earthquake of February 15, 1887. It is the extreme complexity of vibration that complicates the engineer's problems of devising earthquake-resistant designs.

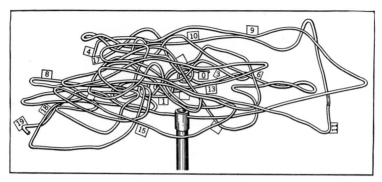

Fig. 136. Model of ground motion during the first 20 seconds of the Japanese earthquake of February 15, 1887. The actual ground motion is greatly exaggerated on the scale shown.

Energy and Amplitude of Earthquake Waves

The actual displacement of the ground is relatively small and seldom exceeds a fraction of an inch. However, in soft ground it may be somewhat greater. According to Gutenberg and Richter,[9] the energy released by a shock of great magnitude is of the order of 10^{25} ergs. This value drops rapidly at lower magnitudes, and it is estimated that at least 10^{11} ergs must be expended in order for a shock originating at a few kilometers in depth to be felt.

Frequency of Earthquakes

It has been estimated that the total number of shocks of all magnitudes occurring each year is well over a million.[9] Most of these are of low intensity. By far the greatest proportion of the energy released is derived from the very small number of major shocks. The energy liberated by earthquakes represents the performance of work at an average rate of 10,000,000 kilowatts. This energy is not released at a uniform rate, and is not regularly distributed over the seismic regions. There are years, or shorter intervals, when activity is either abnormally high or low. Moreover, for a period of weeks significant activity may be concentrated in a limited region. Gutenberg and Richter [9] believe that these variations are chiefly within the limits of statistical fluctuation, but that they may exceed them in certain regions.

Magnitude of Earthquakes

Measurement of the magnitude of earthquake shocks by the seismograph is complicated by the loss of energy which takes place during passage of the waves through the earth's crust. Obviously, a strong shock near at hand and one 180° from the recording station will not register the same amplitude. Gutenberg and Richter [8] have developed a method of estimating the intensity of earthquakes at any distance from the recording station. This method is based on the logarithm of the maximum recorded trace amplitude measured on the seismogram. These investigators have arrived at the following scale of magnitudes.

Class	a	b	c	d	e
Magnitude	7.8–8.5	7–7.7	6–7	5.3–6	<5.3

Increase in the magnitude by half a unit corresponds to an increase by a factor of 10 in the energy released. There is thus a ratio of about 10^{17} between the energies released by the largest and the smallest earthquakes. Shocks of classes a and b are recorded at all stations; class c is recorded up to 90° away; class d up to about 45°; and class e usually up to only about 10° from the recording station. Shocks of magnitude 3 are usually felt; those of 4.5 are capable of causing damage; major shocks range from 7 to 8.5 in magnitude. It is estimated that on the average there is 1 shock of intensity 8 or more per year; 10 of intensity 7–7.9; 100 of intensity 6–6.7; 1000 ranging from 5 to 5.9; 10,000 between 4 and 4.9; and 100,000 of intensity 3–3.9.

Depth of Focus

Earthquakes may originate at almost any depth between the surface and about 700 kilometers below. Shocks originating above 60 kilometers in depth are classified as shallow; from 60 to 300 kilometers as intermediate; and in excess of 300 kilometers as deep. It follows from the laws of propagation of elastic waves that, given equal intensities, shallow shocks will be more violent at the surface, but will affect a smaller area than deep disturbances. On the other hand, a very violent deep shock may cause equal, or even greater, surface damage over a much larger area.

Distribution of Earthquakes

Figure 137 shows the distribution of earthquakes during the period 1904 to 1940. It will be observed that the seismically active regions of the earth fall into four fairly well-marked zones:

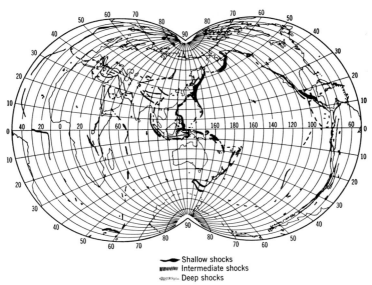

Shallow shocks
Intermediate shocks
Deep shocks

Fig. 137. Map showing distribution of earthquakes throughout the world. (After Beno Gutenberg and C. F. Richter, Seismicity of the Earth, *Geol. Soc. Amer., Spec. Paper* 34, 1941.)

(1) the circum-Pacific belt with many branches and subdivisions; (2) the Mediterranean and trans-Asiatic zone; (3) narrow belts extending through the Arctic and Atlantic Oceans, the western Indian Ocean, and the East-African rift valleys; and (4) scattered belts of only local significance. The circum-Pacific belt includes most shallow shocks, a still larger fraction of the intermediate shocks, and practically all the deep shocks.[9] The Mediterranean and trans-Asiatic zone includes the remainder of the largest shallow shocks (class *a*) and all the remaining intermediate shocks. The last two belts experience only shallow shocks. Activity is not of equal intensity throughout a given belt. In the Pacific coast region of North America activity is greatest,

and the shocks are most intense in the Alaska and southern Mexico areas; the western part of the United States experiences only comparatively moderate activity.

Stable Areas. If Fig. 137 is superimposed on a geologic map, a striking correlation is disclosed between areas of very low seismicity and the so-called shield areas, i.e., regions of very ancient rocks such as those which underlie most of central and eastern Canada, northeastern South America, northern Europe and Asia, and central Australia. Furthermore, with the exception of a few earthquakes near volcanic islands, such as the Hawaiian group, the entire Pacific Ocean appears to be a region of almost complete seismic calm. Earthquakes of great violence have been known to occur in the quiet areas, such as the New Madrid, Missouri, quake of 1811 [6] and the Charleston, South Carolina, disturbance of 1886.[3] The earthquakes of the Saint Lawrence River Valley area and New England regions may be related to isostatic readjustments of the earth's crust incident to the retreat of the ice sheet which formerly covered the region.

CAUSES OF EARTHQUAKES

For an earthquake to occur there must be a local movement of the earth's crust sufficient to impart a measurable acceleration to the adjacent rock masses. Based on the mechanisms involved, earthquakes may be classified as: (1) shocks caused by collapse of underground cavities, (2) vibrations generated by volcanic activity, and (3) tectonic earthquakes caused by movements along fault planes. It is probable that the first two classes have been quantitatively greatly overestimated, and by far the greatest number of earthquakes are tectonic in origin. Mechanisms underlying the first two categories are simple and easily understandable, but nothing is known of the nature of forces responsible for tectonic shocks, except that they are related to mountain building, or orogeny. Fortunately, it is not necessary to understand the ultimate origin of the applied forces in order to arrive at a reasonably sound view of the mechanics of tectonic earthquakes.

The Elastic-Rebound Theory

This explanation of the mechanics of tectonic earthquakes was advanced by H. F. Reid [18] in 1911. The theory postulates that when a force is applied to two adjacent fault blocks friction is at first sufficient to prevent movement parallel to the fault plane.

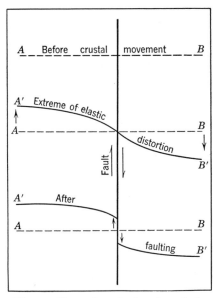

Fig. 138.   Diagram illustrating elastic rebound along a fault.

As a result of continued application of the force, a zone on either side of the fault undergoes elastic distortion. When the distorting forces reach a magnitude which exceeds the frictional resistance of the fault surface, the elastically stored energy is suddenly released, and the disturbed zones recoil to a position of less strain (Fig. 138). According to this view, earthquakes result from a more or less continuous deformational process, the energy liberated being stored elastically in the rocks adjacent to a fault surface and released almost instantaneously. This theory has gained wide acceptance and has many important engineering implications which are outlined in the following paragraphs.

Implications of the Elastic-Rebound Theory. Perhaps the most important by-product of the elastic-rebound theory is that it affords a tempting basis for the prediction of earthquakes. This arises from the consideration that, in any given area where elastic distortion is in progress, the amount of deformation must attain a certain critical value (determined by the strength of the rocks involved, dimensions of the zone affected, and frictional resistance of the fault surface) before an earthquake can take place. Furthermore, the magnitude of a disturbance, other factors being equal, should be proportional to the amount of distortion taking place before elastic rebound. Thus, in Japan attempts have been made to forecast earthquakes by measuring the accumulation of earth tilt in affected areas.[5, 11, 21] In the United States, bench marks have been established on both sides of the San Andreas fault, along which the movements causing the San Francisco quake of 1906 originated, with a view to measuring the accumulation of distorting forces presaging another shock. Resurveys made to date are reported to show an offset of approximately 10 feet. In Japan, experiments have been undertaken with electrodes located along fault planes, on the assumption that a major shock is preceded by minor slips and disturbances. Although these methods offer some hope of foretelling when a shock is imminent, the exact instant depends on numerous extraneous factors such as phases of the moon, tide, heavy rainfall, and barometric pressure. In fact, there is a certain tendency for earthquakes to occur during times of low air pressure. In view of the complexity of the factors involved, exact prediction may never be possible, and, until absolutely infallible results are assured, public announcements of anticipated earthquakes should be discouraged.

Another important deduction from the theory is that at a certain distance from the fault danger from earthquakes declines rapidly. Freeman[4] estimates the distance of serious danger at approximately 25 miles from the center of the disturbance. Small earthquakes may occur far from the main fault, but it has been demonstrated that in southern California at least major disturbances are largely limited to major fault zones (Fig. 139).[9]

According to the elastic-rebound theory there must be an absolute upper limit to the violence which earthquake shocks may attain. This limit is fixed primarily by the elastic properties of the rock and dimensions of the block affected. It is probable that the strength of the earth's crust is such that the energy of the largest earthquake cannot exceed 5.6×10^{24} ergs.[22] This estimate agrees within the limits of error with the energy of the largest known earthquakes. In other words, earthquakes of greater

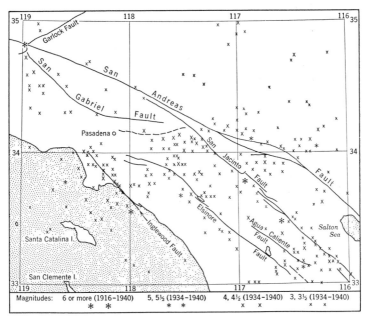

Fig. 139. Map showing distribution of earthquake epicenters and faults in Southern California. (After Beno Gutenberg and C. F. Richter, Seismicity of the Earth, *Geol. Soc. Amer., Spec. Paper* 34, 1941.)

magnitude than those known in the past are exceedingly improbable.

Another implication is that an earthquake may be preceded by foreshocks and followed by aftershocks. Attempts have been made to use the former as a means of predicting the imminence of important disturbances, but reference to Fig. 139 will show that the method has its difficulties. Most small shocks are not centered along important faults, and, unless the epicenters are very accurately known, it is impossible to distinguish between a random shock and a "foreshock." Much remains to be done before the significance of foreshocks can be evaluated. After-shocks, on the other hand, stand in a somewhat clearer light. It has been demonstrated that in the case of certain Japanese earth-quakes the number of aftershocks gradually decreases with time, and when the aftershocks are plotted against time the graph ap-proximates an hyperbola. Investigations indicate that the after-shocks of the Long Beach earthquake of 1933 took place in the neighborhood of the fault along which the main shock originated. In general, the intensity of aftershocks decreases with time.

Another important deduction is that locations of earthquake epicenters are determined by forces that have long been in opera-tion, and sudden shifts of earthquake zones are improbable. Al-though a few tens of thousands of years is thought to be ample time for extensive and significant changes in distribution of local stresses, there seems to be little evidence of change in the distri-bution of earthquakes within historic time. Gutenberg and Richter [9] summarize the problem in the following statement.

> Few definite changes in seismicity have occurred during historical time. Chronologically long histories for such active regions as Japan, China, the Near East, and Italy indicate activity of about the same character as in very recent years, with shocks of the same range of magnitudes occurring in the same areas, apart from a few individually exceptional events. For less active regions, the best available history is that of Great Britain, extending over about a thousand years with no sign of secular change.

It should be borne in mind, however, that there is nothing in the elastic-rebound theory or the above statement to preclude the possibility of an earthquake occurring in an area of pre-sumed seismic quiet. It is entirely a matter of relative proba-bilities.

It has been inferred that frequent small shocks along a major fault may help to relieve strain, and thus act as a sort of "safety valve" delaying, or even preventing, the occurrence of a major earthquake. Studies of the energies liberated by minor and major earthquakes have led to the conclusion that the energy released by the former is so small in comparison with the latter as to make this contingency highly unlikely.[9] On the contrary, it is probable that minor shocks are symptoms of a regional strain, only a small part of which is being transferred away from the major fracture along which it eventually will be released as a major earthquake.

EARTHQUAKES AND OTHER PHENOMENA

The general association of earthquake and volcanically active belts is stressed in many books and treatises. However, this association is rather general and loose.[9] Both are related to weak zones of the earth's crust, and consequently show a parallel distribution when studied on small-scale maps. However, volcanic vents are usually some hundreds of kilometers from the principal tectonically active faults and structures giving rise to shallow earthquakes. Intermediate shocks, on the other hand, frequently occur directly under the structural lines marked by volcanoes. Leading investigators are of the opinion that there is probably no causal connection between such shocks and volcanic activity.

Association with Gravity Anomalies and Foredeeps. The association of gravity anomalies and exceptionally deep oceanic depressions (foredeeps) has been known for some time. However, it has been pointed out that, although negative anomalies * are generally found associated with foredeeps, the greatest anomaly is usually not over the trough, but adjacent to it.[9] The adjacent belt is frequently marked by epicenters of shallow earthquakes. The usual relationship of foredeeps, rises, gravity anomalies, volcanic chains, and seismic belts is illustrated in Fig. 140. The association of these phenomena has been summarized [9] as follows:

> Many authors have correlated deep and shallow earthquakes with oceanic troughs and deeps. This is a general correlation like that with volcanic activity; it applies to small-scale maps, but requires modification in detail. Epicenters usually do not fall in the deep troughs themselves, but on their marginal slopes or along the crests of adjacent submarine ridges. Frequently, as occurs south of Sumatra and Java, the ridge adjacent to the deep is not seismically active, but becomes active in another part of its course where the adjacent depths are less marked. However, most of the greater deeps are in regions where seismicity is at least moderately high.

* Defined as observed values of gravitational acceleration that are lower than the calculated values, and thus implying a deficiency of mass.

This, of course, applies to the true structural troughs, and not to the irregular areas of great depth which occur in the oceanic regions, as in the Atlantic and Indian Oceans, where there are no associated deeps; the seismic belts then follow the ridges.

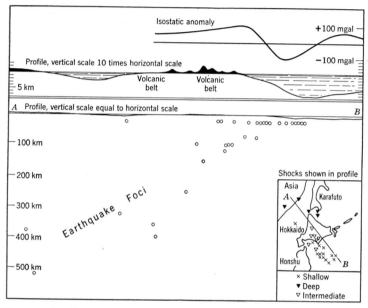

Fig. 140. Profile showing relationships of earthquake epicenters, relief, and isostatic gravity anomalies, North Japanese region. (After Beno Gutenberg and C. F. Richter, Seismicity of the Earth, *Geol. Soc. Amer., Spec. Paper* 34, 1941.)

Association with Landslides. The association of violent earthquakes and landslides both on land and under the sea has long been known, and there is no doubt that earthquakes may cause slides, especially on unstable slopes. The San Francisco earthquake of 1906 caused several landslides along high bluffs or steep hillsides near the coast. At Cape Fortunas a mass approximately a mile long and one-fourth to one-half mile in width moved down a slope nearly 500 feet high. Landslides were also a spectacular feature, especially in loess deposits, during the New Madrid quake of 1811. Perhaps the greatest earth movements known occurred as the result of the earthquake of December 16, 1920 in Kansu

Province, China. This area is underlain by hundreds of feet of loess deposits. Torrential rains saturated the material shortly before the disturbance, and as a result when the structure of the loess was destroyed by the shock, it formed a yielding mass on which the overlying material slid as far as three miles. Visitors to the scene shortly after the disaster described it in the following words: [2]

> ··· Two sections of the ancient, well-packed highway, accompanied by the tall trees which bordered it, were cut from the line of the road following the side hill, swept hundreds of yards over the stream-bed, and set, intact, upon an angle on top of the heap of loose loess.
>
> At the junction of the valley, stands Swen Family Gap, a town of several thousand souls, in which one-tenth were killed by collapse of buildings and cave dwellings; and the other nine-tenths were saved by the miraculous stoppage of two bodies of earth shaken loose from the mother hill and left hanging above the village, lacking only another half-second's tremor to send them down. A third avalanche, having flowed from the hills on the opposite side of the valley across the valley floor and the stream bed, is piled up in a young mountain, near enough to the village to over-shadow the wall.
>
> In each case the earth which came down bore the appearance of having shaken loose, clod from clod, and grain from grain, and then cascaded like water, forming vortices, swirls, and all the convolutions into which a torrent might shape itself.

Submarine landslides, often on a large scale, accompany near-shore and submarine earthquakes. In the Newfoundland earthquake of 1929, submarine cables were broken many miles from the epicenter and known lines of faulting. It is presumed that the breaks were caused by landslides.

"Tidal" Waves and Tsunamis. On the ocean, two events may transpire which are often confused, but which should be differentiated. It has been pointed out that if an earthquake occurs under the ocean, the elastic waves travel through the water. If they strike a ship, the impression is created that the vessel has hit a rock.[7] Only energy is propagated. On the other hand, if an area of the ocean bottom is displaced, a large mass of water forms a wave, usually not very high, which has been misnamed a "tidal" wave. Since it has nothing to do with tides, the Japanese term, *tsunami*, is preferred. Tsunamis are usually harmless in the open ocean, but may produce extensive damage along low-lying coasts. A great portion of the loss of life during the Japanese earthquakes has been caused by tsunamis.

Other Phenomena. A great variety of unusual occurrences has been attributed to earthquakes. Frequent references are found to "visible" waves that pass slowly along the surface. Since the amplitudes described seem to be entirely too large for elastic waves, it may be that they are only indirectly related to earthquakes. It has been suggested that these waves are produced by changes in air pressure which cause a periodic alteration in the indices of refraction of air layers with different densities. A spurious impression of visible ground waves might thus be produced.

Audible sounds are known to be associated with earthquakes. Frequently, the earthquake is heard before it is felt, and at times it is heard but not felt.

Luminous phenomena have been reported during earthquakes, but it is not known whether they result from disruption of power lines, accidental electrical discharges in the atmosphere, or for some unknown reason are actually related to the earthquake.

It has long been known that at times an earthquake may be felt very strongly at the surface, whereas in mines of considerable depth the shock is heard but not felt. This is probably to be attributed to an exponential decrease in the intensity of the surface waves at depth. In any event, it is the surface waves that cause the large vibrations felt in the upper zones of the crust.

ENGINEERING CONSIDERATIONS

The civil engineer is concerned not only with the energy and frequency of earthquakes, but even more with their effects on structures. The relationship between energy and intensity is not simple.[7] Energy does not depend on maximum intensity alone, but, in the shaken area, on extent of faulting, depth range of displacement, and other factors. Freeman [4] has compiled a detailed record of earthquake damage which furnishes a valuable guide for design engineers. Although much still remains to be done, the lessons of the past, if applied with discrimination, are probably sufficiently representative to furnish a sound basis for present practice.

Rossi-Forrel Scale of Earthquakes. Scales devised on the basis of visible damage have long been in use, and, despite their numerous shortcomings, are probably more useful to civil engineers than those based on seismograph records. Perhaps the best scale of this type is that devised by Rossi and Forrel, given in Table IX. It should be observed that, in this and all scales con-

TABLE IX. THE ROSSI-FORREL SCALE OF EARTHQUAKE INTENSITY

Class	Effects	Brief Description
I	Recorded by a single seismograph. May be felt by an experienced observer.	Almost imperceptible
II	Recorded by seismographs of different kinds. May be felt by a number of persons at rest.	Feeble
III	Felt by a number of persons at rest. Strong enough for the duration or direction to be appreciable.	Very slight
IV	Felt by several persons in motion. Movable objects disturbed; doors, windows, creaking of floors.	Slight
V	Generally felt. Disturbance of furniture and beds, ringing of bells.	Weak
VI	General awakening of those asleep. General ringing of bells, oscillation of chandeliers, stopping of clocks, visible disturbance of trees and shrubs, some persons leave their dwellings.	Moderate
VII	Overthrow of movable objects, fall of plaster, ringing of church bells, general panic, without serious damage to buildings.	Strong
VIII	Fall of chimneys, cracks in the walls of buildings.	Very strong
IX	Partial or total destruction of some buildings.	Severe
X	Great disasters, ruins, disturbance of strata, fissures in the earth's crust, rockfalls from mountains, landslides, etc.	Violent or destructive

structed on a similar basis, the design and construction of the buildings, their foundations, etc., enter directly into the measurement of earthquake intensity. They are, in fact, scales of good and poor construction as well as of earthquake intensity. However, when used with care, they are very useful in earthquake studies.

Structural Safety and Local Stability of the Ground. A study of past earthquakes indicates that damage is usually much greater on soft sands and other loose materials than on hard rock. For example, in the San Francisco quake of 1906 it was found that damage was generally five to ten times greater on soft moist sands, loose sediments, and filled land over old swamps than on hard ground.[14] The mechanics underlying the damage on soft ground can be easily demonstrated by a simple experiment. If a quantity of loose sand is poured onto a thin sheet of stiff cardboard and a light tap applied to the latter, the sand grains will bounce perceptibly although no visible disturbance takes place in the rigid cardboard. In addition to the greater vibrations occurring in loose material than in hard rock, if a structure is placed on yielding materials, in the event of an earthquake, the inertia of the structure may cause it to settle into the foundation medium. In certain types of structures, serious damage may result from this cause. The process is very similar to vibrational methods of foundation stabilization in use by engineers. In Japanese earthquakes it has been found that rigid structures on loose materials fare better than non-rigid types.

Area of Damage. Freeman concluded from the record of past earthquakes that about 2000 to 2500 square miles is the maximum area of serious damage, even in severe shocks. The large area over which the shock may be felt has often been confused with the much smaller area of significant damage. Furthermore, even the greatest earthquakes seldom cause damage over 25 miles from the epicenter. Like almost every other general statement, it would be easy to cite exceptions to this rule. The most conspicuous exception is the Assam, India, earthquake of June 12, 1897, in which damage of class X Rossi-Forrel extended on an average of 70 miles from the center of the disturbance. This earthquake ranks with the most violent known, however.

Evaluation of the Earthquake Hazard. Earthquake-resisting designs are described by Freeman[4] and in other engineering treatises, and need not be discussed here. Because precautionary measures entail additional expense, the criteria for determining whether they should be adopted are of considerable importance. Like all statistical methods, data derived from a study of earthquake frequencies are not infallible. It would have been small comfort to the survivors of the New Madrid disaster to learn that the disturbance was exceptional, and yet no engineer would have been justified in suggesting that precautionary measures should be taken. In other words, earthquakes are possible in almost any area, but the range of probability varies enormously. When the probability exceeds a certain critical value, precautionary measures should be taken. This probability factor varies with the purpose, type, and useful life of the structure concerned. The various methods of estimating the earthquake hazard are summarized below.

In Chapter 4 it has been demonstrated how geologic mapping and physiographic studies may disclose the presence of active faults. If such features exist, reasonable precautionary measures should be seriously considered. However, lack of geological and physiographic evidence does not prove that the risk is negligible.

If seismograph records of sufficient accuracy and period of observation are available, the earthquake hazard may be roughly evaluated by this means. Displacement of bench marks and earth tilt are very positive indications of danger, but they are seldom obtainable. Correlation of geological and geophysical evidence furnishes a reasonably sound basis of estimation.

In many regions reasonably long historical records are available. If earthquakes have been known to occur in the region, it should be assumed without question that history will repeat itself. The survival of ancient buildings, of which the flat arch of the church of Santo Domingo in Panama City is a classic example, may be significant, but it must be interpreted with caution. "Freak" survivals are features of many earthquakes. Much depends on unknown details of design, foundation conditions, bed rock conditions, the depth, distance, and direction from which the shock originated. Generally speaking, however, the survival of ancient structures, especially those of poor construction and/or design, at least rules out the occurrence of violent disturbances in the near vicinity. These data cannot, however, be extended farther than 25 miles or so from the site in question.

When a careful study of all available data indicates the earthquake risk to be serious, all possible precautionary measures consistent with sound economics should be taken.

REFERENCES

1. W. H. Bucher, Cryptovolcanic Structures in the United States, *16th Intern. Geol. Congr. Rept.*, Vol. 2, 1933.
2. Upton Close and Elsie McCormick, Where the Mountains Walked, *Natl. Geog. Mag.*, Vol. 41: 461–464, 1922.
3. C. E. Dutton, The Charleston Earthquake of August 13, 1886, *9th Ann. Rept. U. S. Geol. Survey*, 1889.
4. J. R. Freeman, *Earthquake Damage and Earthquake Insurance*, McGraw-Hill Book Co., New York, 1932.
5. T. Fukutomi, Crustal Deformation in the South Idu Peninsula, Japan. *J. Astron. Geophys.*, Vol. 12, 1935.
6. M. L. Fuller, The New Madrid Earthquake, *U. S. Geol. Survey, Bull.* 494, 1912.
7. Beno Gutenberg, Seismology, *Geol. Soc. Amer., 50th Anniversary Vol.*, 1941.
8. Beno Gutenberg and C. F. Richter, on Seismic Waves (3rd Paper), *Gerlands Beitr. Geophys.*, Vol. 47, 1936.
9. Beno Gutenberg and C. F. Richter, Seismicity of the Earth, *Geol. Soc. Amer., Special Paper* 34, 1941. Also published by Princeton University Press, 1954.
10. Arthur Holmes, Radioactivity and the Earth's Thermal History, *Geol. Mag.*, Vol. 52, 1915 and Vol. 62, 1925.
11. A. Imamura, On Chronic and Acute Earth Tilting in the Southern Part of Sikoku, Japan. *J. Astron. Geophys.*, Vol. 8, 1930.
12. J. Joly, *The Surface History of the Earth*, Oxford Univ. Press, 1925.
13. W. D. Keller and Adriano Valduga, The Natural Steam at Larderello, Italy, *J. Geol.*, Vol. 54, 1946.
14. A. C. Lawson et al., The California Earthquake of April 18, 1906, *Carnegie Inst. Wash. Publ.* 87, 1908.
15. L. D. Leet, *Practical Seismology and Seismic Prospecting*, D. Appleton-Century Co., New York, 1938.
16. L. D. Leet, *Causes of Catastrophe*, Whittlesey House, New York, 1948.
17. F. H. Pough, Parícutin Is Born, *Natural History*, Vol. 52, October 1943.
18. H. F. Reid, The Elastic-Rebound Theory of Earthquakes, *Univ. Calif. Publs., Bull. Dept. Geol. Sci.*, Vol. 6, No. 19, 1911.
19. A. Rittmann, *Vulkane und Ihre Tätigkeit*, Ferdinand Enke, Stuttgart, 1936.
20. R. A. Sonder, Zur Theorie und Klassification der Eruptiven Vulkanischen Vorgänge, *Geol. Rundschau*, Vol. 27, 1937.

21. C. Tsuboi, Investigation of Deformation of the Earth's Crust by Means of Precise Geodetic Measurements, *Japan. J. Astron. Geophys.*, Vol. 10, 1932.
22. C. Tsuboi, Isotasy and Maximum Earthquake Energy, *Proc. Tokyo Imp. Acad.*, Vol. 16, 1940.
23. Howel Williams, Volcanology, *Geol. Soc. Amer., 50th Anniversary Vol.*, 1941.

Historical Geology

At first thought nothing seems more remote from civil engineering than historical geology, which deals with the dating of rocks and changes in the earth's surface in times long before the human race existed. Although it is true that whether a structure rests on Pre-Cambrian or Pliocene strata is of no engineering importance, the methods of historical geology are basic to the working out of folds and faults, the presence or absence of which may be of great engineering significance. They also afford a system for cataloging and arranging vast masses of data and are necessary to a proper understanding of the origin and development of land forms on which the aerial photographic interpretation of soils, described in Chapter 16, is based. The methods of historical geology are also basic to a proper understanding of geologic maps and sections. It follows then that civil engineers should know the basic elements of historical geology. The subject is protean in scope, but the underlying principles are very simple. Once these principles are understood, details may be filled in as the occasion requires.

This section would be misleading if it gave the impression that historical geology is as important to engineers as to geologists, for the lengthy elaboration of paleontological and historical data in geologic reports intended primarily for civil engineers has perhaps done more than anything else to discourage the application of geology to civil engineering. The geologist employed in civil engineering work should bear in mind that the civil engineer's interest is usually limited to the land surface and a relatively shallow zone below. Therefore, discussion should be concentrated on explaining events leading to the development of the present land surface. Supporting data are seldom required in the text of the report and, if necessary to the project, should be relegated to appendices. The geologist should also remember that the engineer is often more interested in the soils than in the bedrock, and merely to mention in passing that alluvium is present in a stream valley is entirely to displace the proper emphasis.

It is impossible in a short chapter to give more than the barest outline of the phases of historical geology that are of greatest interest and use to the civil engineer. The subject is intrinsically interesting, however, and it is hoped that some will be motivated to read at least parts of the textbooks cited in the references.

PRINCIPLES OF GEOLOGICAL CHRONOLOGY

Geological chronology is simply a method of bookkeeping. The origin of the earth is lost in remote antiquity, but for billions of years geological processes have been shaping and modifying the planet. Rocks have been alternately deposited and destroyed by erosion. The distribution of lands and seas, mountain ranges and plains, and living things has changed greatly in the past, and is still undergoing slow but sure alteration. Indeed, it is the study of changes now in progress that furnishes the keys to deciphering the past. If the story of the physical and biological changes that have transpired is to be interpreted correctly, it is obviously necessary to determine their time sequence. Otherwise, the geological record might be compared with the scattered pages of an uncollated manuscript. Geological chronology differs from the astronomical in that there is as yet no absolute scale of reference such as that furnished by the motions of astronomical bodies. The geologic time scale is purely relative, but despite its defects it is sufficient for the purpose.

The basis of geological chronology is simply common sense. It rests on the self-evident truth that, in a series of sedimentary strata that have not been overturned by subsequent crustal movements, the oldest stratum must lie on the bottom and in turn be overlain by successively younger beds. Thus, in any area where outcrops are continuous the time sequence of the strata is readily determinable. However, by far the greater portion of the earth's surface is covered with soil, loose rubble, glacial drift, etc. The field geologist sees only isolated outcrops, and the problem of time relations of rock outcrops, or correlation, arises. It will be recalled from Chapter 4 that geologic structures cannot always be determined from purely geometric considerations. A series of uniformly dipping sedimentary beds may be either a monocline or a series of tightly compressed folds (Fig. 141). Of the various alternatives available, how is the correct interpretation to be found? The answer lies in correlating two lines of evidence: the physical and the biological. It should also be mentioned that in many instances faults cannot be detected except from the disturbance of the normal sedimentary sequence brought about by the fault movements.

Physical Methods of Correlation

Consider the situation depicted in Fig. 141*a*. The structure may be: (*a*) a monocline, (*b*) a tightly compressed anticline, (*c*) a tightly compressed syncline, or (*d*) almost any combina-

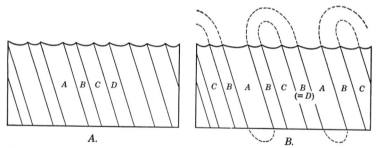

Fig. 141. Diagrams illustrating the necessity of knowing the age relationships of rocks in working out the structure of tightly compressed strata.

tion of (*b*) and (*c*). Let us suppose that the field geologist finds a less highly folded sequence a few miles from the line along which the profile has been drawn, and in this sequence a bed very similar in character to C in Fig. 141*a* is overlain by a stratum very similar in lithology to *B*, which in turn is overlain by a stratum closely resembling *A*. If he assumes that lithological similarity indicates spatial and chronological identity, the structure illustrated must consist of alternating anticlines and synclines, the restoration of which is shown in Fig. 141*b*. In other words, our hypothetical fieldworker has followed the method known as correlation by lithological similarity. The defects of this method will become more apparent as the discussion proceeds.

The above example also illustrates a point of great importance in civil engineering work; namely, the necessity of conducting geological work on a reasonably broad scale. It is apparent that, had our fictitious geologist merely concentrated along the line of the profile, it would have been impossible for him to arrive at a correct solution of the structural problem. Failure to appreciate the methods and complications of geological field work has been responsible for a large proportion of the shortcomings of geologic-engineering studies. As a rule, the best geologist in the world cannot construct a reasonably accurate geologic profile along a proposed construction route merely by walking the center line. Exposures miles away on either side may be equally, if not more, important than those along the center line.

In dealing with igneous rocks the method of superposition cannot be used. Consequently, reliance must be placed on other criteria. In order to establish the time sequence of igneous rocks in contact with sediments, it is always necessary to determine whether the relationships are intrusive or depositional; i.e., whether the igneous body forced its way into the pre-existing sediments or if the latter were deposited on the cooled and eroded surface of the igneous mass. The criteria for distinguishing intrusive relations are: (1) baking and other contact metamorphic effects; (2) presence of apophyses from the intrusive body in the intruded mass; and (3) inclusions of the intruded body in the intrusive, for example, sandstone fragments in a granite. Depositional relations are indicated by: (1) presence of a weathered zone or soil horizon on top of the underlying igneous body; (2) absence of metamorphic effects and apophyses (far from infallible); and (3) presence of igneous boulders and/or pebbles in the overlying sediments. These criteria may also be used to determine the relative ages of intrusive bodies. For example, as shown in Fig. 142, if bed *B* has been intruded by an igneous body *A*, the intrusive is obviously younger than *B*. If bed *D* rests in normal depositional contact on both *A* and *B*, *A* is obviously older than *D*.

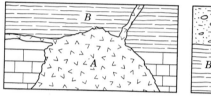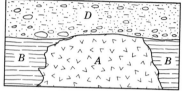

Fig. 142. Diagrams illustrating method for working out age relationships of intrusive rocks.

The relative ages of dikes may be determined by metamorphic effects and spatial relationships. If a dike intersects another dike and one shows baking and the other exhibits chilling at the contact (evidenced by a border of more finely crystalline character), it is apparent that the latter has intruded the former. In the absence of metamorphic effects, the relative ages can often be determined from spatial relationships. For example, the

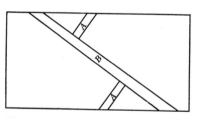

Fig. 143. Diagram illustrating method of working out age relationships of dikes.

situation shown in Fig. 143 is best interpreted as follows. Dike *A* was intruded first and later offset by a small fault. Another dike labeled *B* was then intruded along the fault.

The relative ages of metamorphic rocks are usually very difficult to determine. The chief criterion available is relative intensity of metamorphism, the supposition being that older rocks have been subjected to geological vicissitudes for a longer time and, therefore, should show more pronounced metamorphism. This criterion is far from infallible, for the degree of metamorphism also varies laterally. In metamorphic terrains it is often impossible to work out a reliable sequence, and such rock assemblages are often mapped as a metamorphic complex.

The relative ages of faults can usually be determined from their relationships to strata of known age. Thus, in Fig. 144, the youngest bed cut by the fault is *C*, and the fault is therefore younger than *C*. Bed *E* was laid down on the faulted and eroded surface and has not

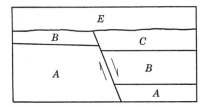

Fig. 144. Diagram illustrating method of determining relative ages of faults.

been displaced by the fault. Therefore, the fault is older than *E*. In dealing with folds, a similar method is applied. In Fig. 145, the youngest stratum affected by the folding is *D*, and the folding must thus postdate *D*. The oldest bed not affected is *F*. Therefore, the folding is older than *F*. The period of erosion which occurred during and after the folding, but before deposition of *F*, has resulted in an unconformity indicated by the discordance in dip between beds *A–D*,

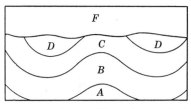

Fig. 145. Diagram illustrating method of determining the age of a fold.

inclusive, and *F*. Its time value represents the lost interval between deposition of beds *D* and *F* and may be assigned a value of *E* in the scale adopted.

Biological Methods of Correlation

In the early 1800's, an English civil engineer, William Smith, discovered that sedimentary beds of the same age contain similar assemblages of fossils, and inferred that these relics of ancient life can be used to establish the time sequence of sedimentary rocks. Smith's discovery was purely empirical, the scientific basis for the application of paleontology to geological correlation being supplied at a later time by Charles Darwin. The theory of evolution holds that all life in existence today has descended from different forms that existed in the past, and the gaps now existing between species and higher categories have come about gradually. If all the organisms that have inhabited the earth could be assembled, a fairly continuous array would emerge (the so-called "missing links"). Species originate only once, and when extinct never reappear. Because the life ranges of most species are short in comparison with geological time, it is apparent that if the life span of a species is known it may be used in geochronology. By fitting together the life spans of many species of overlapping life ranges, a fairly accurate time determination can be made. In other words, a formation can contain as fossils only those species that happened to be living at the time that it was deposited. It should be noted, however, that fossils cannot be used independent of field relationships. The fundamental datum is the stratigraphic sequence (younger rocks rest on older beds), rather than evolutionary sequences, no matter how logical or convincing. However, in most instances it is possible to correlate the physical and biological approaches. In correlating widely separated exposures it is unfortunately necessary to rely almost exclusively on biological data.

ELEMENTS OF STRATIGRAPHY

Stratigraphy is the branch of geology involving the time and spatial relationships of rocks, principally sedimentary rocks. It makes use of all available lines of information gathered from such varied fields as sedimentation, physiography, oceanography, biology, botany, and paleontology to work out the succession and distribution of rocks in the earth's crust. Geological time is divided into eras, periods, epochs, and smaller divisions of unequal duration. The rocks deposited during a period constitute a system; those formed during an epoch are known as a series. No generally accepted term has yet been proposed for the rocks formed during a geologic era. Conversely, the rocks formed during an epoch are subdivided into formations, members, and lentils, but no generally accepted terms have been proposed for the time represented by these units. Eras are separated by world-wide crustal disturbances, resulting in widespread unconformities. Periods are separated by less extensive, but still generally recognizable, unconformities. Epochs are marked by more or less local disturbances. The crustal movements punctuating geologic time not only have affected the spatial relationships of rock units but have exercised a marked influence on the composition and distribution of life assemblages. For example, the widespread folding and mountain making that occurred at the end of the Mesozoic resulted in extinction of the dinosaurs and other characteristic Mesozoic forms of life, and in the introduction of mammals and other modernized types. Thus, every division of geological time has been marked by certain events in the physical and organic worlds, and it is these events that furnish the ultimate basis of geologic correlation.

The Geologic Time Scale

The most generally accepted time scale used in the United States is given in Table X. In making use of this table it should not be inferred that the geologic time scale as now interpreted is complete and accepted by all authorities. The general sequence is well established, but the details are open to dispute and much still remains to be done. It should also be noted that the age of 2,000,000,000 years given for the oldest known rocks does not, as has so often been implied, represent the age of the earth. Our planet probably existed as an astronomical body for a vast length of time preceding the formation of the earliest rocks.

At first reading the geologic time scale may appear a little forbidding and even confused. For example, the term Quaternary is applied to a two-fold division and Tertiary to five-fold grouping. These discrepancies arose historically, and have become so well established in the literature that to replace them by a more consistent nomenclature would cause more, rather than less, confusion. If it is remembered that the time scale is a thing of "shreds and patches" which has been pieced out by generations of geologists who often labored under the handicaps of poor communications, unhealthy climates, and even public hostility, it appears remarkable for its accuracy and consistency. Certainly, the geologic time scale is no more irrational than the English and American systems of weights and measures, the defects of which have a similar historical origin. In order to un-

TABLE X. THE GEOLOGIC TIME SCALE

Era	Period	Epoch	Duration in Millions of Years *
Cenozoic	Quaternary	Recent Pleistocene	0.025 2
	Tertiary	Pliocene Miocene Oligocene Eocene Paleocene	68
Mesozoic	Cretaceous Jurassic Triassic	Too numerous to mention	180
Paleozoic	Permian Pennsylvanian Mississippian Devonian Silurian Ordovician Cambrian	Too numerous to mention	250
Pre-Cambrian	Late Pre-Cambrian Middle Pre-Cambrian Early Pre-Cambrian	Very numerous	1500

* Based on radioactive disintegration.

derstand the table, the following review of its background may be helpful.

At the time when the scale was proposed a knowledge of Latin and Greek was thought to be the mark of educated men, and classical allusions were sought wherever possible. Thus, the names of the three oldest periods of the Paleozoic were derived from the Latin names referring to the country or people of Wales, the region where rocks of this age were first studied.

An earlier classification grouped all rocks as primary (thought to be a part of the original crust of the earth), secondary (deposited on the primary rocks), tertiary, and quaternary. At a later time the classification was revised into eras in which the Paleozoic was thought to represent the most ancient life, the Mesozoic the mediaeval life, and the Cenozoic the modern life. Continued investigation disclosed, however, that life existed long before the Paleozoic, and some authorities include two eras: the Proterozoic (former life) and Archaeozoic (archaic life) in the Pre-Cambrian. Thus, the terms Paleozoic and Mesozoic are in a sense misnomers, for neither represents the oldest or middle life as the names imply. Similarly, the terms Tertiary and Quaternary do not represent the third and fourth eras of earth history, but are merely survivals of an older and now discredited system of classification.

A similar situation pertains with regard to the epochs of the Cenozoic. It was originally believed that the Eocene was the oldest epoch, and it was therefore named to indicate the "dawn of the Recent." Miocene means "less Recent" and was thought to represent the middle Tertiary life. Pliocene indicates "more Recent" and Pleistocene "most Recent." It was subsequently found necessary to interpolate the Paleocene ("ancient Recent") and Oligocene ("a little of the Recent") with the result that most of the names of Tertiary epochs possess misleading connotations.

With few exceptions, the period names of the Mesozoic and Paleozoic eras have been derived from geographic localities. Cambrian was derived from the Roman word for Wales. Ordovician and Silurian are from the Latin names for two Welsh tribes. Devonian is from Devonshire, England. Mississippian is from the river of that name; Pennsylvanian is an allusion to the Quaker state. The Permian takes its name from the province of Perm in Russia. The term Triassic refers to the threefold lithological make-up of the type section in Germany. Jurassic is from the Jura Mountains of Switzerland, and Cretaceous is from *creta*, the Latin for chalk, suggested by the predominately calcareous nature of the rocks formed during this period.

It is now an almost general practice to derive epochs and terms of lower values from the geographic region where the rocks were first recognized. Thus, the Niagara group (Silurian) takes its name from Niagara Falls.

Formation, Member, and Lentil. From an engineering point of view the units, formation, member, and lentil, are very important, and unfortunately they are probably the least understood. It is a popular error to refer to almost any rock, or even physiographic feature, such as a hogback or a glaciated surface, as a "geologic formation." This usage is absolutely indefensible and confusing. A formation is a series of rocks deposited during a particular portion of geological time, and may consist of a single rock type, or represent several types of rock deposited during a sedimentary cycle. A sedimentary cycle is the change in depositional sequence resulting from the advance and retreat of the sea or other water body over a specified area. For example, an advancing sea will at first be shallow, and at this stage conglomerates and other coarse detrital deposits will be laid down. As the sea advances the depth will increase with the result that finer sediments such as shale and siltstone will be deposited on the coarser basal beds. If the water becomes sufficiently deep, limestones may be formed. Such a series constitutes a sedimentary cycle, and the various rock types deposited may be grouped as a formation.

Subdivisions of formations are called members, for example, the basal conglomerate of the above-mentioned sequence. Subdivisions of members are known as lentils, an example of which might be a sandstone lens in the basal conglomerate member. Formations and larger units are often given a short title referring to the predominate rock type. Thus, the term "Monterey shale" (a well-known formation of the California Coast Ranges) should not be construed to mean that the rock is 100 per cent shale, but merely that shale is the predominate constituent. Tuff, limestone, chert, and siltstone also occur in the Monterey. Although the formation is usually the smallest unit shown on geologic maps, for many engineering purposes it is desirable to delineate members or even smaller units. For this purpose unusually large-scale maps must be employed.

Facies

Let us consider an advancing sea from a horizontal rather than a vertical plane of reference. In general, depth of the water will increase gradually from the shore in a seaward direction. Because coarseness of the sediments deposited varies inversely with the depth of water, coarse deposits such as conglomerate and sandstone will be deposited near shore, shales and silts farther offshore, and limestones in deeper water. In other words, normal to the shore line conglomerate will grade into sandstone, sandstone into siltstone, siltstone into shale, and shale into limestone, but parallel to the shore line the same rock type will persist for a much greater distance. A similar situation pertains to organisms. Many types of living things are exclusively terrestrial, others live only in shallow marine waters, and still others only at greater depths. Some may be adapted to a wide range of conditions, however. The variations in rock and fossil assemblages described above which result from purely environmental factors are known as facies. Similarly, rock facies must be taken into account, and it is this factor that makes correlation by lithological similarity alone so very uncertain.

A GENERAL VIEW OF EARTH HISTORY

The most striking feature of earth history is its cyclic character. In any given area mountains have been formed repeatedly only to be destroyed by erosion; seas have alternately advanced and retreated from the continental areas; periods of volcanic activity have alternated with periods of quiescence. It is the recurrence of physical events that makes possible the subdivision of geologic time into eras, periods, etc. The only progressive principle operating through the complex sequence of past changes is organic evolution, for, although the organic world has its cycles of sorts, extinct forms never reappear. In general, life has tended to become increasingly highly organized with the passage of time. Therefore, the more highly developed are the fossil organisms preserved in the strata, the younger their geologic age. With these principles in mind we are prepared to undertake a brief survey of the major divisions of geologic time.

Pre-Cambrian. Distribution of Pre-Cambrian rocks in North America is limited in outcrop to areas where profound crustal disturbances followed by deep erosion have laid bare the deeper portions of the crust. Such areas may be conveniently classified into two types: the cores of folded mountain ranges such as the Appalachians, and the so-called shields, typified by approximately 2,000,000 square miles of central and eastern Canada which geologists refer to as the Canadian shield. The shields are areas which also represent regions of profound crustal disturbance of very ancient date, and which furnished the source for much of the sedimentary materials deposited during later geologic eras. Except for minor warping caused by Pleistocene glaciation, the shield areas of the earth may be regarded as having passed through so many cycles of folding and erosion that isostatic equilibrium has been approximately attained. In other words, they have been relatively stable for a considerable portion of later geologic history, and have furnished the "buttresses" against which younger rocks have been thrust and deformed. Other shield areas exist in central Africa, northern Europe, northern Asia, Arabia, India, and Australia.

An important characteristic of Pre-Cambrian rocks is the almost complete absence of fossils, and at one time it was thought that life originated in the Cambrian. Although the abrupt appearance of highly developed types of life in the Cambrian still constitutes a major geological problem, it is certain that life existed in the Pre-Cambrian, but fossils are too scarce to be useful in correlation. Consequently, the succession of Pre-Cambrian rocks is very imperfectly known. As indicated by Table X, Pre-Cambrian time is immensely long, being estimated to comprise about 75 per cent of all geologic time. Pre-Cambrian rocks have undergone many cycles of mountain making, and as a result are commonly highly metamorphosed. Very deep erosion has exposed the granitic cores of these ancient mountains; consequently, Pre-Cambrian areas comprise a large proportion of the known granites of the world. Rocks of this age are usually "hard rocks," and unless weakened locally by structural features such as joints and faults, are generally satisfactory for most types of engineering work.

Paleozoic. As might be expected, Paleozoic rocks are generally found in belts encircling the Pre-Cambrian shield areas. Rocks of this age are usually well lithified, but many exceptions might be named; for example, the Saint Peter sandstone of Ordovician age is very friable. Except where locally involved in intense crustal movements, Paleozoic rocks tend to be less metamorphosed than Pre-Cambrian. Fossils are often abundant, and are very useful in correlation. Consequently, the structural relationships of Paleozoic rocks can usually be deciphered in great detail. Rocks of this age produce large quantities of oil and gas, and by far the greater proportion of high-rank coals. Except where locally deformed or weakened by solution, Paleozoic rocks are usually satisfactory for a wide range of engineering purposes.

Mesozoic. Mesozoic rocks are usually found in broad belts between folded mountains of Paleozoic age. Except for areas of intense local deformation, Mesozoic rocks are less metamorphosed than the Paleozoic. In contrast with the life of the Paleozoic, which is largely a record of marine invertebrates, in Mesozoic rocks land life, chiefly reptiles, becomes important for the first time. Rocks of this age produce some rather low-rank coal and considerable petroleum.

Cenozoic. Marine Cenozoic rocks tend to form narrow and discontinuous belts fringing the continental borders. Continental deposits, formed by erosion of Mesozoic and Cenozoic mountains, occur in isolated interior basins. In western North and South America Cenozoic beds are characterized by rapid thickening and thinning, abrupt facies changes, and complicated structure. Volcanic and pyroclastic rocks are very common, as are shales, sandstones, and coarser detrital materials. Limestones are relatively rare, as are slates and schists. Marine beds of this age produce a large percentage of the world's oil, whereas continental beds contain vast quantities of lignite which constitute a prospective source of synthetic petroleum products. As a rule, large-scale engineering operations in Cenozoic terrains involve numerous geological problems.

Glacial drift deposited during the Pleistocene covers a large part of northern North America, Europe, and Asia. Pleistocene glacial drift is unconsolidated, and is classified as "soil" by civil engineers. Many of the world's large cities are founded on these deposits, a circumstance which partially explains the importance of soil mechanics in modern civil engineering.

REFERENCES

1. Carl O. Dunbar, *Historical Geology*, John Wiley & Sons, New York, 1949.
2. Hollis D. Hedberg, Time-Stratigraphic Classification of Sedimentary Rocks, *Bull. Geol. Soc. Amer.*, Vol. 59, 1948.
3. Adolph Knopf et al., The Age of the Earth, *Bull. Natl. Research Council*, No. 80, 1931.
4. R. C. Moore, *Introduction to Historical Geology*, McGraw-Hill Book Co., New York, 1949.
5. M. Grace Wilmarth, The Geologic Time Classification of the United States Geological Survey Compared with Other Classifications, *U. S. Geol. Survey, Bull.* 769, 1925.
6. M. Grace Wilmarth, Lexicon of Geologic Names of the United States, *U. S. Geol. Survey, Bull.* 896, 1938.

CHAPTER 15

Geologic Maps and Sections

Geologic maps and sections are the two chief means of conveying geologic information and, therefore, the civil engineer should know how to use them. The reading of geologic maps and sections is not difficult and requires the application of only a few easily understood principles, but their intelligent use in engineering work requires a geologic background sufficient to appreciate how the data were obtained in the field. The primary purposes of this chapter are to explain the principles on which the reading of geologic maps and sections depends and to outline briefly the methods used in geologic mapping and the problems involved. Topographic maps may also convey a certain amount of geologic information, and, since geologic maps are often not available for areas that have been mapped topographically, a brief discussion of geologic interpretation of topographic maps is included. The use of aerial photographs in geologic and soil mapping is discussed in the following chapter.

GEOLOGIC INTERPRETATION OF TOPOGRAPHIC MAPS

Civil engineers are familiar with the construction and reading of topographic maps, in fact, such maps are usually made by men who have been trained as engineers. However, many engineers do not know that reasonably accurate geologic inferences can often be made from the topography of an area. This is possible because during the erosion cycle topography and underlying rock structure are often closely related. In other words, the topography tends to reflect the structure and composition of the underlying rocks.

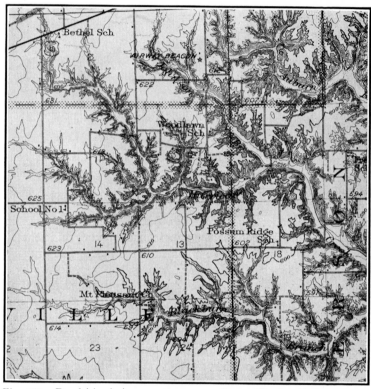

Fig. 146. Dendritic drainage pattern near Casey, Illinois. Scale approximately 1 mile to the inch, contour interval 10 feet. (U. S. Geological Survey, Casey, Illinois, Quadrangle.)

The simplest case that can be selected for illustration is that of a region underlain by rocks of uniform resistance to erosion. If such a region is subject to stream erosion, the drainage pattern developed will tend to resemble the veins of a leaf, the so-called dendritic drainage pattern described in Chapter 7. Any type of homogeneous rock may give rise to dendritic drainage; horizontally bedded sediments and large bodies of igneous rocks are examples.

A more general case may be illustrated by beds of alternating hard and soft rocks that have been folded into alternate anticlines

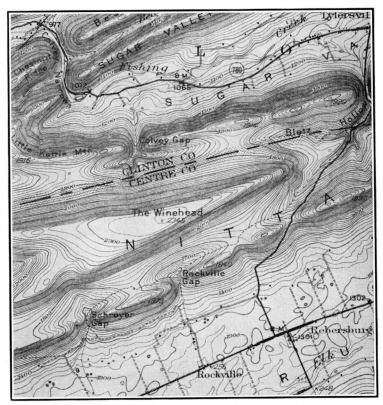

Fig. 147. Topographic expression of anticlines and synclines near Millheim, Pennsylvania. Scale approximately 1 mile to the inch, contour interval 20 feet. (U. S. Geological Survey, Millheim, Pennsylvania, Quadrangle.)

and synclines. As is indicated in Fig. 147, this type of structure is characterized by a topography consisting of arcuate asymmetrical ridges (hogbacks). Since the bedding planes of sedimentary rocks represent horizons of equal resistance to erosion, it is to be expected that on outcropping layers of tilted sediments an erosional surface will develop which approximately parallels the plane formed by intersection of the dip and strike, the so-called dip slope. The side opposite the dip slope, or the erosion slope as it is called, is not influenced by the bedding planes, and is usually steeper than the dip slope. Thus, the dip is normally

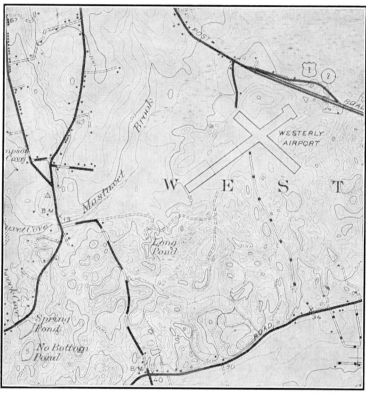

Fig. 148. Topographic expression of moraine near Westerly, Rhode Island. Scale approximately 2 inches to the mile, contour interval 10 feet. (U. S. Geological Survey, Watch Hill, Rhode Island-Connecticut, Quadrangle.)

away from the direction toward which the steeper side of the escarpment faces. Applying this principle to the case in point, it will be seen that in passing from top to bottom of the example illustrated the structure consists of an anticline followed by a syncline which in turn is followed by an anticline. The folds plunge uniformly to the left.

Faults are usually difficult to recognize from topographic expression and on topographic maps are easily confused with features of entirely different origin. For example, uplifted marine terraces, cliffs, hogbacks, dikes, sills, and walls of glaciated valleys superficially resemble faults. Although the above features can usually be distinguished by a careful study of the map, it does not follow that a topographic feature that cannot be accounted for otherwise is necessarily a fault. In fact the only really reliable criterion is actually to locate displaced beds on the ground. However, under favorable circumstances it is possible to detect suspected faults. On the other hand, absence of evidence of faulting by no means implies that faults do not exist.

Under favorable conditions glacial topography, especially the morainal type, is readily distinguishable on a topographic map. The chief criteria for its recognition are: (1) lack of a systematic drainage pattern, (2) presence of numerous small lakes and undrained depressions, (3) presence of elongated ridges, or eskers, and (4) presence of elliptical-shaped hills known as drumlins. Mountain glaciation may be detected by the presence of: (1) moraines, both recessional and lateral, (2) U-shaped valleys, (3) cirques and cirque lakes, (4) faceted spurs, and (5) hanging valleys.

The interpretation of topographic maps has been the subject of numerous books and treatises, a few of which are cited in the references.

GEOLOGIC MAPS

A geologic map is a two-dimensional representation of the distribution and structure of bedrock in an area with as much detail as is commensurate with the scale on which the map is constructed. Areal geology maps show distribution of bedrock as it would appear if the overlying soil and unconsolidated surficial materials were removed. In glaciated areas, distribution of various types of drift is shown on so-called surficial geology maps. In practice the smallest rock unit that can be delineated on the scale adopted is usually the formation. A formation may be defined as a rock unit, usually of sedimentary origin, composed of strata that have been deposited in uninterrupted sequence under the same set of conditions, or under conditions that changed gradually. The term is often used as a synonym for rock and when so used is not correct. Because a formation usually consists of more than one type of rock, the average areal geology map does not as a rule show the distribution of distinct rock types. In other words, the uses to which a geologic map is adaptable are governed by the scale to which it is constructed. Small-scale maps (one or more miles to the inch) are sufficient to establish the principal rock types present in the area, the broader structural features, and the main outlines of the history through which the region has passed. However, they do not show details of structure or distribution of members and lentils. They are, therefore, inadequate for many engineering purposes for which scales of 500 feet to the inch, or even less, are sometimes necessary.

Methods of Construction. Geologic maps may be constructed on a base furnished by topographic maps or aerial photographs, or they may be made simultaneously with topographic surveys by plane-table and alidade. Although the last method is the most accurate, and therefore preferable, it is seldom used, especially where topographic maps are available. The techniques of plotting outcrops on a topographic map are very simple, and, if this were the only task involved, geological mapping would be a simple matter indeed. However, the interpretation of structure depends largely on the ages assigned to the various outcrops encountered in the field, i.e., on the correlation of isolated rock exposures. A geologic map is, in fact, a systematic body of inferences based on interpretation of the rock exposures present. Consequently, its accuracy depends mainly on the accuracy of the inferences involved. In making the necessary inferences the geologist is guided by his past training and experience as well as by his knowledge of the area in question. He is therefore in a position to check, test, and compare these inferences during various stages of the work and to arrive at an over-all picture of considerable accuracy no matter how inaccurate the details may be. Despite their inevitable shortcomings, geologic maps have a wide range of application in civil engineering work, and they furnish an indispensable basis for all subsequent subsurface explorations.

It should be noted that accuracy of a geologic map is not entirely a function of the abundance of outcrops and the abilities of the geologist responsible for the work. It also depends on the complexity of the area in question. In many parts of the world (especially in areas underlain by Paleozoic rocks), highly fossiliferous formations of uniform character and thickness occur over very wide areas. These formations lie nearly level, or are only slightly disturbed. Given even comparatively few exposures, the geology of such areas can usually be worked out with considerable accuracy. In other regions, such as the Tertiary areas of western North and Central America, outcrops are relatively abundant, but the sedimentary formations are characterized by abrupt changes in lithology and thickness, and extensive folding and faulting; they are still further complicated by a variety of intrusive and pyroclastic rocks. In such areas, absolute accuracy in a geologic map can hardly be expected.

Whereas the accuracy of geological inferences tends to vary directly with the geologist's knowledge of the area, it follows that the larger the area studied (thoroughness of investigation held constant) the more accurate the map is likely to be. Generally speaking, it is impossible to base accurate inferences on experience gained from a study of a narrow strip of territory; and this accounts for many of the shortcomings of geologic maps made especially for engineering purposes, where considerations of expense seem to have dictated limiting the scope of the work. Because accuracy is highly desirable, the money spent for mapping a wider area than at first seemed necessary is very much justified, for geologic mapping is very cheap in comparison with other methods of engineering exploration.

Methods of Presentation. The geologic maps published by the United States Geological Survey use ordinary topographic maps as a base. The areal extent of the various formations is shown by an overlay composed of a combined color and line pattern along with a letter symbol corresponding to the geologic age and name of the formation [Sc, for example, means Silurian (Clinton formation)]. A columnar section on the margin indicates the relative ages of the formations involved, the oldest unit being at the bottom with younger formations occupying successively higher positions. Formational contacts, faults, dip and strike, etc., are shown by conventional symbols which are explained in the legend. Although many engineering organizations lack the facilities for producing such an elaborate type of

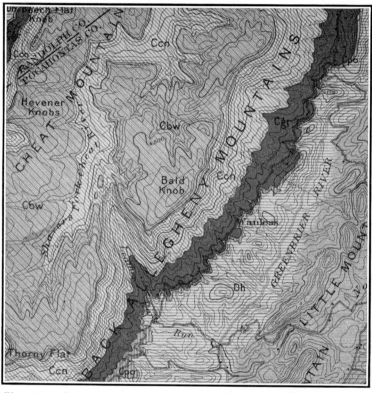

Fig. 149. Geologic map showing nearly flat-lying sedimentary rocks. Symbols beginning with D indicate Devonian strata; those beginning with C indicate Carboniferous (Mississippian and Pennsylvanian) strata. Formations are indicated by line patterns. Scale approximately 2 miles to the inch. (U. S. Geological Survey, Monterey, Virginia-West Virginia, Quadrangle.)

map, they do well to follow the methods of the Geological Survey as closely as conditions permit. Figure 149 presents a portion of a typical Geological Survey geologic map minus the color overlay and legend.

In studies involving accurate and detailed structural mapping, subsurface water supplies, oil and gas fields, for example, the use of structure contours is very convenient and often obligatory. Structure contours are similar to topographic contours in all respects except the reference surface, which is an identifiable layer of distinctive lithology and/or fossil content (marker bed) rather than the land surface. As indicated in Fig. 150, directions parallel

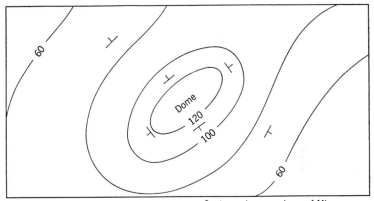

Contours drawn on base of Miocene

Fig. 150. Hypothetical structure contour map of a dome in Tertiary strata. Contour interval 20 feet; elevation datum, mean sea level.

to the contours correspond to the strike, whereas those normal thereto correspond to the dip of the strata. The direction and amount of dip may be found by comparing elevations of adjacent contour lines.

Interpretation of Geologic Maps

The interpretation of geologic maps requires a certain ability to visualize spatial relationships in three dimensions, but it is otherwise very simple. Because visualization cannot be taught, but only illustrated, the following discussion will deal only with the principles involved.

Relation of Outcrop Pattern to Topography. In strata of fairly uniform thickness the underlying structure is revealed by the relationship of formational contacts to topographic contours. If the beds are horizontal, it follows from the nature of topo-

Geologic Maps and Sections

graphic contours that both the upper and lower contacts of a formation will parallel the contour lines. Gently dipping strata of uniform thickness can be detected by the relationships of the

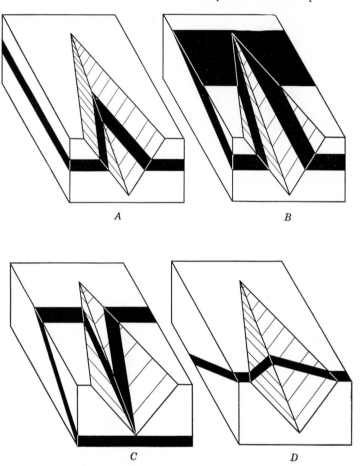

A

B

C

D

Fig. 151. Diagrams illustrating relationship between dip and outcrop pattern. *A*, horizontal strata; *B*, strata dipping gently downstream; *C*, strata dipping steeply downstream; *D*, strata dipping upstream.

contacts and contours. Both the upper and lower contacts will cut across the contours. The direction and amount of dip can be estimated by measuring the change in elevation of the formational boundaries per unit distance.

Examination of Fig. 151 discloses that, when formational contacts strike across stream valleys, V-shaped indentations of the formational boundaries result. If the indentations follow the contour lines, the strata are horizontal; but, if they cut across the contours, the strata are inclined, and, therefore, possess some measure of dip. A little reflection will disclose that the configuration of the formational boundaries intersected by a stream valley is dependent on both the direction of dip and gradient of the drainage channel. When the dip is in an upstream direction the apex of the V will point upstream regardless of whether the angle of dip is higher or lower than the stream gradient. In strata that dip in the opposite direction, however, the V will point downstream when the angle of dip exceeds the stream gradient. If the dip is lower than the valley gradient the V will point upstream. Topography has no effect on the outcrop pattern of vertical strata.

In horizontal beds of non-uniform thickness either the upper or lower formational contact, or both, cut across the contour lines, and thus simulate gently dipping strata. However, wedge-like formations can always be distinguished from dipping beds of uniform thickness by an increase or decrease in the number of contour intervals included within the upper and lower limits of the formation.

Folds. In folded areas, the various types of folds can be distinguished by application of the simple rule that in synclines the youngest stratum will always occupy a central, or axial, position; in anticlines the opposite is the case. Application of this rule is aided by the corollary that older rocks dip toward younger strata. This corollary may also be found useful in determining the direction of plunge along the canoe-shaped ends of folds (Fig. 147). However, when the section has been overturned the above rules are, of course, reversed.

Unconformities. Many problems of interpretation arise from the fact that folding followed by erosion and renewed deposition gives rise to marked discontinuities in the distribution of strata. Figure 152 illustrates a type of unconformity known as an overlap. Because unconformities obscure the underlying rocks, in most instances an areal geologic map cannot be used to determine the character of underlying beds. This can only be established by subsurface explorations. To determine the exact nature of the rocks below an unconformity in any other way would require not a geologist, but a geomancer.

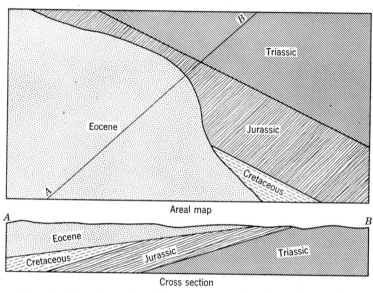

Fig. 152. Hypothetical geologic map and section illustrating overlap.

Faults. On many geologic maps the interpretation of faults is aided by symbols showing the type of fault (normal or thrust) and the direction of the last recognizable displacement. In the lack of symbols, however, it is always possible to determine the relative displacement and type of fault in question by application of the simple rule that the oldest rocks involved outcrop on the upthrown side of the fault. When a fault intersects dipping strata a considerable amount of visualization is necessary in order to determine the direction of displacement. The example illustrated in Fig. 153 shows strata dipping uniformly to the right. Because the contact between formations is offset to the right on the upper block, it is apparent that this block has been displaced upward relative to the lower block. Consequently, if the dip of the strata is known, the relative displacement of faults can be determined; and, conversely, if the relative displacement of the fault is known, the dip of the dislocated strata can always be worked out. It should also be observed that the presence of a

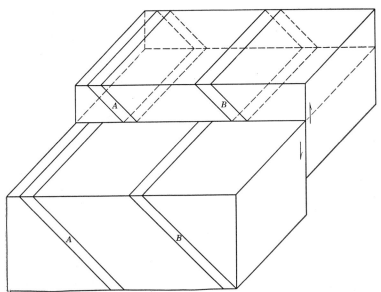

Fig. 153. Block diagram showing dipping strata offset by a fault. If the upthrown side of the fault is eroded to the level indicated by the dashed lines, movement along the fault will appear to have been entirely in a horizontal direction.

fault may often be detected by the omission or duplication of beds (Fig. 154). Determination of the ages of folds and faults has been discussed in the preceding chapter.

Soil Maps

As previously mentioned, areal geology maps tell little or nothing about the soils of a region. Surficial geology maps contain considerable soil information, but it is seldom presented from an engineering point of view. Hence, the importance of a geologic background to the engineer engaged in soil work.

To be of greatest value to the civil engineer, a soil map should convey the following information, where possible: (1) range of thickness; (2) range of grain size; (3) stratification, if any; (4)

plasticity and plasticity index, where applicable; and (5) weathering of the material and soil profile development. The last can usually be obtained from agricultural soil maps, where available. Classification of soils should correspond to accepted engineering standards, for example, fat clay, lean clay, clay sand, well-graded gravel, etc. This subject is too involved to be discussed here and is treated in Chapter 17. Where possible, alluvial soils should

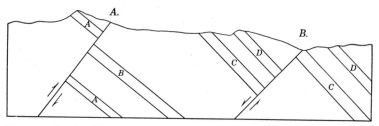

Fig. 154. Section illustrating omission (*A*) and duplication (*B*) of strata by faulting.

be broken down into the natural units described in Chapter 7, and the soils of these units should be described in terms of their engineering properties. The above considerations apply to all types of generalized engineering soil maps, and it should be added that maps made for a specific engineering purpose should stress the soil properties that have the greatest influence on the project contemplated. For example, maps made for selection of sites for earth dams should emphasize factors bearing on permeability; maps made for locating construction materials should emphasize grain size, gradation, mineral composition, and weathering of the materials.

GEOLOGIC SECTIONS

A geologic section shows the inferred relationships of rock units as they would occur in the sides of a vertical cut made along the line of the section. As a rule, the sections published by the United States Geological Survey and the various state geological surveys have the same horizontal and vertical scale. However, in engineering work it is often necessary to exaggerate the vertical scale. In such cases the angle of dip component cannot be read directly from the section. Figure 152 shows a rather conventionalized example of a geologic section.

Although no particular training is necessary in order to read geologic sections, it is well for engineers to keep in mind that they are constructed on a rather inferential basis. As previously mentioned, the accuracy of a geologic map or section rests ultimately on the matching of age relations, or correlation, of surface outcrops. Consequently, as is true of geologic maps, a geologic section is as accurate as the correlations on which it is based.

ENGINEERING USES AND ABUSES

Civil engineers are usually more interested in the subsurface geology than in surface exposures, and are prone to overlook the value of geologic maps and sections as a guide to planning subsurface investigations. On the other hand, examples might be cited of the opposite extreme where surface geologic maps were relied on exclusively for subsurface information with none too satisfactory results.

Uses. The shortcomings of geologic maps have been stressed sufficiently to indicate that they cannot be used as a source of subsurface information of accuracy sufficient for most engineering purposes. However, their merits far outweigh their defects, and in any event the first step in engineering investigations should be to obtain an adequate geologic map. If such a map cannot be obtained from published sources, qualified members of the geologic staff should be assigned to mapping the area. Even when published maps are available, it is always wise to check their accuracy, especially in critical locations.

After the geologic map has been completed, it may be found very useful in selecting the most favorable site for more intensive exploration. For example, alternate locations for a dam may be about equally favorable topographically. However, the geologic map may show that one site is crossed by a fault which brings a soft formation into contact with a very hard and competent one. Geologically this site is far inferior to the other, and, at least at first, geological explorations should be confined to the unfaulted location.

Geologic maps may also be very useful in making preliminary estimates of cost. Alternate highway alignments may be about equally favorable from a topographic point of view. However, one alignment may involve cuts through a very soft formation requiring slopes averaging 1 vertical on 2 horizontal; the second alignment may involve excavation of harder rock at a higher unit

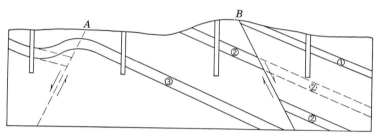

Fig. 155. Section illustrating shortcomings of arbitrary spacing and depth of borings located without reference to the surface geology. True situation indicated by solid lines; inferred conditions indicated by dashed lines. Not only is it likely that faults will be erroneously inferred (*A*), but actual faults are likely to remain undetected (*B*).

cost, but with side slopes averaging 2 vertical on 1 horizontal. When due allowance for differences in quantities is made, the second may prove to be the more economical.

Perhaps the most important use of geologic maps is in the laying out of core-boring locations. If data furnished by such maps are lacking an arbitrary spacing must be adopted. Figure 155 illustrates the shortcomings of a purely arbitrary method of spacing. If this method is adopted two consequences may follow: (1) important geologic features will remain undetected or be misinterpreted, or (2) a much larger number of borings will be required to disclose the underlying rock conditions than would have been necessary had the surface geology been considered.

In addition to the above applications, surface geologic maps may also be useful in laying out exploratory programs for building materials, coal, oil, gas, and mineral deposits.

Geologic sections are of considerable and obvious importance in problems involving excavation, tunneling, water supplies, and dams. Most sections contained in published reports are to too small a scale and too few in number to be of much use in other than a reconnaissance way. In most instances, their accuracy should be checked by core drilling.

Abuses. Misuse of geologic information arises from two widely different causes. The first is overconfidence in the accuracy and thoroughness of the information with the consequent failure to secure sufficient corroboration from drill holes or other sources. The second stems from the opposite attitude. It may be argued: because geologic maps and sections are seldom accurate when checked by subsurface work, why not disregard them entirely? The answer to this view has already been made in preceding paragraphs, but it might be well to emphasize once more that, although no geologic map or section is so accurate that it can be relied on to the exclusion of subsurface exploration, a map or section is seldom so poor as to be valueless in the planning stages at least.

REFERENCES

1. C. L. Dake and J. S. Brown, *Interpretation of Topographic and Geologic Maps*, McGraw-Hill Book Co., New York, 1925.
2. Edwin B. Eckel, Interpreting Geologic Maps for Engineers, *Am. Soc. Testing Materials, Spec. Tech. Publ.* 122, 1951.
3. Samuel L. Greitzer, *Elementary Topography and Map Reading*, McGraw-Hill Book Co., New York, 1944.
4. Charles B. Hunt, Military Geology, in *Application of Geology to Engineering Practice* (Berkey Volume), Geol. Soc. Amer., 1950.
5. F. H. Lahee, *Field Geology*, 4th Ed., McGraw-Hill Book Co., New York, 1941.
6. Norman Maclean and Averett C. Olson, *Manual for Instruction in Military Maps and Aerial Photographs*, Harper & Brothers, New York, 1943.
7. William C. Putnam, *Map Interpretation with Military Applications*, McGraw-Hill Book Co., New York, 1943.
8. R. D. Salisbury and Wallace W. Atwood, The Interpretation of Topographic Maps, *U. S. Geol. Survey, Prof. Paper* 60, 1908.

C H A P T E R   16

. *Aerial Photographic Interpretation*
of Soils

Aerial photographs have long been used by geologists in mapping bedrock, but it remained largely for civil engineers to apply geologic principles to the aerial photographic interpretation of soils. This comparatively new technique fills a long-standing need for a rapid and inexpensive method of soil reconnaissance. It probably has done more to bring geology and civil engineering together than any other single development since the turn of the century, for aerial photographic interpretation of soils requires both a thorough understanding of the engineering characteristics of soils and the geologic principles discussed in the preceding chapters of this book. It is, in fact, a form of applied geomorphology, as the scientific study of the surface features of the earth is called.

Aerial photographic interpretation of soils was developed originally to meet the needs of highway and airfield work, but it has expanded very rapidly to cover nearly all types of civil engineering exploration. Its value in the recognition of permafrost is generally recognized,[11] as are its uses in flood control and navigation work. Aerial photographs are also indispensable in laying out soil borings and interpreting the results. To put it briefly, planning and exploration for large engineering projects should almost never be undertaken without a thorough study of aerial photographs of the region by a competent interpreter.

Although the principles underlying aerial photographic interpretation of soils are relatively simple, their application requires long familiarity with soils and associated geologic features as they occur in nature and a thorough training in laboratory techniques. Therefore, the actual work is best left to specialists, but civil engineers in general, and particularly the soil mechanics specialists, should know something about the method in order to apply the results in an intelligent manner. Geologists may also gain considerably from a knowledge of the subject. It is, of course, impossible in a short chapter to do more than give the barest outline of the principles and procedures involved.

BASIC PRINCIPLES

The interpretation of soils by means of aerial photographs rests largely on the interrelationships of soils and geomorphic features, of which the land form is the basic unit. A given land form is produced by only one set of natural agencies. Therefore, to identify the land form is equivalent to identifying the processes that produced it. These processes can form only a more or less restricted range of soil types; and in some instances, sand dunes, for example (Fig. 89), only a single type of soil. Consequently, the first two steps are to identify the land forms and list the range of soils that may be associated with them.

To choose a concrete example, let us suppose that aerial photographs of a region show only three land forms: a glacial moraine, an outwash plain, and a till plain. The moraine may consist almost entirely of clay, silt, sand, gravel, boulders, or almost any combination of them; the outwash plain may consist entirely of gravel or sand, with minor amounts of silt and clay, or a combination of gravel, sand, silt, and clay; the till plain may consist largely of clay or silt with scattered deposits of sand, gravel, and boulders; peat and organic soils may fill former surface depressions. In view of the rather wide range of possibilities, it is apparent that the next steps must be to narrow this range. This is done by a study of secondary geomorphic features, such as drainage pattern and stream cross sections, vegetation, land use, and color tones appearing on the photographs.

In each consecutive step the range of possibilities is narrowed, and under favorable conditions it is often possible to plot with considerable accuracy the distribution of the soils according to grain size. The over-all procedure may be compared with mechanical analysis of soil samples. The land form furnishes a coarse sieve; secondary geomorphic features are comparable to sieves of intermediate mesh; land use and distribution of vegetation provide still finer sieves; color tones and other special markings that are sometimes seen on aerial photographs may furnish even finer sieves capable of separating sand from silt and clay.

It should not be supposed, however, that the procedure is as cut and dried as the mechanical analysis of a soil sample. In some instances various lines of evidence may conflict, and in others essential data may be lacking. Much may also depend on climate and other more or less obscure factors, and mistakes may be made if experience gained in one region is extrapolated too far. The loess deposits of the United States, for example, are characterized by undulating ridges and a herringbone drainage pattern (see Fig. 156); those of western Europe lack these features. Hence, the importance of experience.

Classification of Land Forms

For purposes of soil interpretation, land forms may be classed as constructional, destructional, and a combination of the two. Constructional land forms are those that owe their present configuration chiefly to the constructional agencies which formed them. Growing volcanic cones and migrating sand dunes are good examples. Destructional land forms are those that owe their present configuration chiefly to erosion; mesas and buttes, for example. Combinations arise when erosion of any constructional land form becomes well advanced, and this class of land form is very common.

Constructional Land Forms. Land forms of the constructional class may be divided into the following major groups, based on the agency which formed them: (1) glacial, (2) fluvial-glacial, (3) alluvial, (4) colluvial, (5) lacustrine, (6) littoral, (7) eolian, (8) volcanic, and (9) tectonic. Glacial land forms have been discussed in Chapter 10. They include moraines, drumlins, till plains, boulder trains, etc. Glacial-fluvial forms include outwash plains, kames, deltas, and eskers. All these features are usually readily recognized on aerial photographs. Alluvial land forms have been described in Chapters 7 and 8, and it is only necessary to point out that they too are usually easily recognized on aerial photographs. Colluvial land forms are caused by down-hill creep and/or slope wash. In some instances, solifluction dis-

Fig. 156. Herringbone drainage pattern in loess, northern Mississippi. Loess drainage pattern in upper right contrasts strongly with flood plain of Yocona River in lower left. Note how land-use pattern follows the drainage pattern. (Corps of Engineers, U. S. Army.)

cussed in Chapters 11 and 12, for example, they are fairly easily recognized, but in others they may present difficulties. Lacustrine land forms are those formed by lakes. They are often associated with littoral (shore line) land forms, discussed in Chapter 9, and when so associated are fairly easily recognized. Eolian forms include sand dunes, sand ridges, and loess topography. They are usually readily identifiable and diagonistic of a fairly narrow range of soil types. In Morocco, lithified sand ridges, called *sahels*, are widely distributed along the coast. On aerial photographs, they look like sand ridges, but their lithified character may often be determined by the presence of quarries.

Volcanic land forms, discussed in Chapter 13, include strato-cones, cinder cones, shield volcanoes, and lava flows. They are usually easily identified, and under favorable conditions it is often possible to make out separate lava flows. Tectonic land forms include various types of folds and faults, which in the uneroded state are extremely rare. Faults are often readily identified, even when concealed under thick blankets of residual or alluvial soils. Derangements of normal surface drainage patterns have been used successfully in locating faults buried under over 100 feet of alluvium, but in such cases the faults must be more or less active. Reference 12 describes application of the method to detection of faults in the lower Mississippi River Alluvial Valley.

Destructional Land Forms. Land forms resulting from erosion are so numerous and varied that a detailed classification would serve no useful purpose. The erosional agencies involved may be streams, waves, glaciers, wind, gravity, or almost any combination thereof. Those caused by streams are the most widespread and important, and it should be recalled that the configuration of land forms caused by stream erosion depends not only on rock and soil types involved but also on the length of time during which erosion has been active, i.e., on the "stage" in the erosion cycle described in Chapter 7.

Interpretation of the types of rock and soil associated with destructional land forms is generally more difficult than interpretation of rocks and soils associated with constructional features. Stream patterns, discussed in the following section, may offer important clues, but in situations where thick blankets of residual soils have been formed determination of their range of grain size may be difficult to impossible. If the parent materials can be identified, it is often possible to make reasonably accurate inferences concerning the types of residual soils resulting. Shales, for example, produce clayey or silty soils, sandstones tend to produce sandy soils, and limestones tend to produce clays. In humid tropical regions, like central Africa, which have been exposed to weathering and erosion for geologic ages, the influence of parent material may be reduced to the vanishing point. In such situations climate and geologic history assume primary importance. In other words, the composition of deep residual soils is largely a function of climate and the length of time during which they have been forming.

Compound Land Forms. Most of the earth's surface has been more or less affected by stream erosion, and compound land forms are thus extremely common. This is fortunate, for it is usually easier to interpret the soils and rocks composing such land forms than either the purely constructional or destructional forms. As in the case of the destructional features, compound land forms are too numerous to be classified here. It should be noted, however, that the methods used in interpreting the associated soil and rock types are a combination of those described in the preceding paragraphs. The extent of erosion may be very useful in distinguishing similar land forms of different geologic age. The older till plains (Illinoian) of central United States, for example, are generally characterized by integrated drainage systems, whereas the younger till plains (Wisconsin) generally lack this feature.

Drainage and Drainage Patterns

The drainage characteristics of a soil or rock are largely a function of its permeability, which in turn is often related to the grain size. Therefore, the drainage of an area may furnish important information concerning the soil and rocks. Sands and gravels are generally permeable, and regions underlain by these materials are generally characterized by comparatively few streams, which in turn have relatively few tributaries, and the general lack of ponds, marshy and swampy areas. Clays and silts, on the other hand, are generally impermeable, and regions underlain by these materials are characterized by numerous streams, with many branching tributaries. Swamps and marshes may be present in varying degrees. It should be added, however, that the position of the water table must also be taken into account, for when found below the water table pervious materials may exhibit ponds and marshy areas.

Drainage Patterns. The spatial arrangement of surface drainage channels constitutes what is known as the drainage pattern. The dendritic pattern developed in uniform materials and the trellis pattern formed in tilted or folded sedimentary rocks have been described in the preceding chapter. There are numerous other patterns, many of which are closely related to specific soil and/or rock types. Space permits mention of only a few of the more important patterns. Limestones and other soluble rocks give rise to numerous sink holes and a generally poorly developed surface drainage. Large streams may disappear into sink holes, flow underground for many miles, and reappear as springs which give rise to new surface streams. In extreme cases, the surface drainage may be almost entirely extinguished, giving rise to a chaotic type of topography in which isolated knobs and hills, known as "haystacks," rise above a jumbled, though reasonably flat, topography. This type of surface drainage is not universally developed on soluble materials, for the extensive chalk deposits of the Paris Basin are characterized by a topography closely resembling that ordinarily associated with shales. Radial and concentric patterns are often associated with domes, laccoliths, and volcanoes. Deltas exhibit a characteristic pattern formed by channels which diverge in a downstream direction (see Fig. 79). Flood plains of alluvial rivers are characterized by a unique type of surface drainage in which the minor streams flow away from, rather than into, the major stream.

Stream Profiles and Cross Sections. The characteristics of stream profiles and cross sections are often closely related to soil and/or rock type, and are thus a valuable adjunct to aerial photographic interpretation. In fine-grained materials like clay and silt, for example, stream cross sections and intervening divides tend to be smoothly rounded; in coarse-grained materials like sand and gravel stream cross sections tend to be sharply V-shaped. Stream cross sections also depend on the distribution of rainfall. In regions where torrential rains are frequent, clays are eroded into a network of deep narrow valleys separated by sharp divides. The herringbone drainage pattern and sharp gullies of North American loess deposits are probably the result of erosion by torrential rains; for in western Europe, where concentrated downpours seldom if ever occur, stream channels in loess deposits are smoothly rounded. Hard intercalations in otherwise soft materials, sandstone ledges in thick shale beds for example, tend to cause local steepenings of the stream profiles.

Vegetation and Land Use

The frequent association of certain types of vegetation with specific types of soil is fairly well known and is often of considerable value in soil interpretation. This is especially true for northern temperate to subarctic climates where the association of vegetation and soil type is often extremely close. Fir trees are generally found on comparatively well-drained soils, whereas spruce, alder, and willow trees tend to grow in less well-drained environments. Swamps and bogs are generally treeless, but may contain scattered bushes. The characteristic swamp and bog vegetation is composed mainly of moss and grass. Vegetation may also give indications relative to the presence or absence of permafrost and the thickness of the active layer; the "drunken forests" mentioned in Chapter 11 are an outstanding example of such an indicator. Tropical forests do not show the same tendency toward concentration of a single species in a certain place, and in the tropics more than a few individuals of the same species are rarely seen growing side by side. The value of tropical vegetation in soil interpretation is, therefore, doubtful, but until a close study has been made it would be premature to discount it completely. In detailed studies of vegetation-covered regions, the services of the botanist and pedologist are extremely valuable, if not essential. It should be noted that tree cover is seldom, if ever, dense enough to obliterate land forms and associated features.

Vegetation may, surprisingly enough, offer clues to valuable mineral deposits. The vegetation of ore-bearing regions generally contains larger concentrations of the elements found in the ore than that of surrounding areas, and this difference in concentration may produce variations in the light reflected from trees and bushes. Some plants are associated with certain minerals, and are said to serve as indicators. Some are indicative of copper and others of lead deposits. The zinc violet, *Viola calaminaria zinci*, is frequently found growing on waste dumps of zinc mines. Large copper deposits in Rhodesia were blocked out in the 1920's by correlating the presence of copper ores with distribution of vegetation.

Land Use. At first thought it might seem that intensive cultivation tends to obscure or obliterate essential clues to soil identification. Although this may be true in extreme cases, such as the extensive rice terraces of southern Asia and the Philippine Islands, the loss (if any) is compensated by the gain, for the works of man are generally very noticeably influenced by soil conditions. In eastern France and western Germany, for example, the rugged topography and associated sandy soils developed on sandstones are generally left in forest; the comparatively level topography and associated clayey soils developed on shales and limestones are cultivated. Orchards and vineyards are generally planted on stony to sandy soils. Peat bogs and other highly organic soils are normally left in pasture, and in countries where peat is used for fuel, peat diggings can be readily recognized on aerial photographs. Highways and railways show a preference for level well-drained tracts such as river terraces; airfields tend to be situated on topographically high comparatively level areas, preferably underlain by coarse-grained soils. Cracks in concrete pavements may be much more numerous where silts and clays form the subgrade than where the pavement rests on sands and gravels. Large cities do not, of course, offer a promising field for aerial photographic interpretation of soils, but even here they may be of value. In some instances data found from a study of surrounding less well-built up areas may be projected with some confidence into the main part of the city.

Color Tones and Special Markings

The varying shades of white to black seen on aerial photographs constitute what may be called color tones and form their most obvious feature. The causes of differences in color tone are many and must be very carefully evaluated. Changes in tone may be caused by shadows cast by clouds or other objects, some result from careless developing and printing, and others reflect farming practices. Therefore, color tone should not be used as the sole basis of soil interpretation, but when combined with other lines of evidence it may aid in separating silts and clays from sands and gravels. The color tone of soils is largely a function of their water content; the wetter the material the darker the tone. Consequently, sands and gravels generally print white to light gray, and clays and silts print gray to almost black. It should be noted that, when thoroughly dry, clays and silts may print as light as sands and gravels (see Fig. 157). Naturally dark-colored materials like peat and highly organic soils generally retain water and tend to print darker than other materials. Color aerial photo-

Fig. 157. Buried tiles in clay disclosed by white streaks caused by drying of the soil. (Production and Marketing Administration, U. S. Department of Agriculture.)

graphs show, of course, nearly the full range of colors observable on the ground. They may be very useful, particularly where there is a close correlation between vegetation and soil type, but they are also rather expensive.

An interesting and important feature of color tones is that they may disclose disturbances of the soil dating from thousands of years ago. The outlines of Celtic fields in the Netherlands, abandoned long before the time of Julius Caesar, can be plainly traced on aerial photographs.[3] Roman roads and fortifications in England have been located by aerial photographs, and application of the method to exploration for Inca and pre-Inca civilizations of South America has furnished archaeologists with enough material to occupy them for at least a generation.

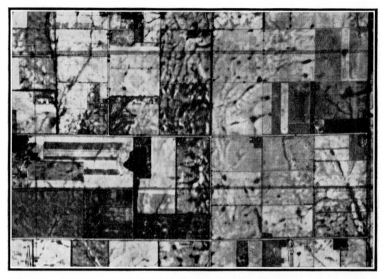

Fig. 158. Gravel outwash deposits, Laporte County, Indiana. The speckled pattern illustrated is often associated with coarse-grained materials. (Production and Marketing Administration, U. S. Department of Agriculture.)

Special Markings. Many aerial photographs show more or less peculiar markings, some of which are important in soil interpretation and some of which are entirely fortuitous. At least two types of special markings have been found to correlate with soil conditions. They are the speckled markings illustrated by Fig. 158 and the intricately banded and spotted markings shown by Fig. 159. The speckled markings are often associated with coarse-grained materials such as sand and gravel, and gravel and sand mixtures. They appear to be formed by the washing of fine-grained moisture-retentive soils into surface depressions. The second type of marking is more or less characteristic of the older till plains of central United States. It is caused by a complicated network of

Fig. 159. Illinoian glacial drift, southern Indiana. The white-fringed gullies illustrated are often associated with deposits of this type. (Production and Marketing Administration, U. S. Department of Agriculture.)

streams which have cut through a soil horizon made up of thin bands of light- and darker-colored materials.

TYPES OF AERIAL PHOTOGRAPHS AND STUDY METHODS

Aerial photographs may represent either vertical or oblique views, and the so-called trimetrogon coverage consists of a horizon-to-horizon coverage of a strip at right angles to the line of flight. Vertical views are indispensable in soil interpretation, and oblique views may also be very useful. Oblique views are particularly useful in areas where the relief features are rather broad and flat, various types of terraces for example, where the relief is not readily distinguished on vertical photographs. To be useful for stereoscopic study, individual vertical photographs should have at least a 50 per cent overlap.

Scales and Coverage Required

Vertical photographs do not, of course, possess a fixed horizontal scale. Scale distortion is at a minimum near the centers of the photographs and a maximum around the edges. An absolutely accurate horizontal scale is seldom required in soil interpretation, but it is good practice to use only the central portions of individual prints in constructing soil maps.

The best horizontal scale for soil interpretation depends somewhat on the purpose of the work and the geomorphic features occurring in the area of interest. In some instances, scales of 1:50,000, or more, may be usable; in others, scales as large as 1:5000 may be required. A scale of 1:20,000 is suitable for a wide range of conditions, and combines economy of cost with generally adequate detail. Dual, or even multiple, coverages, one to a larger and the other to a smaller scale possess considerable advantages and are often highly desirable.

Coverage. Aerial photographic interpretation of soils is similar to geologic mapping, explained in Chapter 15, and it is generally as impossible to interpret correctly the soils of a small area as it is to map the geology. In many instances, the area of aerial photograph coverage should be several times greater than the area of primary interest. For example, a group of aerial photographs showing only a uniformly level plain generally cannot be used successfully for soil interpretation, for it does not tell if the plain in question is a till plain, a river terrace, a mesa, or some other similar geomorphic feature. If the area of coverage is extended to include surrounding land forms, however, the origin of the plain in question and the type of soils to be expected can usually be determined. If the plain lies behind a glacial moraine, it is very probably a till plain, and the soils are very likely to be mainly silts and clays. If it borders a river, it is likely to be a river terrace, and the soils are probably largely sands and gravels. Therefore, the area of aerial photograph coverage should always be large enough to include a fair sample of the land forms of the region, and in practically no situation should less than two distinct land forms be represented.

Study Methods

Overlapping vertical photographs should always be obtained in duplicate. One set is used to construct a roughly assembled mosaic of the entire area of coverage; the other is used in stereoscopic study. The mosaic is best constructed by stapling the prints to a Celotex board, and after the mosaic has served its purpose the photographs can be readily detached. Stereoscopic study is accomplished most expeditiously by placing a duplicate print beside the proper print in the mosaic.

Study of the Mosaic. After the mosaic has been assembled, the first step should be to delineate the major land forms shown thereon. It is in this step that aerial photographs disclose one of their most important and useful features, which for want of a better term may be referred to as aerial perspective. An aerial mosaic enables the observer to see at a glance geomorphic features that could otherwise be seen only in sequence, if at all, and thus emphasizes in a striking manner the mutual relationships which are so important in soil interpretation. Paradoxical as it may seem, in many regions a trained observer can tell more about the surface features and associated soils of a region from an aerial mosaic than he could from a ground traverse. Figure 78 illustrates such a situation. Any one familiar with alluvial geology would have no difficulty in recognizing the natural levees and associated crevasses, clay plug, point bar, and backswamp areas on the aerial mosaic, but the same person might spend several uncomfortable days on the ground without forming even a reasonably adequate idea of the over-all situation, let alone the details of clay-filled swale and sandy ridge distribution. The ground observer almost literally cannot see the trees for the forest. As this illustration tends to emphasize, however, an appreciation of the advantages of aerial perspective requires a thorough grasp of geomorphology and related branches of geology.

Study of the aerial mosaic may be aided by examination from various angles and under various conditions of lighting. Very important features may be detectable only from certain angles and under exactly the proper lighting conditions. If he is to extract the greatest possible information from the mosaic, the interpreter should have about the same appreciation of perspective and lighting effects as the artist. Published descriptions and oblique aerial views of the region may also be used to advantage during this stage. After the interpreter has satisfied himself concerning the land forms and associated features, he should sketch their boundaries lightly with a china marking pencil. This type of pencil has the advantage of easy erasure without damage to the prints, and erasures there will be.

Stereoscopic Study. Stereoscopic examination adds the third dimension to aerial photographic interpretation, and is very important in fixing the precise boundaries of land forms and the detection of minor geomorphic features, many of which may not be distinguishable without it. It is best accomplished by placing a separate print beside the proper print in the mosaic. Various types of stereoscopes are available, but for all-purpose use the small pocket type of instrument is the most satisfactory. Proper use of the stereoscope can be found only through trial and error, and detailed directions concerning it would be rather useless. It might be pointed out, however, that when viewed through the stereoscope relief features are often highly exaggerated. Shallow gullies less than a foot deep can often be recognized on large-scale photographs.

As in the case of the two-dimensional study of the mosaic, proper application of the information gained from the stereoscopic study demands a keen appreciation of geomorphology and the significance of slopes. Boundaries between land forms are generally marked by more or less abrupt changes in slope, but not all slope changes represent boundaries of land forms. They may merely indicate changes in rock hardness, old stream channels, etc. Slope changes often involve changes in soil type, and this possibility should be borne in mind throughout this and later phases of the study.

Information gained from the stereoscopic study should then be correlated with the tentative boundaries already drawn on the mosaic. Corrections and details should be added to these boundaries and preliminary soil interpretations should be made. After these preliminary interpretations have been checked against vegetation, land use, color tones, and special markings, the office study may be regarded as complete. Unless field checks are to be made, the soil map may then be constructed.

Field Checks. It is always desirable and for some purposes essential to check the soil interpretations in the field. The interpreter will know long before where the most critical points are and these should be checked first. Less doubtful points may then be checked, and, if time and money permit, what are believed to be well-established determinations should be verified. Checking is best done on the photographs used in the stereoscopic study, and field notes and revisions can usually be placed directly on the prints. If field checks are properly planned and executed, it should be possible to construct a very accurate soil map from the combined data.

Construction of Soil Maps. Soil maps of various degrees of refinement may be drawn on the basis of aerial photographs. Where a great deal of first-hand information is available, either from published accounts or later field checks, it is often possible to map the soils on the basis of a fairly narrow range of grain size, but it is never wise to go into more detail than the information warrants. In situations where direct information is scarce or completely lacking, land-form boundaries may be used as the basis of a very effective generalized soil map. A generalized map of a glaciated region, for example, may show the boundaries between several glacial land forms such as moraines and till plains with a legend describing the range of soils and the predominant soil which may be expected to be associated with them. One of the signs by which a good interpreter is known is that he is always conscious of the limitations of the method and, therefore, never pushes his inferences too far.

USE OF AERIAL PHOTOGRAPHS IN SUBSURFACE EXPLORATION

It has been well said that every well-located exploratory boring is put down to test a geological hypothesis. Aerial photographs are extremely useful, and often indispensable, in framing geological hypotheses, and thus are very valuable in laying out programs for subsurface exploration. Under favorable conditions, a study of aerial photographs may greatly reduce or even eliminate the need for preliminary borings. Even under unfavorable conditions, they may still be useful in locating critical points, many of which might otherwise be missed completely even with very closely spaced borings. Aerial photographs are, in fact, so useful in this branch of civil engineering work that the laying out of borings on aerial photographs should be regarded as standard practice.

Aerial photographs are equally important in interpreting the results of borings. Their use in this phase of the work is well illustrated by Fig. 160. Section *A* shows the best interpretation possible without the aid of aerial photographs; section *B* shows the results obtained by combining the boring data with information obtained from a study of the aerial mosaic. Boring No. 1 is located on a thin natural levee deposit of silts which covers alternate sandy ridge and clay-filled swale point bar deposits. Boring No. 2 is located in a clay-filled cutoff channel (clay plug), and between these two borings there are at least three clay-filled swales. Swales are seldom deeper than 40 feet, and three clay-filled swales less than 40 feet deep are, therefore, drawn between borings 1 and 2. There is also evidence that two swales occur to the left of Boring No. 1. Clay plugs extend to the bottom of the former river channel, and, since the present channel is about 125 feet deep and the boring is known to be to the side, the clay plug is shown 125 feet thick. The width is made to equal that indicated by the mosaic. Boring No. 3 is located on silty natural levee deposits overlying numerous clay-filled swales and sandy ridges, and a number of swales and ridges are, therefore, indicated below the natural levee silts. Boring No. 4 is located just beyond the opposite arm of the clay plug, but the data supplied by Boring No. 2 and the mosaic are sufficient to permit drawing it in with considerable confidence.

Section *B* is still lacking in details, such as the exact number and depth of clay-filled swales. These could be determined, if necessary, by a number of carefully located shallow borings, but section *B* is obviously a vast improvement over section *A*, which is so poor as to be worse than nothing at all. These sections also illustrate the importance of locating borings according to geologic hypotheses, for if a purely arbitrary spacing had been used the clay plug, which is the most important feature present, might have

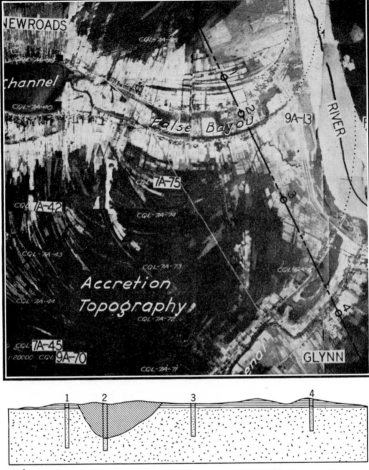

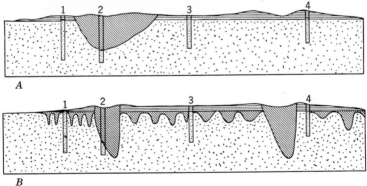

Fig. 160. A portion of False River Cutoff, Louisiana, with hypothetical sections illustrating uses of aerial photographs in soil exploration. Section *A* constructed without the aid of aerial photographs; section *B* constructed with their aid. Horizontal rulings indicate silts; slant rulings indicate clays; dots indicate sands. (Aerial photograph, Corps of Engineers, U. S. Army.)

been missed completely. It might also be added that no matter how closely spaced the borings may be, proper interpretation would be difficult to impossible without the assistance of aerial photographs.

REFERENCES

1. Airphoto Interpretation Laboratory, Joint Highway Research Project, *A Manual of the Airphoto Interpretation of Soils and Rocks for Engineering Purposes*, Purdue University, August, 1953.
2. R. E. Frost and K. B. Woods, Aerial Photographs Used for an Engineering Evaluation of Soil Materials, *Proc. 2nd Intern. Conf. Soil Mech. and Foundation Eng.*, Vol. 1, Rotterdam, 1948.
3. C. A. J. von Frytag Drabbe, Photo-Interpretation in the Netherlands, *Archives Internationale de Photogrammetrie*, Tome X, Fascicule 2, 1950.
4. D. S. Jenkins, D. J. Belcher, L. E. Gregg, and K. B. Woods, The Origin, Distribution, and Airphoto Identification of United States Soils, *U. S. Dept. Commerce, Civil Aeronautics Admin., Tech. Devel. Report* 52, May, 1946.
5. R. D. Leuder, The Preparation of an Engineering Soil Map of New Jersey, *Am. Soc. Testing Materials, Special Tech. Publ.* 122, June, 1951.
6. Robert D. Miles, Preparation of Engineering Soils and Drainage Survey Strip Maps from Aerial Photographs, *Proc. 36th Annual Road School*, Series 71, Purdue Univ., 1950.
7. Robert D. Miles, Application of Aerial Photographs to Preliminary Engineering Soil Surveys, *Am. Soc. Testing Materials, Special Tech. Publ.* 122, June, 1951.
8. M. Parvis, Drainage Pattern Significance in Airphoto Interpretation of Soils and Bedrocks, *Highway Research Board, Bull.* 28, Washington, D. C., November, 1950.
9. R. J. Russell, Louisiana Stream Patterns, *Bull. Amer. Assoc. Petroleum Geologists*, Vol. 23, 1939.
10. R. J. Russell, Quaternary History of Louisiana, *Bull. Geol. Soc. Amer.*, Vol. 51, 1940.
11. K. B. Woods, Jean E. Hittle, and R. E. Frost, Correlation between Permafrost and Soils As Indicated by Aerial Photographs, *Proc. 2nd Intern. Conf. Soil Mech. and Foundation Eng.*, Vol. 1, Rotterdam, 1948.
12. *Waterways Experiment Station*, Geological Investigation of Faulting in the Lower Mississippi Valley, *Tech. Memo.* 3–311, Vicksburg, Miss., May, 1950.

CHAPTER 17

Soil Mechanics *

As mentioned in Chapter 6, soils may be defined and studied from various points of view—geology, pedology, physics, chemistry, and civil engineering. The geologist is primarily interested in the origin of soils and the time sequences of natural deposits, whereas the pedologist is concerned with soils from a basically agricultural point of view. The civil engineer is chiefly concerned with the behavior of soil and rock under load and their potentialities as construction materials. The study of the engineering properties of soils comprises the field of soil mechanics, which in the early stages of its development was considered an exclusively engineering subject related mainly to strength of materials. It is now fairly generally recognized that soil mechanics has many points of contact with geology and pedology, and few important investigations are attempted without reference to these natural sciences. It is perhaps less widely realized that soil mechanics has much to contribute to the field of pure geology.

Geology enters into soil mechanics studies in many ways. It is a basic consideration in planning field investigations, locating borings, and interpreting the results. It also aids in establishing the origin and history of the materials; neglect of these factors may have a very adverse influence on the engineer's conclusions. Moreover, an engineer who is thoroughly acquainted with geol-

* This chapter was prepared by E. J. Yoder of Purdue University.

ogy is often able to infer the engineering properties of soils from geologic data, for soils developed under similar conditions and from similar materials tend to behave similarly under load, even though situated many miles apart. To put it briefly, correlation of geologic and soil mechanics data generally results in considerable reduction of the number of field samples and laboratory tests that would otherwise be required. It follows, therefore, that the soil engineer should be conversant with geology, and the geologist engaged in civil engineering work should have at least a basic knowledge of soil mechanics. To correlate fully the fields of soil mechanics and geology is beyond the scope of a single chapter, and the material in the following pages has been selected mainly for the use of geologists and civil engineers who are not soil mechanics specialists.

COMPONENTS AND CLASSIFICATION OF SOILS

For the most part the basic components of soils are differentiated on a grain size basis. However, the upper and lower grain size limits used for each component have not been too well standardized, with the result that engineers use various size classifications. The broad groups that have come into common use include boulders, gravel, sand, silt, and clay. Table XI contains a summary of the soil components as defined in the Unified Soil Classification System of the Corps of Engineers, U. S. Army.[3]

TABLE XI.   DEFINITIONS OF SOIL COMPONENTS

Component	Size Range
Cobbles	Above 3 in.
Gravel	3 in. to No. 4 sieve (4.76 mm)
coarse gravel	3 in. to $\frac{3}{4}$ in.
fine gravel	$\frac{3}{4}$ in. to No. 4 sieve (4.76 mm)
Sand	No. 4 sieve (4.76 mm) to No. 200 sieve (0.074 mm)
coarse sand	No. 4 sieve (4.76 mm) to No. 10 sieve (2.0 mm)
medium sand	No. 10 sieve (2.0 mm) to No. 40 sieve (0.42 mm)
fine sand	No. 40 sieve (0.42 mm) to No. 200 sieve (0.074 mm)
Fines (silt or clay)	Below No. 200 sieve (0.074 mm)

The size limitations are arbitrary; they are proposed in an attempt to standardize the definitions. Though for many years the difference between silt and clay was considered one of grain size only, the difference is not one of grain size alone but is also re-

Fig. 161. A portion of the Soil Mechanics Laboratory of the Waterways Experiment Station. *A*, consolidation testing device; *B*, torsion shear testing device; *C*, direct shear testing device; *D*, Baldwin-Southwark aircell testing machine, used mainly for triaxial testing; *E*, triaxial testing machines; *F*, Baldwin-Southwark standard testing machine, used mainly for unconfined compression and CBR testing; *G*, humid room for storage and preparation of samples; *H*, Tinius-Olsen testing machine, used mainly for CBR testing. Classification tests conducted in alcove in far left; compaction carried out in separate room at rear. Electron microscope and differential thermal analyzer, not shown in photograph, are also part of the equipment.

lated to the plasticity of the soil. For this reason, most classification systems in use at the present time define silt and clay in terms of plasticity, the clay soil being the more plastic. Here again, the difference is strictly arbitrary.

Plasticity may be defined as the ability of a soil mass to be deformed by an external force without rupture and without regaining its shape after removal of the force. The liquid and plastic limit tests, often referred to as the Atterberg limits, have come into common use for detecting soil plasticity.[1] The plastic limit is defined as the moisture content at which the soil will just begin to crumble when rolled into threads ⅛ inch in diameter. The liquid limit is the moisture content at which the soil will just begin to flow when lightly jarred in a standard manner. Because the liquid limit test is strictly arbitrary, it has been standardized, and an identical procedure must be followed at all times in order to obtain reproducible results. In the standard test a soil-water slurry is placed in a metal cup and a groove cut through the soil with a small tool. The liquid limit is then defined as the moisture content at which this groove will close for a distance of ½ inch when the cup is dropped from a height of 1 centimeter, 25 times on a hard rubber base (see reference 1 for details).

The numerical difference between the liquid and plastic limits, termed the plasticity index, represents the range in moisture contents through which the soil is plastic. High liquid limits and high plasticity indices indicate highly plastic soils; soils of low plasticity produce low test values.

Soil Classification

In the design of a structure, the engineer follows a sequence of procedures in which he first decides upon the type of material to be used, and then makes a decision concerning the allowable working stresses. Soils, unlike other engineering materials, must in many cases be used as they occur in nature. Investigation may indicate means by which the soil may be improved, such as compaction, chemical stabilization, or mixing with soils of different properties, but, in any event, the first step is to identify the soil encountered. It is often possible to infer design data if the soil has been adequately classified.

To be classified adequately, a soil must be described in sufficient detail to enable other engineers to recognize it and to form an opinion regarding its probable behavior. Many classification systems, based primarily on size of the soil components, have been used in the past. This type of classification may be very misleading, because it does not take into account topographic position, climate, mineral composition, and many other factors. In the classification of a soil, a logical procedure must be adopted that takes into account, so far as possible, all the factors that may affect structures built on the soil, or structures such as dams and fills that will be constructed with the soil.

Soils derived from similar parent materials under identical climatic conditions will, within limits, have the same physical properties. Performance surveys of highway and airport pavements indicate that soils deposited under similar geological conditions have similar performance records. Therefore, for adequate soil identification, geological data regarding type of parent material, climatic conditions, and topographic position should be included in the description.

At the present time three soil classification systems for engineering purposes are in widespread use in the United States. These include the Unified System adopted by the Corps of Engineers, U. S. Army,[3] the Civil Aeronautics Administration System,[2] and the Highway Research Board System.[4] Only the first will be discussed. Table XII contains a summary of this system.

The coarse-grained soils are identified by a sieve analysis. Sands and gravels are materials more than 50 per cent of which is retained on the No. 200 sieve. The GM, SM, GC, and SC groups are made up of granular materials containing more than 12 per cent by weight passing the No. 200 sieve. Poorly graded sands and gravels are those in which all the material is of one size, or in which some size is missing.

The plasticity chart (Fig. 162) is used in the classification of fine-grained soils. Test values from liquid and plastic limit tests made on the portion of the soil sample smaller than the No. 40 sieve (0.42 mm) are entered in Fig. 162, and the appropriate classification symbol is assigned to the soil.

Several factors must be kept in mind relative to classifying a soil by this means. The identification tests are made on disturbed samples, and test results are greatly influenced by the testing procedure. For example, the liquid limit obtained after air-drying the soil may be vastly different from that of the soil in its natural state. Soil structure may also greatly influence soil behavior in an undisturbed state. A soil may appear to have good bearing capacity on the basis of its classification, but it may be sensitive and lose practically all its strength upon slight disturbance.

TABLE XII. CORPS OF ENGINEERS UNIFIED SOIL CLASSIFICATION

Major Divisions		Letter	Name	Characteristics Pertaining to Embankments and Foundations		Value as Foundation When Not Subject to Frost Action	Value as Base Directly under Bituminous Pavement	Potential Frost Action
				Value for Embankments	Value for Foundations			
Coarse-grained soils	Gravel and gravelly soils	GW	Well-graded gravels or gravel-sand mixtures, little or no fines	Very stable, pervious shells of dikes and dams	Good bearing value	Excellent	Good	None to very slight
		GP	Poorly graded gravels or gravel-sand mixtures, little or no fines	Reasonably stable, pervious shells of dikes and dams	Good bearing value	Good to excellent	Poor to fair	None to very slight
		GM	Silty gravels, gravel-sand-silt mixtures	Reasonably stable, not particularly suited to shells, but may be used for impervious cores or blankets	Good bearing value	Good to excellent	Poor to good	Slight to medium
	Sand and sandy soils	GC	Clayey gravels, gravel-sand-clay mixtures	Fairly stable, may be used for impervious core	Good bearing value	Good	Poor	Slight to medium
		SW	Well-graded sands or gravelly sands, little or no fines	Very stable, pervious sections, slope protection required	Good bearing value	Good	Poor	None to very slight
		SP	Poorly graded sands or gravelly sands, little or no fines	Reasonably stable, may be used in dike section with flat slopes	Good to poor bearing value depending on density	Fair to good	Poor to not suitable	None to very slight

Major division	Symbol	Name	Value for embankments	Value as foundation	Compaction characteristics	Requirements for seepage control	Compressibility and expansion
Fine-grained soils	SM	Silty sands, sand-silt mixtures	Fairly stable, not particularly suited to shells, but may be used for impervious cores or dikes	Good to poor bearing value depending on density	Fair to good	Poor to not suitable	Slight to high
	SC	Clayey sands, sand-silt mixtures	Fairly stable, use for impervious core for flood control structures	Good to poor bearing value	Fair to good	Not suitable	Slight to high
Silts and clays LL < 50	ML	Inorganic silts and very fine sands, rock flour, silty or clayey silts with slight plasticity	Poor stability, may be used for embankments with proper control	Very poor, susceptible to liquefaction	Fair to poor	Not suitable	Medium to very high
	CL	Inorganic clays of low to medium plasticity, gravelly clays, sandy clays, silty clays, lean clays	Stable, impervious cores and blankets	Good to poor bearing	Fair to poor	Not suitable	Medium to high
	OL	Organic silts and organic silt-clays of low plasticity	Not suitable for embankments	Good to poor bearing, may have excessive settlements	Poor	Not suitable	Medium to high
Silts and clays LL > 50	MH	Inorganic silts, micaceous or diatomaceous fine sandy or silty soils, clastic silts	Poor stability, core of hydraulic fill dam, not desirable in rolled filled construction	Poor bearing	Poor	Not suitable	Medium to very high
	CH	Inorganic clays of high plasticity, fat clays	Fair stability with flat slopes, thin cores, blankets and dike sections	Fair to poor bearing	Poor to very poor	Not suitable	Medium
	OH	Organic clays of medium to high plasticity, organic silts	Not suitable for embankments	Very poor bearing	Poor to very poor	Not suitable	Medium
Highly organic soils	Pt	Peat and other highly organic soils	Not used for construction	Remove from foundations	Not suitable	Not suitable	Slight

The sensitivity of clay soils is most important. If an undisturbed sample of clay is tested in simple compression applied so rapidly that no moisture changes occur during the test, it will have a certain compressive strength. If the same soil is reworked and recompacted to the same moisture content and dry density and then tested, it may have a much lower compressive strength. This loss in strength is attributed to several factors, including breakdown of the soil structure. The ratio of the unconfined

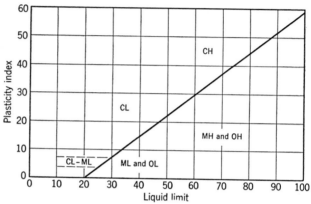

Fig. 162. Plasticity chart. (After Corps of Engineers, U. S. Army.)

compressive strength in an undisturbed state to that after remolding is known as sensitivity, S in the equation below.

$$S = \frac{\text{Unconfined strength, undisturbed}}{\text{Unconfined strength, remolded}} \qquad (1)$$

A clay is said to be sensitive when values of S range between 4 and 8, and to be extra sensitive when values of 8 and more are encountered.[11] It is thus apparent that tests on disturbed samples should be supplemented with information concerning the consistency and structure of the soil.

The use of classification tests has become a routine matter for most engineers, and it is important enough to warrant this type of testing by all persons dealing with soils from an engineering point of view. Many useful correlations have been evolved from the classification, as indicated in Table XII, but the greatest benefits of this method are realized after the engineer has personally run many classification tests and has formed an opinion regarding the behavior of the various types of soils. Furthermore, the wide-spread use of classification systems enables engineers from widely scattered geographic locations to speak in common terms.

DEFINITIONS OF RELATIONSHIPS PERTAINING TO UNIT WEIGHT

A mass of soil can be either in a dry state, partially saturated, or completely saturated. A sketch illustrating a partially saturated soil is shown in Fig. 163. For purposes of clarity, the soil

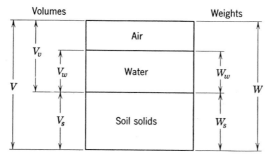

Fig. 163. Diagram illustrating a partially saturated soil.

solids, water, and air are shown separately, although in actuality the soil mass consists of an intimate mixture of the three.

The true specific gravity of the soil grains is equal to the weight of the solids divided by the weight of an equal volume of water,

$$S_s = \frac{W_s}{V_s \gamma_w} \tag{2}$$

where γ_w is equal to the unit weight of water.

Mass unit weight is defined as total weight divided by the volume. Generally this is expressed in pounds per cubic foot.

$$\gamma_m = \frac{W}{V} \tag{3}$$

Dry unit weight is equal to the weight of the soil solids divided by the total volume. This represents the unit weight of the soil mass after removal of all water.

$$\gamma_d = \frac{W_s}{V} \qquad (4)$$

The moisture content is the ratio of the weight of water contained in the soil to that of the soil solids. This value is expressed as a per cent.

$$w = \frac{W_w}{W_s} (100) \qquad (5)$$

Void ratio is the ratio of volume of voids to volume of solids.

$$e = \frac{V_v}{V_s} \qquad (6)$$

Density, or unit weight, and moisture values are highly significant, particularly in airfield, highway, and embankment construction. Generally, a certain density is specified to which the soil is to be compacted, and the moisture content at time of compaction is most important for many soils. Density measurements are generally made by weighing a portion of the soil from the construction area and measuring the volume of the hole from which it was taken. Moisture content is conveniently measured by drying a small portion of the soil to oven dry weight and determining its weight before and after drying.

Void ratio is generally obtained by measuring the volume V, either by direct measurement, immersion in mercury, or as indicated above, and determining the moisture content and dry weight. If the specific gravity of the soil solids is known, V_s is calculated by equation 2, and the volume of voids is determined by subtracting V_s from the total volume.

METHODS OF EVALUATING SOIL STRENGTH IN THE LABORATORY

Strength tests for estimating shearing resistance may be made either on undisturbed samples or on remolded samples, depending on the data desired. If a structure is to be built on soil in situ, the so-called undisturbed test is made. Undisturbed soil samples may be obtained by means of sampling pits, thin-walled tube samplers, or other procedures wherein disturbance of the soil is kept at a minimum. Such a procedure might be indicated in building, bridge, dam, or earth embankment foundations. The aim of undisturbed sampling procedures is to obtain samples at the moisture content, density, and structure in which the soil exists in the field. This necessitates coating the sample with paraffin, or similar substance, to insure that no soil water is lost previous to testing. It is also necessary to insure that the soil sample is not jarred in handling. Undisturbed samples are relieved of their overburden pressures upon being removed from the ground and are always disturbed to at least a minor degree.

Remolded samples are tested for design of embankments, pavements, etc., where the soil will be reworked and recompacted. The procedure followed is to compact the soil to the density and moisture content expected during construction and then subject it to weathering and/or loading conditions that duplicate, so far as possible, actual conditions as they will exist at some future date.

Direct Shear Test

A soil mass will support a load only so long as its shearing resistance is not exceeded, or consolidation of the soil does not

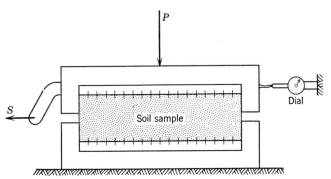

Fig. 164. Diagram illustrating direct shear apparatus.

take place. In the analysis of the stability of foundations these two conditions must be investigated. The oldest, and perhaps simplest, method of determining the shearing resistance of soil is the direct shear test. This consists of placing a sample of the soil in a split box and applying a shearing force (S) to the sample as indicated in Fig. 164. A normal force (P) is applied perpendicular to the shearing force. As the strain is increased, the force required to cause further strain is increased until it approaches a peak value. The shearing forces may then abruptly or gradually decrease with further strain, depending on the type of

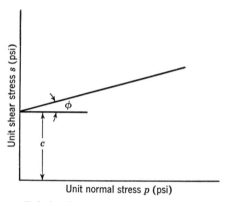

Fig. 165. Relation between unit shear and normal stress.

material being tested. The test is repeated under conditions of various normal loads, and the results are plotted as shown in Fig. 165. Unit shear stress (s) at failure is plotted against the corresponding value of unit normal stress (p), and a straight line is fitted through the series of points thus obtained. The unit shearing stress (s) and the unit normal stress (p) are calculated by dividing S and P by the cross-sectional area.

Normally the data fit a straight line, but, owing to testing techniques, etc., there may be some scattering. The intersection on the y axis of the line drawn through the test values is termed the value of unit cohesion (c). The angle that this line makes is known as the angle of internal friction (ϕ). For cohesionless sands, the value of cohesion will be zero; for soft-saturated clays tested under conditions of constant moisture content the angle of internal friction is, for all practical purposes, equal to zero. The equation of the straight line for the general case is

$$s = c + p \tan \phi \tag{7}$$

which is Coulomb's equation for friction. Shearing resistance is thus made up of two components, one value being the resistance to shear under zero normal load, and the other consisting of friction caused by the normal load. In actuality, it is most difficult to determine the true values for any one soil, for the shearing resistance is greatly affected by other factors such as preconsolidation and rate of load application.

Triaxial Compression Test

In the triaxial test, confining pressures are applied to the sample. Figure 166 is a sketch illustrating this type of test. The sample is generally encased in a thin rubber membrane, fitted snugly and bound to porous stones at the top and bottom of the sample. The sample is then placed in a transparent cylinder, and confining pressures are applied through valve A by means of air, water, or more often glycerine. Vertical loads are applied by a piston passing through the cylinder head. Valve B may be open or closed during the test, depending on whether a drained or undrained test is desired. The loads may be applied so that either a constant rate of loading or a constant rate of strain is maintained.

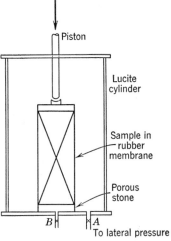

Fig. 166. Diagram illustrating triaxial apparatus.

An abrupt failure of the sample is sometimes obtained. At other times, however, the stress-strain curve does not bend down, but increases gradually with increasing strain. In this case the ultimate stress is taken at some arbitrary strain, generally 20 per cent. The peak or ultimate stress is termed the deviator stress p_d, and the confining stress is designated p_3. Thus, the total vertical stress at failure (p_1) is equal to $p_d + p_3$.

Triaxial tests may be made with the drainage valve *B* open or closed during the testing period. If the sample is permitted to consolidate fully (with the valve open) under the confining stress before application of the deviator load, the test is termed either the slow test or quick-consolidated test. In the quick-consolidated test, the valve is closed before application of the deviator load so that no further drainage can take place. If the valve is kept open and the deviator load is applied at a very slow rate, so that the moisture in the sample may drain and readjust itself, the test is called a slow test. The quick test is one that is made before consolidation under lateral stress is permitted, with valve *B* closed, and at a relatively fast rate of load application. In this test, the water cannot drain from the specimen with increased strain, with the result that the water is placed under pressure, which in turn greatly affects the stress condition at failure.

Mohr Diagram. A detailed treatment of the application of the results of the triaxial test is beyond the scope of this book, but the use of the Mohr diagram plays such an important role in the analysis of the test that some mention of it is warranted. The data are plotted in the form of Mohr's circles of stress as indicated in Fig. 167. Stresses p_1 and p_3 are plotted as shown in Fig. 167a, and a circle with its center on the x axis is drawn through the points. It can be proved that the circle represents the state of stress on any plane of the sample under the system of loading p_1 and p_3. Thus, in Fig. 167b the normal stress (p) and the shearing stress (s) acting on a plane of angle α to the horizontal are represented on the circumference of the circle in Fig. 167a at an angle of 2α from the center of the circle. If a series of tests is made, each test at a different lateral pressure, and the corresponding circles plotted, a line drawn tangent to the circles results in Mohr's rupture envelope for the general case as shown in Fig. 167c. This line will intersect the y axis at a value of c and be inclined at an angle equal to ϕ. The equation of this line can be expressed by equation 7.

California Bearing Ratio Test

Many other tests have been devised for measuring various strength properties of soils. The unconfined and triaxial tests described above are rational tests in which it is possible to determine the state of stress on any plane at failure. Other tests which

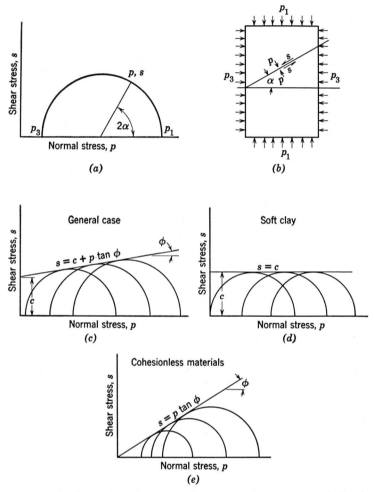

Fig. 167. Mohr circles of stress, and stresses acting on a sample in the triaxial test.

are performed to evaluate soil strength are for the most part empirical, and their usefulness depends on correlation of the test results with field performance.

The California Bearing Ratio test (usually shortened to CBR) falls in the latter group and has come into widespread use for the design of flexible airport and highway pavements. This test is of the penetration type, wherein a standardized piston having an end area of 3 square inches is pushed into the soil at a standard rate of 0.05 inch per minute (Fig. 168). The unit load at each $\frac{1}{10}$-inch penetration up to 0.5 inch is recorded, and the California Bearing Ratio is computed as the ratio of an arbitrarily selected unit load to that of a standard. The standard values were obtained by testing a high-quality crushed stone material and are as follows: 0.1-inch penetration, 1000 psi; 0.2-inch penetration, 1500 psi; 0.3-inch penetration, 1900 psi; 0.4-inch penetration, 2300 psi; 0.5-inch penetration, 2600 psi. The unit load generally taken for design is at 0.1-inch penetration. Thus, if the unit load on the piston at 0.1-inch penetration is 100 psi, the California Bearing Ratio is 10 per cent. The California Bearing Ratio has been correlated with pavement performance, and design curves have been formulated giving pavement thickness for various wheel loads and CBR values.

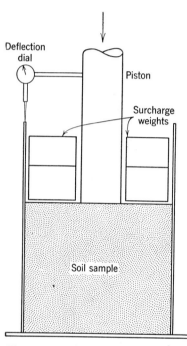

Fig. 168. Diagram illustrating the California Bearing Ratio test.

All features of this test have been standardized, such as soaking the sample in water for a period of 4 days to saturate the soil, method of compaction, etc. The cylinder in which the soil is tested has a diameter of 6 inches; the piston has a diameter under 2 inches. Surcharge weights estimated to exert a pressure equal to that of the pavement are placed on the soil during soaking and testing in order to prevent the soil from swelling during soaking and heaving around the piston during testing. Because this test is an empirical one, it is absolutely essential that the standard procedures be followed throughout. Deviations from prescribed rate of loading, size of piston, etc., completely invalidate the results.

EFFECTIVE AND NEUTRAL STRESSES

References to effective and neutral stresses are common in soil mechanics discussions. Effective stress (or pressure) is transmitted from grain to grain in a soil mass; neutral stresses are pressures in the soil water. Effective stresses cause deformation of the soil, and it is important that they be clearly distinguished from the neutral stresses. Effective stresses are always obtained by subtracting the neutral, or pore-water, pressures from the total. Pore-water pressures may be caused by hydrostatic head, flow of water, or deformation of the soil which tends to squeeze water from the pores, and other causes. If the ground water table is at the ground surface, the effective stress at a depth Z due to the hydrostatic condition is:

$$p = Z\gamma_m - Z\gamma_w = Z(\gamma_m - \gamma_w) \qquad (8)$$

where $Z\gamma_m$ represents the total stress in the soil and $Z\gamma_w$ the stress in the pore-water. The quantity $\gamma_m - \gamma_w$ is the submerged (or effective) unit weight.

SHEARING RESISTANCE OF SOIL DEPOSITS

The subject of shearing resistance of soils is complex, and for a thorough understanding of its significance the reader should consult textbooks on soil mechanics. It is too important, however, to be omitted entirely.

The angle of internal friction of dry sand is nearly equal to its angle of repose. Determination of the angle of repose is simple: an oven-dried sample is poured onto a flat surface, and the angle of the cone-shaped mass with the horizontal is measured. The shearing resistance of sand is dependent on the relative density. That is, a dense sand tends to increase in volume, owing to the tendency of the grains to slide over one another.

On the other hand, if it was originally in a loose state, the sand when subjected to shear tends to decrease in volume. If a triaxial test is made with full drainage permitted, and if water is free to move into or out of the pores, the shearing resistance can be expressed as

$$s = p \tan \phi \qquad (9)$$

This is the equation of Mohr's rupture line that passes through the origin (Fig. 167e). However, if water cannot move freely through the sand, the shearing resistance is either reduced or increased, depending on whether the volume tends to increase or decrease. For the condition where the pore-water is under pressure owing to decrease in volume during shearing, the effective grain-to-grain stress is reduced by an amount equal to the pore-water pressure. The shearing resistance can thus be expressed as

$$s = (p - u) \tan \phi \qquad (10)$$

where u is equal to the pore-water pressure. If a sand is very loose, it may lose its shear strength upon being loaded. This is popularly known as a quick sand. Loss of shear strength may result from pore-water pressures reducing the effective, or intergranular, pressures to zero, or may be caused by an upward flow of water.

The angle of internal friction of a granular material depends upon both the density of the mass and the angularity of the grains. Angles of internal friction for a dry sand are commonly about 30°; those for dense well-graded crushed stone with sharp angular grains may run as high as 40° to 50°.

The shearing resistance of silty sands can be expressed by equation 9, but ϕ will be smaller than for clean sands. Values of ϕ depend on the type of test performed, as well as on the degree of saturation of the soil. For purposes of estimating shearing resistance of these materials, a series of quick-consolidated triaxial tests may be employed.

The shearing resistance of natural saturated clay deposits can be approximated by unconfined compression tests. The permeability of a clay is very low, and if a load is applied rapidly a condition arises that is the same as that resulting from the quick undrained laboratory test. Inasmuch as the angle of internal friction (ϕ) for a saturated undrained clay is for all practical purposes equal to zero, the shearing resistance may be given as

$$s = c \tag{11}$$

Referring to Fig. 167d, it will be seen that for the assumption $\phi = 0$ the shearing resistance is numerically equal to the radius of the Mohr circle, which in turn is equal to one-half the unconfined compressive strength.

Should the clay mentioned above consolidate during the loading process, it will behave as if it possessed some internal friction. For this reason, quick-consolidated tests may at times be made. However, owing to the time and expense involved in laboratory testing and the uncertainty of a similar condition existing in the field, determination of more than the unconfined compressive strength is seldom warranted.

Partially saturated clays placed in embankments and under roadways behave as indicated by equation 7. The values of c and ϕ will depend on the soil texture, density, and moisture content. Even so, for embankment design it is rather common to perform unconfined compressive tests on remolded samples. The shearing resistance is then given by equation 11, which is an approximation on the safe side.

CONSOLIDATION OF SOILS

It is necessary at the outset to distinguish between compaction and consolidation. Compaction refers to increasing the density of a soil by mechanical means, such as rolling or vibrating. The stability of earth dams, embankments, and other soil structures is largely dependent on soil density. Compaction of soil will be discussed more fully in later paragraphs. It should be mentioned that if a micaceous soil or an organic soil is reworked and compacted it will be highly compressible but at the same time subject to rebound upon removal of load. This happens because the compression is due in part to the elastic nature of the solid particles.

A soil may also deform, upon being loaded, because of expulsion of water and a rearrangement of the soil grains. This process is termed consolidation. The rate at which consolidation takes place depends on the permeability of the soil. If the permeability is low, consolidation takes place over a long period of time; if the permeability is relatively high, consolidation takes place at a much faster rate.

The amount of settlement for any given case will depend on the magnitude of the load as well as on the compressibility of the soil layers themselves. Soil compressibility is measured in the laboratory by means of the consolidation test. A sample is placed in a rigid metal ring, and loads of varying intensity are applied to the upper face as indicated in Fig. 169. Loads are

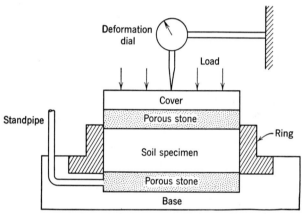

Fig. 169. Diagram of consolidation ring.

transmitted through porous stones placed at both the top and bottom of the sample. Each increment of load is permitted to remain on the sample until nearly complete consolidation takes place for that particular load. During the consolidation process, water is forced from the specimen through the disks. Since water itself is incompressible, the rate at which consolidation takes place depends on the rate at which the water escapes from the specimen. As an increment of load is applied to the soil the initial rate of deformation is relatively high, but the rate decreases rapidly until after a period of time no additional deformation is apparent.

In making a settlement analysis it is necessary to consider

whether the clay is a normally loaded or a preconsolidated material. A normally loaded clay is one on which the present overburden pressure is at least as great as any pressure that ever existed in the past. A preconsolidated clay is one which at some

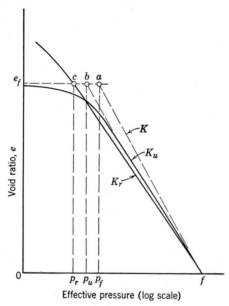

Fig. 170. Void ratio-pressure curves for normally loaded clay of ordinary sensitivity. (After Terzaghi and Peck, *Soil Mechanics in Engineering Practice*, copyrighted by John Wiley & Sons, 1948.)

time during its geologic history was subjected to pressures in excess of the overburden pressure at the time of the analysis. The settlement resulting from loading a normally loaded clay may be considerable; the settlement resulting from loading a highly preconsolidated clay is generally insignificant. Preconsolidation may be caused by desiccation or by loading by glaciers or superincumbent deposits which may be subsequently removed by melting or erosion.

Pressures used in laboratory consolidation tests are plotted on a logarithmic scale against the void ratio (Fig. 170). Because the soil is disturbed at least to a minor degree during sampling, the e-log p curve obtained in the laboratory is not the same as that of the soil in nature. To determine the field e-log p curve the following steps are taken. The void ratio for each increment of load in the laboratory test is plotted against the log of the corresponding pressure as indicated in Fig. 170. The curve labeled K_u represents the data obtained from testing an undisturbed sample in the laboratory. K_r is the curve obtained from testing a disturbed sample of soil which was mixed into thick paste before testing. This latter curve intersects the zero void ratio axis at f. Below c this curve is nearly a straight line. The initial condition in the field is represented by point a, with a void ratio e_f and the effective pressure p_f of the overburden, which is calculated from the depth times the effective unit weight. The field curve must pass through point a, and, if the curves K_r and K_u intersect at point f, the field curve should logically pass through f also. Therefore, a straight line connecting points a and f represents the field consolidation curve. Point f is based on the results of consolidation tests on undisturbed samples, and to determine the point the curve is extended downward until it intersects the x axis.

It will be noted that point a lies to the right of point b. The latter point is determined by the intersection of the extension of the K_u curve and e_f. In a preconsolidated clay, point a will lie to the left of point b. However, undue disturbance of a preconsolidated clay will sometimes result in point a being to the right of b, and in this event it may be difficult to determine whether the clay has been preconsolidated. Therefore, extreme caution in sampling is essential. Procedures for estimating the magnitude of preconsolidation pressures are described in textbooks on soil mechanics. It should also be mentioned that the e-log p curves for sensitive clays are somewhat different in shape from those for clays of ordinary sensitivity.

The total amount of settlement which will occur when a structure is placed over a compressible layer can be estimated from the field e-log p curve. The initial effective stress (p_f) and the final effective stress after loading (p_2) are calculated, and the amount of settlement is found from the relationship

$$\Delta h = H \left(\frac{e_f - e_2}{1 + e_f} \right) \qquad (12)$$

where H is the original height of the compressible layer, e_f is the initial void ratio corresponding to p_f, and e_2 is the final void ratio corresponding to p_2. Equation 12 can be written as

$$\Delta h = H \frac{C_c}{1 + e_f} \log_{10} \frac{p_f + \Delta p}{p_f} \tag{13}$$

where C_c is termed the compression index and is equal to the tangent of the angle of the straight portion of curve K (Fig. 170), and Δp is equal to the change in effective stress on the soil. Statistical correlations have been found which give values of C_c in terms of the liquid limit for normally loaded clays of ordinary sensitivity.[11]

$$C_c = 0.009(L_w - 10\%) \tag{14}$$

where L_w is equal to the liquid limit expressed as a percentage. Thus it is often possible to estimate settlement, without performing consolidation tests, by substituting values of C_c from equation 14 into equation 13.

Because the effective stress in a soil layer varies with depth, the compressible layer is generally broken down into smaller layers for calculation purposes, and Δh is calculated for each layer. The effective stress is calculated for the midpoint of each layer, and the summation of the Δh's gives the total settlement.

The time rate of consolidation for any per cent consolidation is given by

$$\frac{T}{t} = \frac{D^2}{d^2} \tag{15}$$

where T is time for foundation consolidation, t is time for consolidation of the test sample, and D and d are equal to the corresponding distances the water must travel during consolidation. If the compressible layer in the field (as well as in the laboratory) can drain from two faces (top and bottom), D and d are equal to one-half the thickness. If drainage takes place at one face only, the distance values are equal to the total thickness.

Inasmuch as the rate of consolidation for a given per cent of consolidation depends only on the distance which the water must travel during the consolidation process, consolidation of soft clays, peat, muck, etc., can be hastened by the use of vertical sand drains or other measures which aid in expelling the water. Moreover, the time varies as the square of the distance; therefore, reducing the distance that the water must travel from H to $H/2$ increases the rate of consolidation by a factor of 4.

It should be noted that equation 15 gives time of consolidation for a given per cent of total consolidation. This does not necessarily imply "total" consolidation of a soil. As a load is applied to a soil, initial consolidation takes place at a relatively fast rate. However, as time progresses the amount of consolidation per unit of time decreases so that total consolidation in the field might take place only after many years.

PLATE LOADING TESTS

Plate loading tests are used to determine allowable bearing pressure for footings and for design of concrete pavements. The test consists of loading a small plate to a predetermined load and measuring settlement of the plate. Interpretation of the data is largely a matter of personal opinion in selecting a limiting deflection, and then taking the allowable unit load as that which caused that particular deflection. This in itself may be misleading because settlement of the plate varies with size of the plate. To eliminate this variable, the plate has been more or less standardized to a 1-foot square for building foundation work, and a 30-inch diameter circular plate for evaluation of foundations of pavements.

There are many limitations to this type of test. If the loads to be placed on a footing are relatively high, the plate bearing test does not take into account settlement which might result in deep clay layers. Furthermore, should a relatively stiff layer of soil occur at shallow depths under the loaded area, the results may indicate unreasonably high bearing values.

ALLOWABLE BEARING PRESSURE FOR FOOTINGS

A footing for a structure may fail because of consolidation of the underlying soil, shear, or a combination of both, and in the design of a foundation both these factors must be considered.

It is general practice in most parts of the United States arbitrarily to set up a code listing allowable bearing pressures for various types of soils. Customary values range as high as 100 tons per square foot or more for rock to as low as $\frac{1}{2}$ ton per square foot for extremely unstable materials such as alluvial deposits. Theoretical procedures beyond the scope of this chapter are available for estimating allowable bearing pressure.[10, 11] These procedures employ the results of the unconfined compression and triaxial compression tests. The allowable bearing pressure is obtained by semi-empirical formulae giving allowable load in terms of cohesion, angle of friction, and depth and size of footing.

The allowable bearing pressure actually selected for design purposes should satisfy both requirements given at the beginning of this section. Because of unknown variations and irregularities in the soil, an ample factor of safety should be provided so that neither settlement nor shear failure will occur. It can be shown from theory that for footings on soft clays allowable bearing pressure at which shear failure will not occur is independent of width of the footing and depends on depth to only a minor extent. This relationship holds true only so long as the moisture content of the clay is not altered during loading and ϕ is equal to zero. Foundations on clay have settled excessively, even though a factor of safety against shear failure was provided.

It can be shown from theory that the allowable bearing pressure at which shear failure will not occur in sandy foundation soils is dependent on cohesion, angle of friction, unit weight of the soil, and depth and width of the footing. The weight of the soil above the footing has a restraining and confining effect tending to overcome heaving and spreading of the soil. If the water table is above the base of the footing, or at a depth less than the width of the footing, the allowable bearing pressure is reduced, for the unit weight of the soil is then equal to its submerged unit weight ($\gamma'_m = \gamma_m - \gamma_w$). The resistance of the soil to failure by shear is increased as depth and width of the footing are increased.

The vertical pressure under a loaded area decreases with increasing depth. Therefore, deep compressible layers may not affect settlement appreciably if the stress in these layers is low enough. However, the vertical stress at a given point in a deep soil layer is practically the same for point loads as for loads distributed over a footing, owing to distribution of stresses throughout the soil mass. Therefore, increasing the width of the footings by any reasonable amount has practically no influence on settlements caused by deep compressible layers.

FOUNDATIONS FOR AIRFIELD AND HIGHWAY PAVEMENTS

The design of pavements to sustain heavy loads has become an important branch of civil engineering and is closely connected with characteristics and strength of the subgrade soils. Soil mechanics is thus an important part of airfield and highway work, and brief accounts of its applications are given in Chapter 21.

COMPACTION OF SOILS

Proper compaction of embankments, levees, earth dams, and foundations for highways and flight strips is essential. Compaction increases density with consequent decrease of moisture content even in the event of subsequent saturation. Both of these factors result in an increase in shearing resistance.

The compaction characteristics of a soil can best be described in terms of a standard compaction test. In the Proctor test, named after R. R. Proctor, who did much of the original compaction research,[8] the soil is compacted at various moisture contents in a cylinder having a volume of $\frac{1}{30}$ of a cubic foot, using 25 blows of a 5.5-pound hammer dropped from a height of 12 inches, on each of three soil layers of equal thickness. The mass unit weight (γ_m) of the soil is determined by weighing the soil in the cylinder and dividing this value by the volume.

If the soil is compacted in the cylinder at a low moisture content, say 6 per cent for a clayey soil, the resulting mass density will be equal to γ_m. The dry density is then obtained by

$$\gamma_d = \frac{\gamma_{m_1}}{1 + w_1} \tag{16}$$

where w_1 is equal to the moisture content expressed as a decimal.

If more water is added to the soil and the test is repeated, a higher dry density results, owing to breaking down of soil clods and other factors influencing workability of the soil. This can be repeated up to a certain moisture content at which the density begins to decrease. This decrease is caused by the voids of the soil

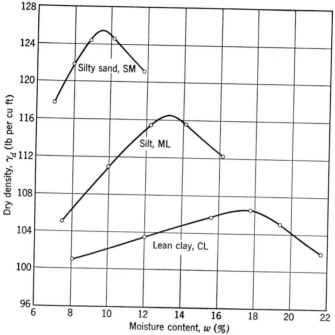

Fig. 171. Standard Proctor compaction curves.

becoming practically filled with water, with the result that with further increases in moisture content, water actually replaces a portion of the soil solids. Figure 171 shows a set of typical compaction curves for several soils. The moisture content at which the maximum dry density is attained is termed the optimum moisture content.

Testing compaction characteristics of soils has become a matter of routine in most large laboratories. Most highway departments employ the standard Proctor test; for airfields and highways the Corps of Engineers, U. S. Army, has adopted a modification of this test but employs the standard procedure for earth dams, levees, etc. The modified Proctor test uses a 10-pound hammer dropped from a height of 18 inches on each of 5 layers of equal thickness. This results in a compactive effort, per unit of volume, roughly equal to $4\frac{1}{2}$ times the standard.

It can be seen from the shape of the curves near optimum that moisture content is extremely critical in the compaction of silts, and that proper compaction of clay soils can be obtained over a rather wide range of moisture contents. Cohesionless sands are best compacted when in a saturated condition.

Compaction of embankments and other structures involving huge quantities of earth is best accomplished by means of rollers. Rollers in use at present are the sheepsfoot, pneumatic, and smooth-wheel types. The first is a drum with feet ranging from 6 to 8 inches in length. The area of the tamping feet and the weight of the roller are adjusted so that the unit contact pressure ranges from about 200 pounds per square inch up to 1000 pounds per square inch or even higher. Clay soils are best compacted by sheepsfoot rollers. The action of the tamping feet compacts the soil from the bottom upward.

Pneumatic rollers are generally made up of a series of rubber-tired wheels, the total weights ranging as high as 50 tons or more. They are widely used for compacting sandy soils. In contrast to the sheepsfoot, the pneumatic roller compacts the soil from the top down. Heavy pneumatic rollers have been used to compact soils in highway cuts to depths of as much as 3 feet.

Smooth-wheel rollers have steel wheels and total weights of about 10 tons. This type of roller is used mainly for compacting bituminous surfaces, but it is also very effective for compacting gravel, and some clay soils.

Thickness of the layer to be compacted is very important, as well as moisture control at the time of compaction. The thickness of the lift generally should not be greater than about 8 to 10 inches, unless a very heavy roller is employed. The number of passes required for adequate compaction depends on size of the roller, thickness of the layer, and soil texture.

Degree of compaction is generally specified as a certain per cent of the peak value of the curve. In construction of earth dams 100 per cent of the standard Proctor test is very often used; for airfields compaction ranging as high as 100 per cent of modified Proctor may be specified in order to minimize failures caused by extremely heavy wheel loads. On many large construction jobs it is economical to construct test sections in order to determine ahead of actual construction the number of passes required for good compaction.

FIELD EXPLORATION

Adequate subsurface explorations are essential if the design engineer is to utilize fully all the tools at his disposal. The type of exploration adopted depends on the particular job and the type of information required for that job. Therefore, generalized statements regarding proper procedures may be misleading. Nevertheless, several factors that should influence the thinking of the soil engineer are nearly always present.

The first step is to make a complete study of all geological and agricultural survey reports on the area under consideration. These reports will indicate the type of parent material; depth of soil horizon development, in many cases; depth to ground water. Literature pertaining to performance of structures built on various types of soil in the vicinity should also be investigated. The use of aerial photographs in soil exploratory work has been outlined in the preceding chapter.

After the broad distribution and general characteristics of the soils of the area are known, the engineer can intelligently plan a preliminary program. This involves borings and the obtaining of soil samples from each distinctive layer for visual inspection and testing purposes. The size of a job will determine whether or not more detailed information is required.

The simplest means of subsurface exploration is the common hand auger. This type of operation is laborious and is generally not employed except for very shallow work, but power augers capable of boring to considerable depth in a relatively short time are used fairly often. Many of these augers, however, cause an excessive amount of mixing of the various soil horizons, which in turn may lead to serious errors in design.

Wash borings are not to be recommended, except where the only information desired is depth to rock. Sampling by means of sample spoons produces reliable results and if thin-walled tubes are employed disturbance may be kept to the minimum necessary

Fig. 172. Taking soil samples with a portable drilling rig. The hinged auger shown in the insert is a rather uncommon device, used mainly for sampling sticky clays. (Corps of Engineers, U. S. Army.)

for most design purposes. Spoon sampling is accomplished by pushing or driving an open tube into the soil. Much useful information may be obtained by measuring the force or number of blows required to push the tube into the ground.

In some situations it is desirable to obtain soil samples from open pits in order to minimize, so far as possible, disturbance caused by sampling. Care is taken not to disturb the pit walls. Samples are then carefully cut from the walls of the pit, coated with paraffin, and sent to the laboratory for testing.

The number and depth of exploratory holes depend on the

job in question. At times a standard spacing and depth of holes is chosen without proper regard to the geology of the area. Chapter 16 discusses ways of locating borings in order to obtain a maximum of information from a minimum number of holes.

Fig. 173. Taking an undisturbed box sample. (Corps of Engineers, U. S. Army.)

For routine highway and airport work, samples for classification tests are obtained from each horizon to a depth of 5 or 10 feet. Additional samples may be obtained for determining design data based on the California Bearing Ratio, and for determining the compaction characteristics of fill materials. In foundations for embankments, dams, and buildings, samples are obtained for the standard classification tests and, if the job warrants, for determining the shearing resistance and consolidation characteristics of each soil layer to a considerable depth.

Each soil layer encountered in the exploration program should be identified in the field as to color, consistency, texture, and, if possible, parent material. The upper elevation of each layer must be very carefully logged as well as the elevation of the surface of the ground. An accurate recording of depths is very essential, for stability computations are dependent on it. Further, accurate records should be kept of depth to the water table, springs in the vicinity, water-bearing strata, and other data that might have a bearing on the final design.

The actual depth to which borings are put down depends on the stresses that will be imposed on the soil and the geology of the area. Preliminary estimates of the stress in the soil after loading should be made, and the borings should be carried at least to depths at which critical stresses will exist. If weak material is encountered, the borings should be carried to firm strata. A sufficient number of borings should be made to define the lateral and vertical extent of all critical strata.

REFERENCES

1. Amer. Soc. for Testing Materials, *Procedures for Testing Soils*, Sponsored by A.S.T.M. Committee D-18 on Soils for Engineering Purposes, July, 1950.
2. Civil Aeronautics Administration, *Airport Paving*, May, 1948.
3. Corps of Engineers, U. S. Army, The Unified Soil Classification System, *Waterways Experiment Station, Tech. Memo.* 3-357, Vicksburg, Miss., March, 1953.
4. Highway Research Board, Report of Committee on Classification of Materials for Subgrades and Granular Type Roads, *Proceedings*, Vol. 25, 1945.
5. Highway Research Board, Compaction of Embankments, Subgrades and Bases, *Bull.* No. 58, 1952.
6. M. J. Hvorslev, *Subsurface Exploration and Sampling of Soils for Civil Engineering Purposes*, Prepared for the Soil Mechanics and Foundations Division of the American Society of Civil Engineers, Distributed by the Engineering Foundation, New York. Printed by Waterways Experiment Station, Vicksburg, Miss., November, 1949.
7. Ralph B. Peck, Walter E. Hanson, and Thomas H. Thornburn, *Foundation Engineering*, John Wiley & Sons, New York, 1953.
8. R. R. Proctor, Fundamental Principles of Soil Compaction, *Eng. News Rec.*, Vol. III, Nos. 9, 10, 12, and 13, Aug.–Sept., 1933.
9. A. W. Skempton, The Bearing Capacity of Clays, *Building Research Congress, Papers Presented in Division* 1, London, 1951.
10. Donald W. Taylor, *Fundamentals of Soil Mechanics*, John Wiley & Sons, New York, 1948.

11. Karl Terzaghi and Ralph B. Peck, *Soil Mechanics in Engineering Practice*, John Wiley & Sons, New York, 1948.
12. Gregory P. Tschebotarioff, *Soil Mechanics, Foundations, and Earth Structures*, McGraw-Hill Book Co., New York, 1951.
13. *Waterways Experiment Station*, Undisturbed Sand Sampling below the Water Table, *Bull.* 35, Vicksburg, Miss., June, 1950.

C H A P T E R   1 8

Subsurface Exploration

Accurate and complete subsurface information is necessary for all types of civil engineering projects, for without this information it is not possible to arrive at a rational design for the structure and the proper construction procedures. Structures have failed because of inadequate or misleading subsurface data, and many so-called successful structures could probably have been completed at much less cost had proper consideration been given to obtaining more complete subsurface information. Engineers are fully aware of the importance of the subject, and a comprehensive study of subsurface exploration and sampling methods, prepared by M. J. Hvorslev for the American Society of Civil Engineers, was published in 1949.[4] This publication should be consulted for details that cannot be discussed in a short chapter.

Subsurface exploration can be divided into three types: (1) soil sampling and exploration; (2) sampling and exploration of rock; and (3) geophysical exploration. Inasmuch as the information sought is essentially geological in nature, geology is important to all three, but in the interest of brevity mention of its role is confined to generalities. Application of subsurface information to design and construction of projects is, of course, the province of the civil engineer. Soil sampling and exploration have been discussed briefly in Chapter 17 and are treated exhaustively in Hvorslev's review of the subject. This phase of subsurface exploration thus needs no further treatment, and in the present chapter emphasis is placed on rock drilling and geophysical exploration.

The type of subsurface exploration required depends in a large measure on what the surface work has revealed and the type of project contemplated. The initial subsurface investigations are carried out for the primary purpose of providing the engineers in charge with the information needed for intelligent selection of site, type, and design of the structure, and the proper construction procedures. Builders are often able to recover substantial damages from the owners because of erroneous and misleading information contained in boring reports and test pit data supplied to them prior to submission of their bids. This is an additional reason for obtaining the best and soundest geological advice and information possible, for where the subsurface data are accurate and complete it is usually possible to demonstrate that the contractor misunderstood the problem and proceeded with methods entirely unsuited to the existing conditions.

Frequently, contracts contain clauses disclaiming all responsibility for the reliability of the subsurface information. This procedure is often prompted by the owner's desire to save a few dollars at the expense of thoroughness and quality of the work. It is better to have no borings at all than data that are so poor as to be completely unreliable. It most certainly is not a fair practice to submit the results of poor or incomplete studies to a builder, at his risk, when his bid often depends in a large measure on the information furnished him. Furthermore, disclaimer clauses have not been upheld invariably by the courts. Such clauses are almost never upheld when there is reason to believe that information has been withheld. In placing the blame for mistaken assumptions concerning subsurface conditions, the boring contractor is presumably at fault. Nevertheless, he is the owner's agent, and it is not equitable to pass financial responsibility to an innocent outsider (the builder) who had no voice in the matter.

ROCK EXPLORATION

It might appear that generally strong rocks, such as granite, are always suitable for almost all engineering purposes, but experience shows that this is not invariably true. Much depends on details of jointing, weathering and other forms of alteration, permeability, ground water conditions, etc. It is thus poor practice

Fig. 174. Bore-hole camera developed by the Corps of Engineers, U. S. Army. *e*, camera proper, stainless steel tube 2¾ inches in diameter, with quartz window inclosing conical mirror near lower end; *a*, power supply; *b*, level-wind control for cable; *c*, cable wheel; *d*, leveling screws; *g*, hand crank for raising and lowering camera, geared into wheel *c*. (Corps of Engineers, U. S. Army.)

to dispense with subsurface investigations merely because surface exposures indicate generally sound and suitable rock, for important hidden flaws have been found at many construction sites.

Although one or more of the geophysical methods described in a later part of this chapter can often be used to advantage, rock exploration generally involves borings. These are of two types: (1) core borings, the primary purpose of which is to obtain samples of the materials penetrated, and (2) large-diameter calyx holes, the primary purpose of which is to afford access to the subsurface in order that an observer may examine the rock in situ. The cores obtained from calyx holes are useful, but not essential. Large-diameter borings are expensive, and the Corps of Engineers, U. S. Army, has devoted considerable effort to the development of a camera capable of photographing the walls of borings as small as 3 inches in diameter.[2] The bore-hole camera has passed several preliminary tests, and its use should be considered in all situations where a truly realistic conception of subsurface conditions is desired.

CORE BORINGS

Core borings can be made in almost any type of rock, and the samples are machined cylinders of the materials penetrated. A good grade of shale is usually the softest rock from which reasonable core recovery can be obtained. At times cleavage planes and fracture systems interfere to the extent that only very poor recovery is possible. Core holes may be vertical, inclined, or horizontal, but the last require special types of drilling rigs. Almost any type of tool may be employed to penetrate the overburden. Casing is always used. It must be seated on the top of sound rock and tightly sealed. Otherwise, control of the drilling fluid may be lost, and abrasive materials may enter and damage the tool. The equipment is driven by a variable-speed machine. The bit makes an annular opening in the rock, and as drilling progresses the core barrel slides down over the central core of rock material. When the core barrel is raised, a catcher holds the core in place and the core is raised with the core barrel.

Types of Drill Bits and Drilling Fluids

Three types of cutting tools are in use: (1) black diamonds (carbon) and bortz, (2) shot, and (3) steel-alloy toothed cutters. The black diamond has no cleavage planes and hence seldom frac-

HANDSET CARBON
A

HANDSET BORTZ
B

CAST BORTZ
C

BORTZ IMPREGNATED
D

BLANK   FINISHED

BORTZ INSERTS
E

HANDSET   INSERTS   CONCAVE   PILOT

REAMING SHELLS   SOLID DIAMOND BITS

F   G

SPRAGUE & HENWOOD CO. – J. K. SMIT & SONS CO. – CARBOLOY CO. – SULLIVAN MACHINERY CO. – E. J. LONGYEAR CO.

Fig. 175. Various types of diamond drill bits. (After M. Juul Hvorslev.[4])

tures. Bortz is a diamond by-product or reject of the jewelry trade. Because of the cleavage planes, bortz is subject to fracturing. Diamond bits will cut even the hardest rocks, but tend to clog in softer materials. Alloy-steel toothed bits (carboloy inserts, for example) are usually better adapted to shales and soft rocks and will operate satisfactorily in rocks as hard as dense unweathered basalt. The diameter of cores taken with diamond or alloy-steel bits ranges from $\frac{7}{8}$ to about 4 inches. Larger-diameter cores are usually taken with shot drills described later in this chapter. Length of the core cut is limited to length of the core barrel, which is usually about 10 feet.

Diamond drilling is expensive, and it is apparent that the cost will be greatly increased if diamonds are lost. A chattering bit, or one improperly fed, seamy rock, or poorly seated casing, which permits sand to enter the hole, can cause loss of stones. A diamond lost in the hole can force abandonment of the location, for unless it is removed it will tear up the drill bit.

Drilling Fluids. During drilling operations, water, or other suitable drilling fluid, is forced down the hollow interior of the drill rods and returned to the surface via the annular space between the casing and the outside of the drill stem. The drilling fluid cools and lubricates the bit and removes the cuttings from the hole. In loose sands and other materials susceptible to caving, it is often desirable to use heavy mud as a drilling fluid. Drilling mud tends to prevent caving, but it may contaminate the samples and close off joints and openings in the walls of the hole. Sealing of the wall rock can be a serious objection to the use of drilling mud in situations where permeability of the rock is to be determined.

Core Recovery

The success of a core-drilling job is measured by the percentage of core recovered. Frequently, a good operator can obtain between 90 and 100 per cent recovery, but bad rock conditions may cause blocking of the core in the barrel and consequent grinding of the core. If blocking occurs, with accompanying chattering of the rods, the drill stem must be pulled and the blocked core removed. In hard rock a single-tube core barrel may prove satisfactory, but in soft rock a double-, or even triple-, tube barrel must be used. If core recovery is not good, the inspector must know why and how the losses occurred, and should estimate the nature of the lost intervals from behavior of the drill rig and character of the returning drilling fluid and cuttings. The lost portions of the core may represent open cavities, or merely seams of softer materials. In either event, it is precisely such conditions that are of most interest to the engineer. Paradoxical as it may seem, proper interpretation of the intervals from which no core was recovered is more important than a precise description of the core itself. In situations where core losses are excessive, consideration should be given to the use of the bore-hole camera mentioned in a preceding paragraph.

Deviation and Caving of Bore Holes

In some instances it is essential to determine the deviation of the hole in both a horizontal and vertical plane. Surveying of a hole in the vertical plane is generally accomplished with a "test tube," or clinometer. This apparatus consists of a metal tube case attached to the end of the drill rods. A glass vial partly filled with hydrofluoric acid is fitted inside the case, and the rods are quickly lowered to the depth where the survey is desired. The vial is allowed to remain in this position for at least 1 hour and then withdrawn. The line etched into the walls of the vial is a measure of the deviation of the hole from the vertical. Surveying a bore hole in the horizontal plane has not proved very satisfactory. A small compass is introduced in place of the acid vial. A liquid that solidifies quickly is placed inside the compass. The theory is that after coming to rest the liquid will congeal and hold the compass needle in place, and upon retraction of the rods the correct bearing may be obtained. This procedure has not worked very well, but in fairly large-diameter holes a small camera has been adapted for photographing the position of the needle. Ordinarily, it is not necessary to survey shallow holes, but if they exceed 1000 feet or so in depth it is often desirable to do so.

When a horizontal boring begins to deviate from the desired direction and completion of the hole within the limits set forth in the specifications is endangered, a change in size of the core taken (either to a smaller or larger diameter) will often cause the boring to straighten out. In view of this contingency, when a horizontal boring is being drilled it is wise to begin with a diameter one or two sizes larger than is absolutely necessary so that reductions can be made. Wedging has been tried, but it is not always successful, and if it is not the hole must be abandoned. Surveys of horizontal borings are usually taken at every 100 feet.

Caving. Many weak beds start to cave during drilling operations. This can be so serious that the drilling tools cannot be withdrawn, and both they and the hole are lost. When the tools can be withdrawn and further drilling progress is necessary, cementing is often resorted to. Usually the driller knows exactly where the caving took place, and the bore hole is cemented to a height considerably above the caving interval. Quick setting cement is always used. The cement is allowed to set overnight, and next day the cement is drilled through. If all goes well, further progress is possible. If not, another attempt may be made to cement the caving zone. Not infrequently, caving cannot be stopped by the use of cement, and the only alternative to abandoning the hole is to set a string of casing of the next smaller size and proceed with a smaller-diameter core bit. For this reason, in order to permit telescoping of the casing, it is well to begin deep borings one or two sizes larger than necessary.

Logging of Cores

Cores are of little or no value unless properly logged. In the past it was common practice to have cores logged by drillers, foremen, or engineering aides. Highly individualistic and confusing results often followed. In an effort to improve the quality of core logs, geologists are often employed, and core logging is usually the first task that the aspiring engineer-geologist undertakes. Many strictly geologically trained men have failed to produce the results expected of them, largely because of their failure to appreciate that more is required than an accurate petrographic description of the samples. In order to write a good log the inspector must understand drilling methods and comprehend the engineering purposes for which the work was undertaken. For example, the percentage of core recovery should be determined after each pull and an effort made to determine where and why the losses occurred. If the drill stem dropped suddenly and circulation of the drilling fluid was temporarily lost, it is very probable that the core loss was caused by an open cavity. If, on the other hand, the drill stem only dropped at a faster rate than normal while the circulation remained constant, and the drilling fluid became more muddy, it is probable that the core loss was caused by grinding or washing away of the core. In order to interpret reasons for core losses it is absolutely necessary that the inspector be present at all times during drilling operations and note carefully operation of the rig.

Log Sheets. Core logs should be written on sheets especially prepared for the purpose. They should contain all the information listed in Fig. 176, and in addition should specify the position of the casing, type of drilling fluid, and any other factors that may influence the amount and quality of core recovery. Action of the drilling rig during the time the core was cut should also be described. If blocking occurred, the depth at which it took place should be noted.

Water Table. The position of the water level in the drill hole at the beginning of each shift should also be noted, and after the hole has been completed the depth to the water table should be carefully determined. Because addition of the drilling fluid causes a temporary rise of the water table, it is well to wait a week or so after completion of the hole before soundings are attempted. In materials of silt size and finer, it may not be possible to determine the water table by these methods, and determination of the water table in such materials may constitute a very real and difficult problem. If artesian water was encountered, the original pressure and pressure decline over a period of time should be determined and stated in the log.

Rock Descriptions. The core should be logged immediately after removal from the core barrel, for exposure to drying may materially alter its hardness and other physical properties. Each rock type encountered, no matter how thin, should be classified according to the best geological nomenclature, but hair-splitting distinctions of no possible engineering importance should be avoided. The hardness should be determined according to a definite scale as in the following scheme. Hardness 0 can be dented

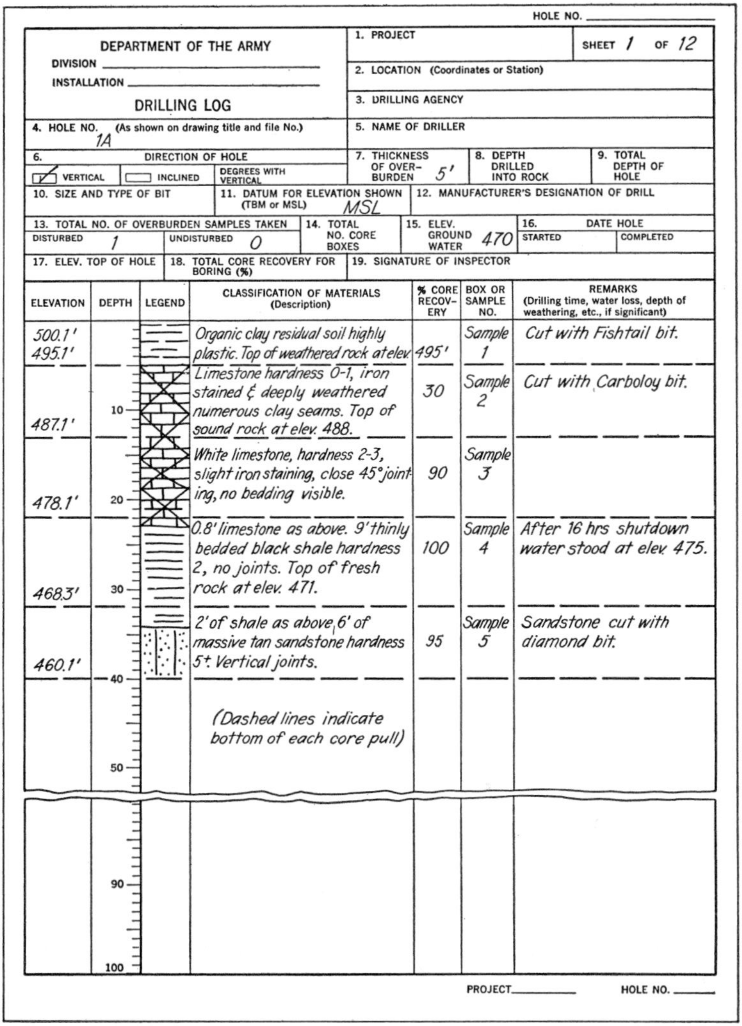

HOLE NO. _____							

DEPARTMENT OF THE ARMY	1. PROJECT	SHEET *1* OF *12*

DIVISION _____
INSTALLATION _____

DRILLING LOG

2. LOCATION (Coordinates or Station)

3. DRILLING AGENCY

4. HOLE NO. (As shown on drawing title and file No.) *1A*	5. NAME OF DRILLER

6. DIRECTION OF HOLE		7. THICKNESS OF OVER-BURDEN *5'*	8. DEPTH DRILLED INTO ROCK	9. TOTAL DEPTH OF HOLE
☑ VERTICAL ☐ INCLINED	DEGREES WITH VERTICAL			

10. SIZE AND TYPE OF BIT	11. DATUM FOR ELEVATION SHOWN (TBM or MSL) *MSL*	12. MANUFACTURER'S DESIGNATION OF DRILL

13. TOTAL NO. OF OVERBURDEN SAMPLES TAKEN		14. TOTAL NO. CORE BOXES	15. ELEV. GROUND WATER *470*	16. DATE HOLE	
DISTURBED *1*	UNDISTURBED *0*			STARTED	COMPLETED

17. ELEV. TOP OF HOLE	18. TOTAL CORE RECOVERY FOR BORING (%)	19. SIGNATURE OF INSPECTOR

ELEVATION	DEPTH	LEGEND	CLASSIFICATION OF MATERIALS (Description)	% CORE RECOVERY	BOX OR SAMPLE NO.	REMARKS (Drilling time, water loss, depth of weathering, etc., if significant)
500.1' 495.1'			Organic clay residual soil highly plastic. Top of weathered rock at elev. 495'		Sample 1	Cut with Fishtail bit.
487.1'	10		Limestone hardness 0-1, iron stained & deeply weathered numerous clay seams. Top of sound rock at elev. 488.	30	Sample 2	Cut with Carboloy bit.
478.1'	20		White limestone, hardness 2-3, slight iron staining, close 45° jointing, no bedding visible.	90	Sample 3	
468.3'	30		0.8' limestone as above. 9' thinly bedded black shale hardness 2, no joints. Top of fresh rock at elev. 471.	100	Sample 4	After 16 hrs shutdown water stood at elev. 475.
460.1'	40		2' of shale as above, 6' of massive tan sandstone hardness 5± Vertical joints.	95	Sample 5	Sandstone cut with diamond bit.
	50		(Dashed lines indicate bottom of each core pull)			
	90					
	100					

PROJECT_____   HOLE NO. _____

Fig. 176. Log sheet used by Corps of Engineers, U. S. Army, for recording core-boring data.

with moderate pressure of the fingers; 1 cannot be dented with moderate finger pressure, but can be dented with the point of a pencil; 2 cannot be dented with a pencil point, but can be cut with a knife; 3 cannot be cut with a knife, but can be dented or broken with light blows of a hammer; 4 can be dented or broken with moderately heavy blows of a hammer; 5 can be broken only with heavy hammer blows. The porosity, jointing, bedding, and weathering of the samples should be accurately and fully described. All changes in rock character should be mentioned, and the depths at which these changes occurred should be noted. In addition, fossils or beds of distinctive lithology which may be useful in correlating drill holes should be adequately described.

Weathering and Weathered Rock. The log should also specify the following: (1) top of weathered rock; (2) top of sound or firm rock; and (3) top of fresh or unweathered rock. Correct determination of these surfaces is a matter of considerable engineering importance and calls for experience and sound judgment. Where soils are transported, it is usually not difficult to determine the top of weathered rock, but in regions where the soils are residual it may be exceedingly difficult. The best over-all guide is to place the weathered rock line at the upper surface of materials that in the judgment of the inspector cannot be excavated without blasting. The top of sound rock, as used by most engineering organizations, is meant to indicate the upper surface of materials suitable for the foundations of heavy structures. Such materials do not necessarily need to be entirely free from weathering, and the term "top of fresh rock" should be used to indicate the complete absence of weathering in compact unjointed materials. Weathering may still be apparent along joints and fissures, however. In some regions, this degree of refinement may not be called for. In glaciated areas, for example, the top of fresh rock may be encountered immediately below the soil or overburden.

Shot, or Calyx, Borings

Shot drilling is conducted with the type of tools used for diamond drilling to the point where casing is seated on the bedrock. The chief difference lies in the type of bit used, which in this case consists of a core barrel of soft steel with one or more throats cut in the bottom of the rim. Chilled steel shot of suitable size, depending in some degree on the nature of the rock to be cored, are fed into the wash water. The shot settle to the bottom of the hole and embed themselves into the soft steel barrel. Others find their way under the tool, some on the outside and some on the inside. Cutting is a combination of wear, abrasion, and scratching. Too many shot tend to form a ball bearing under the barrel, and too few tend to slow or stop drilling progress. The stresses set up in the rock are severe, and attempts to take cores less than 2 inches in diameter should not be made. The size of cores obtainable ranges from about 2 inches to 54 inches in diameter. The larger-diameter holes permit the lowering of an inspector in a boatswain's chair or cage and make possible examination of the rock in place. In these borings core recovery is usually a secondary consideration. It should be noted, however, that large-diameter borings cannot be used for visual examination below the water table unless large capacity pumps for lowering the water level are available, or the rock has been thoroughly grouted in advance. In drilling large-diameter holes the core is lifted out by a cable attached to a jack hammer hole drilled in the center, or by a special type of core catcher. In smaller-diameter holes, the core is reclaimed by packing the barrel with sand or gravel, usually termed grout. Shot-cut cores of necessity have a rough surface.

GEOPHYSICAL EXPLORATION

Application of physical laws and measurements to the earth is known as geophysics; it includes such diverse subjects as meteorology, oceanography, seismology, geodesy, and terrestrial magnetism. Geophysical prospecting constitutes the practical side of the subject, and may be defined as "prospecting for mineral deposits and geologic structure by surface measurements of their physical quantities." [3] Engineering applications are of two types: (1) geological, having to do with foundation problems, construction materials, etc., and (2) dynamic vibration tests of structures, pipe and metal location, and other problems of a non-geological character. Geological applications are, in effect, forms of subsurface exploration, but they differ from core boring and other standard engineering exploratory methods in that the information obtained is always indirect and not subject to direct visual verification. Because of this distinction, geophysical exploration is treated separately. The following discussions are intended merely to acquaint the reader with the principal geophysical methods in use and their potentialities and limitations. A quantitative and rigorous treatment is beyond the scope of this book.

It has been indicated in Chapter 15 that, although surface geological maps are seldom accurate enough to stand unsupported by subsurface exploration, geological mapping is the least expensive of all available methods and furnishes an indispensable basis for later studies. Geophysical exploration is most effective when carried out in conjunction with accurate geological knowledge, and thus cannot be regarded as a substitute for the latter. Moreover, seldom, if ever, can it be used as a substitute for core boring or other more direct methods of subsurface exploration. Compared with core borings or test pits, the cost of geophysical work is low, and it is often possible to effect considerable savings by a judicious use of geophysical methods. When adopted, they should always be considered something additional to surface geological studies and direct methods of subsurface investigation rather than a substitute for one or both.

GRAVITY METHODS

If the earth were stationary and a perfect sphere of uniform density, the value of gravity at all points on its surface would be the same. In actuality it is an oblate spheroid with an equatorial diameter approximately 27 miles greater than the polar diameter, rotating at a rate varying from about 1000 miles per hour at the equator to zero at the poles. It follows that under these conditions the value of gravity should be a maximum at the poles and a minimum at the equator. Furthermore, the surface is marked by considerable local variation in altitude. Assuming uniform density distribution, by application of physical laws it is possible to correct for the above factors and arrive at a computed value of gravity for any point on the earth's surface. If, after these corrections have been made, the computed value of gravity does not equal the observed value the difference between the two constitutes a gravity anomaly. Such anomalies, of course, must arise from irregularities in the distribution of mass in the earth's crust, and are thus closely connected with the local geology. For this reason, gravity methods can be used in subsurface geological exploration.

Newton's law of gravitation states that the attraction between two masses m_1 and m_2 is proportional to their product and inversely proportional to the square of the distance between them. Thus, when rocks of different densities are in contact, there will exist a corresponding gravitational field the configuration of which may be shown by surfaces of equal gravity potential. In the simple case of a body of dense rock intruded into lighter materials the surfaces of equal gravity potential form concentric hemispheres centered around the intrusive body.

Units and Instruments Employed

The force of gravity is measured in dynes, 1 dyne being the force required to impart an acceleration of 1 centimeter per second per second to a mass of 1 gram. The acceleration of gravity is measured in gals, 1 gal being equal to an acceleration of 1 centimeter per second per second. In geophysical work the milligal, equal to $1/1000$ of a gal, is the unit most frequently employed. As might be expected, the difference in density between very light and very heavy rocks is not great enough to produce marked differences in gravity potentials. Those encountered in practice are usually only a few milligals, and seldom exceed 100 milligals. To measure such small quantities, very sensitive instruments are necessary.

Instruments. Various types of instruments are possible, but in practice pendulums, gravimeters, and torsion balances are the devices most frequently employed. All these types of instruments actually measure the force of gravity, but the results are usually expressed in milligals. Both pendulums and gravimeters measure relative values of gravity, whereas the torsion balance measures the gravity gradient, i.e., the variation in gravity per unit horizontal distance. To obtain the necessary accuracy with a pendulum the period must be timed to within $1/10,000,000$ of a second, and various ingenious methods of timing have been devised. The simplest type of gravimeter consists of a standard mass suspended from a spring, the extension of which is amplified by various means. The simplest type of torsion balance, shown in Fig. 177, consists essentially of a horizontal rod, or beam, supporting equal masses at each end suspended from a vertical torsion wire. If differences exist in the horizontal components of gravity, which are at right angles to the surfaces of equipotential, the beam will be deflected in the direction of minimum

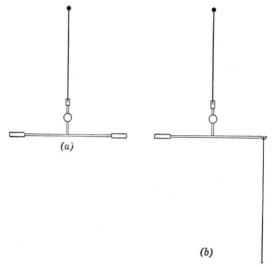

(a)

(b)

Fig. 177. Two types of torsion balance. (After C. A. Heiland, *Geophysical Exploration*, copyrighted by Prentice-Hall, 1940.)

curvature of the equipotential surface. In the instruments used in practice the beam is ⌐-shaped rather than straight, and thus utilizes the effect of convergence of surfaces of equipotential with depth as well as the horizontal components of gravity.

Survey Methods and Data Obtained

Gravity instruments are usually mounted on trucks, but are portable if the nature of the terrain so demands. Regardless of the type of instrument used, both the latitude and the elevation of the observation points must be known. Because the local topographic relief also influences results, a topographic map is desirable. Usually, surveying work is necessary and constitutes a considerable item of the total cost. The number of readings per unit area necessary to detect a concealed geologic body depends chiefly on its size. The smaller the size, the greater is the number of observations per unit area required. The time required for an observation depends chiefly on the type of instrument employed. As a rule the torsion balance requires a considerable time to come to rest, and for this reason fewer readings per unit time are usually possible than with other types of instruments, especially the gravimeter. Furthermore, torsion balance readings may be disturbed by the presence of a large boulder or other object near by.

Corrections Applied. All gravity measurements must be corrected for latitude and elevation above sea level. In precise work the effect of the surrounding topography must also be taken into account. This is calculated from selected elevations along radial lines and concentric circles centered on the station. Torsion balance readings are particularly influenced by topography, and in rugged terrains the corrections may be very complicated and inaccurate. Consequently, this instrument is not well adapted for use in such regions. When pendulums or gravimeters are used, it is necessary to correct for the influence of the rocks between the station and sea level, the so-called Bouger correction. Under certain conditions it may also be necessary to consider the effects of what are known as regional gravity gradients caused by the geologic structure of the area, for example, a sloping surface of dense Pre-Cambrian rocks overlain by lighter materials. Because the gravitational effect of the underlying rocks tends to mask anomalies caused by structure of the overlying materials, if shallow structures are to be detected it is obviously necessary to eliminate the gravitational effects of the underlying basement. This can readily be done when the depth, density, and configuration of the basement is known. The configuration of the upper surface of the underlying rocks should be determined by a method other than the gravitational. The magnetic method is well adapted to this purpose.

Plotting of Data. In pendulum and gravimeter surveys the corrected values of gravity are plotted beside the appropriate stations. Gravity anomalies are usually shown by contours drawn at a suitable interval through or around the observation points. An observed value or gravity greater than the computed value constitutes a positive anomaly; an observed value less than the computed value constitutes a negative anomaly. Figure 178 illustrates a region in which there is an unusually close correlation between gravity anomalies and geologic structure.

In torsion balance work the results are plotted as vectors or as curves. In plan view, gravity gradients are shown by arrows

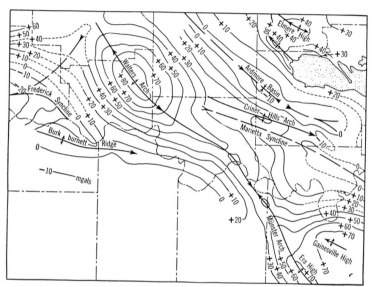

Fig. 178. Gravimetric map of the Wichita and Arbuckle Mountain region. (After A. Van Weelden, *World Petroleum Congr., Proc.*, 1934.)

pointing in the direction in which the gradient is a maximum. Length of the arrow is proportional to the gravity gradient. Curvature values are plotted as straight lines passing through the station. Length of the line is proportional to the deviation of the equipotential surface from the spherical. The data may also be plotted as curves on profiles taken at right angles to the strike. Relative gravity may be calculated from gradients, and the results may be plotted similarly to those obtained by gravimeter surveys. Figure 179 shows the results of a torsion balance survey of a salt dome. Because the salt is less dense than the surrounding rock, the dome is indicated by a negative anomaly.

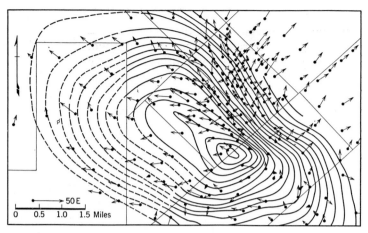

Fig. 179.   Reflection seismograph (contours) and torsion balance survey (arrows) of the Shepards-Mott salt dome.   (After J. B. Eby and R. P. Clark, Relation of Geophysics to Salt-Dome Structure, *Bull. Am. Assoc. Petroleum Geol.*, Vol. 19, No. 3, March 1935.)

Interpretation of Results

The object of geophysical work is to determine the outlines and depths of concealed geologic bodies.   Inasmuch as gravity methods depend on measurements of density distribution in a gravitational field, no depth control (lower limit of measurement) is possible.   In theory an infinite number of mass arrangements may cause any given anomaly, and it might appear that geologic interpretation of gravity measurements is impossible.   However, if the number of unknowns is reduced by making tentative and plausible assumptions regarding the geology, a qualitative interpretation can always be made.   In civil engineering work it is usually possible to test these assumptions by core drilling, and with the data so obtained arrive at a reasonably reliable quantitative interpretation.   To choose a concrete example, a strong positive anomaly may be indicated by gravity data.   This anomaly may be caused by an infinite number of mass arrangements, only a few of which are geologically probable.   They are: (1) a relatively thin body of very dense rock near the surface, (2) a very thick body of moderately dense rock near the surface, and (3) an extensive body of dense rock far below the surface. A core boring is put down at the point where the denser rock should approach nearest the surface and encounters andesite (average density 2.84) at a depth of 400 feet. The second alternative is thus indicated, and with the data so obtained it is possible to plot with some assurance the upper configuration of the andesitic material. The lower limit must be left as undetermined, but by inference may be assumed to lie far below the surface. Thus, with a single drill hole and at considerably less expense than if borings had been used throughout, a reasonably reliable estimate of subsurface conditions may be obtained.

Under certain conditions concealed geologic bodies will not produce a gravity anomaly. A deep-seated salt dome may be overlain by a zone of compacted and denser material resulting from the pressure of the rising salt mass. If the compacted zone exactly compensates for the lighter salt that has displaced uncompacted rock of intermediate density (differences in density of the two materials and their distances from the measuring instrument being taken into account), no anomaly will result. If the compacted zone overbalances the salt mass in gravitational effect, the anomaly will be positive. Under all other conditions, a salt dome will produce a negative anomaly.

Engineering Applications

To date, gravity methods do not appear to have been much used in civil engineering investigations, but because of their low cost a number of applications seem possible. Surveys of the bottoms of frozen lakes have been made with considerable success, a technique of possible engineering importance. Another possibility is in preliminary surveys of reservoirs in limestone terrains similar to those of the Tennessee Valley. At least the larger underground caverns should produce recognizable negative anomalies, and, although the limits might still require detailing by core borings, application of gravity methods might entail less expense than exploration by core boring alone. Other applications are found in the detection of buried intrusive bodies, faults, and folds.

MAGNETIC METHODS

The law governing magnetic attraction is very similar to the law of gravitation, and in a sense the magnetic method may be regarded as a variation of the gravitational. As in gravity surveys, it is impossible to control the working depth. However, the magnetic properties of geologic bodies depend in some measure on the magnitude and direction of the earth's magnetic fields. In addition, magnetism is a polar quantity; i.e., the sign depends on the direction of measurement.

Units and Instruments Employed

The ultimate cause of the earth's magnetism is unknown,* but the intensity, direction, and variations have been measured with considerable accuracy. In practical work it is convenient to distinguish between the vertical and the horizontal fields, the former being the one more frequently measured. If a magnetic body is present, its existence will be disclosed by a change in the normal value of the magnetic fields for the area in question (magnetic anomaly). The unit of magnetic measurement is the gauss, which is defined as 1 line per square centimeter (the field intensity which will exert a force of 1 dyne on a unit magnetic pole). Physicists refer to this quantity as the oersted, but geophysicists continue to employ the term gauss. In geophysical work units of $1/100,000$ of a gauss known as the gamma (γ) are normally used. Unlike the instruments used in gravity measurements, commercial magnetometers are designed with relative ease.

Instruments. A great variety of instruments is available, the one most frequently used being the vertical intensity magnetometer. In the Schmidt-type instrument a magnetic system is suspended on an axis oriented at right angles to the magnetic meridian. Deflections from the horizontal are measured in scale divi-

---

* It has been suggested that magnetic fields are fundamental properties of rotating bodies. In any event, it is doubtful that the earth's magnetism is solely the result of the presence of magnetic substances. It should be noted, however, that magnetic methods are almost solely concerned with the detection of magnetic bodies.

sions and then multiplied by a suitable factor to give relative vertical intensities.

Survey Methods and Data Obtained

Magnetometers are portable, and thus are easily adaptable to work in all types of terrain. Furthermore, the effect of topography is comparatively slight, but for accurate work the locations of observation points must be determined by surveying. Strong anomalies are produced by wires, rails, pipes, etc., causing the magnetic method to be best suited to poorly developed regions. Inasmuch as the response of the instrument may change as the result of minor jarring, or other causes, it is good practice to tie in a traverse with a previously established point. An instrument has also been developed for use in airplanes, and aeromagnetic surveys offer a cheap and rapid reconnaissance method.[13]

Corrections Applied. A number of corrections must be applied to magnetic observations. The magnetism of materials decreases with temperature, and, unless the instrument is well insulated, a temperature correction should be applied. Diurnal changes in the earth's magnetism must be taken into account. The influence of magnetic storms and secular variations must also be considered. When iron objects are present it is necessary to stay at a distance in keeping with accuracy of the survey. Allowances must also be made for terrain anomalies. A correction for errors in closure disclosed by checking on a base station is made in the normal manner, unless the observer has accurate information concerning the cause of the error (jarring of the instrument, etc.).

Plotting of Data. The corrected results of magnetic surveys may be presented as lines of equal magnetic anomaly (isanomalic lines) or as curves along profiles at right angles to the assumed strike. Peg models and anomaly vectors may also be employed. The latter are tangent to the lines of force radiating from the poles of disturbing bodies, and their usefulness depends very largely on the proper selection of the normal value.

Interpretation of Results

Quantitative interpretation of the results of magnetic surveys is unusually difficult for the following reasons: (1) there is lack

largely of insoluble minerals and thus tend to have relatively low concentrations of dissolved electrolytes in the interstices. As a result, their electrical conductivity tends to be low and their resistance high. Weathered rocks tend to be somewhat more porous than sound materials, and their resistances are therefore lower. It follows that stratified beds of unequal porosities will tend to possess rather uniform resistances parallel to the bedding planes, or a given layer, but at right angles to the bedding their resistances will be variable. For accurate interpretation it is necessary that the electrical properties measured be continuous in at least one plane; hence resistivity methods are well adapted to use in horizontally stratified materials.

Methods and Instruments Employed

Electrical measurements may be carried out in a number of ways among which are the self-potential, surface potential, and electromagnetic methods. The first is used mainly in ore prospecting where certain ore bodies tend to act as wet-cell batteries and thus generate a measureable current in their vicinity. Surface potential methods operate on the principle that when an electric circuit is grounded at two points an electrical field is produced in the earth. Lines of equal potential (along which no current flows) or lines of equal resistivity may be determined. Electromagnetic methods utilize the electromagnetic field of the ground currents rather than their electrical fields. The resistivity method is the one most applicable to civil engineering problems, and is thus the only method which needs be treated in detail.

The principles underlying the resistivity method have been well described by Schlumberger.[11] If two electrodes * are placed in the ground and joined in a circuit to a source of electrical energy (usually a dry cell), a current will flow from one electrode, pass through the ground, and leave by the other electrode. Experience has shown that the average depth of penetration of the current is one fourth of the distance separating the two electrodes. Consequently, the greater the spacing of the electrodes, the greater the depth of penetration. As shown by Fig. 181, the difference in potential between two points spaced equal

* Actually, four are generally used.

distances from the electrodes is measured by a potentiometer. The resistivity thus measured constitutes an average of the resistivities of the ground, and is called the "apparent resistivity." Hence, if the spacing of the electrodes is held constant, the average resistivities of a constant thickness of earth materials may be explored over any desired area. If the location of the potentiometer is held constant and the spacing of the electrodes is progressively increased, a series of measurements showing the

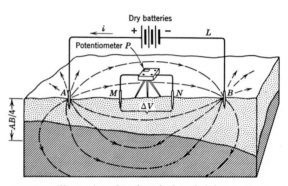

Fig. 181.   Diagram illustrating the electrical resistivity method.   (After C. and M. Schlumberger, *Second Congress on Large Dams*, Vol. IV, 1936.)

changes of resistivity with depth may be made.   In civil engineering work both types of measurements are important.

Instruments. The equipment used in resistivity work consists of two spools of insulated cable equipped with folding stands, copper pegs for earth contacts, dry-cell batteries, a potentiometer, and a tripod. This equipment is easily portable, and a field party consists of one technician and two or three unskilled helpers.

Survey Methods and Data Obtained

Four pegs are driven into the ground with uniform separation in a straight line and connected in a circuit to a source of electrical energy. In making resistivity maps a spacing of 400 feet between electrodes is usually employed. About 40 measurements per day can ordinarily be made. If the changes of resistivity with depth are measured (electrical sounding), about 4 stations can be occupied in a day.[11]

Corrections Applied. In practical work no corrections are made. The effect of topography cannot be estimated, and, be-

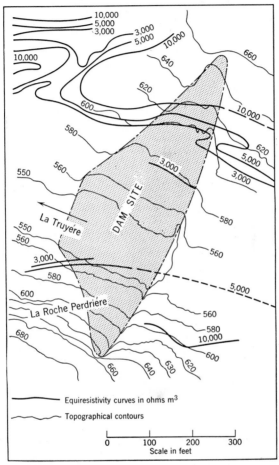

Fig. 182. Results of electrical resistivity measurements at the Sarrans Dam site, Truyére River, France. (After C. and M. Schlumberger, *Second Congress on Large Dams*, Vol. IV, 1936.)

cause the theoretical curves of electrical measurements can be calculated only by assuming a horizontal surface and stratification, in uneven topography resistivity measurements lose most

of their meaning. Depth of the water table also influences results, and in practice adds much to the difficulty of interpretation.

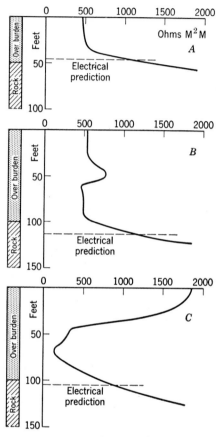

Fig. 183. Electrical determination of depth of overburden on the Connecticut River. (After C. and M. Schlumberger, *Second Congress on Large
Dams*, Vol. IV, 1936.)

Plotting of Data. The results of measurements at a constant
depth (resistivity mapping) are usually presented in the form
of equiresistivity curves plotted on a topographic map (Fig. 182).
The results of electrical soundings are presented as curves showing the change of resistivity with depth (Fig. 183).

Interpretation of Results

Interpretation of resistivity maps is based on the assumption that the resistance of a given bed is constant over the area investigated. Thus, if a bed of uniform character is shattered by a fault, the greater porosity resulting will give rise to a lower resistance value. The presence of unweathered protuberances surrounded by more porous weathered material should, of course, be disclosed by the higher porosity and lower resistivity of the latter. These principles are well illustrated in the interpretation of Fig. 182. The bedrock consists of granite overlain by weathered derivatives. Resistances of 5000 ohms, or more, were estimated for the fresh granite, resistivities of the weathered material being below that value. On the left bank of the Truyére River the 3000-ohms resistivity contour shows a deflection in the direction of the slope near the downstream portion of the dam site. Construction work disclosed the presence of a narrow zone of crushed and moist granite paralleling the 3000-ohms contour.[11] On the right bank, the abnormal configuration of the contours downstream from the dam site suggests the presence of granite surrounded by weathered zones. This inference was verified during construction operations.

The curves obtained from electrical soundings are interpreted in a similar manner. Figure 183 shows some typical examples. It will be noted that lack of sharp discontinuities at rock contacts makes precise interpretation very difficult. Lack of sharply defined breaks results from the fact that average, rather than specific, resistivities are measured.

Engineering Applications

The chief civil engineering applications of the resistivity method are in estimating the depth of overburden at dam sites (depth to bedrock), estimation of the physical characteristics of the rocks underlying construction sites, location of construction materials, and location of faults and water-bearing formations. When applied to underground water location, it is necessary to know in considerable detail the stratigraphic conditions in the area in question. When used for exploration of construction sites, the resistivity method does not do away with the necessity of core boring, but it does serve to guide the location of borings and reduce the number of holes required.

SEISMIC METHODS

Seismic methods depend on the properties of the elastic waves propagated in the earth's crust for calculation of the depths and properties of the strata through which they have been transmitted. Seismic exploration differs from seismology in that the energy generating the elastic waves is obtained from explosive charges rather than from earthquakes. The distance between the point of disturbance and the receiving station can be controlled, and subsurface conditions can be estimated in considerable detail. As mentioned in Chapter 13, in the non-homogeneous earth seismic waves are both reflected and refracted. Seismic exploration methods may therefore be classified as reflection and refraction, depending on the type of wave utilized.

Methods and Instruments Employed

The refraction method is based on the variable velocity of wave propagation in the different materials composing the earth's crust. Granular and plastic materials transmit seismic waves at velocities ranging from approximately 600 to 6000 feet per second, whereas rigid rocks such as shale and granite exhibit velocities of about 7000 to 20,000 feet per second. As a rule, velocity increases with depth, and the laws of wave propagation are such that a wave is always refracted, or bent, away from the medium of higher velocity. In instances where the velocity is greatest at or near the surface, the refracted wave is bent away from the surface and the refraction technique cannot be used. It has been suggested by Shepard [12] that the wide range of velocities exhibited by natural materials opens possibilities of enabling civil engineers to measure the rigidity of subsurface materials by seismic methods. However, many factors such as water content, consolidation, porosity, and cementation enter into speed of propagation, and the extent to which such physical properties as bearing strength, rigidity, etc., may be estimated from velocity measurements is not generally known. As research develops along these lines, however, it may be possible eventually to make reasonably reliable estimates of these physical properties from velocity measurements.

Figure 184 shows the paths of the refracted waves in horizontal beds of constant, although different, velocities. The depth of penetration is proportional to the distance between the source

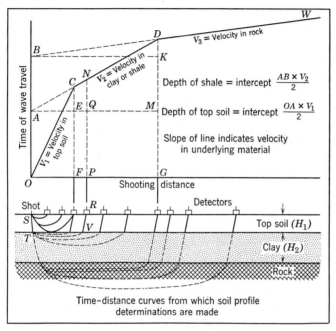

Fig. 184. Diagrams illustrating principles underlying the seismic refraction method. (After E. R. Shepard, *Military Engineer*, Vol. 31, No. 179, 1939.)

of the wave and the receiving station. It follows that in a homogeneous medium having a constant velocity (V_1) the time of arrival will be proportional to the distance separating the source and receiver. If the times of arrival at a series of equally spaced receivers are plotted against distance, the points will fall along a straight line. As the distance between the receiver and source reaches a certain critical value, represented by point F, the wave which travels through the upper low-speed layer and the wave that has entered the underlying higher-speed medium (velocity V_2) will arrive simultaneously. From this point outward the arrival of waves that have penetrated the underlying high-speed layer will be marked by a break in slope of the travel-time curve. In other words, from this point on the velocity will be higher. In shallow work, where the wave paths may be assumed to be essentially vertical, depths may be calculated as indicated in Fig. 184. More precise formulae, beyond the scope of this book, are required for the comparatively great depths involved in explorations for dams and other large engineering structures.

In the case of dipping beds the travel-time curves do not give true, but only apparent, velocities. As contrasted with horizontal layers, the up-dip profile exhibits a greater apparent velocity in the high-speed layer, and the down-dip profile shows a lower apparent velocity. Thus, both the dip and depth may be found from two profiles, one up- and the other down-dip, both being perpendicular to the strike.

In addition to refraction impulses, reflected waves are also received. The reflection technique is based on the latter type of wave. In order to separate the two types of impulses, multiple receivers are employed, the usual number being 6 or 12. On the multiple records thus obtained, reflections may be identified by their almost simultaneous arrival. Paths of the reflected waves are shown in Fig. 185. It is apparent that depth of the reflecting surface must equal one-half the speed of wave travel multiplied by the time elapsing between the instant of the explosion and the arrival of the reflected wave.* Similar to the refraction method, if the beds are not horizontal, up- and down-dip profiles must be taken. Relative depth may be determined from the travel time, but actual depth must be measured from the velocity of wave propagation. This may be determined in various ways, the most direct of which is actually to measure the velocity in the field. This can be done by exploding a charge in a well which penetrates identical strata and measuring the time of arrival of the waves a known distance away from the shot point.

Instruments. The energy generating the elastic waves is derived from an explosive charge, which is usually dynamite. If the charge is placed on the surface, the greater part of the energy

---

* This is strictly true only when the angle of incidence is very nearly vertical.

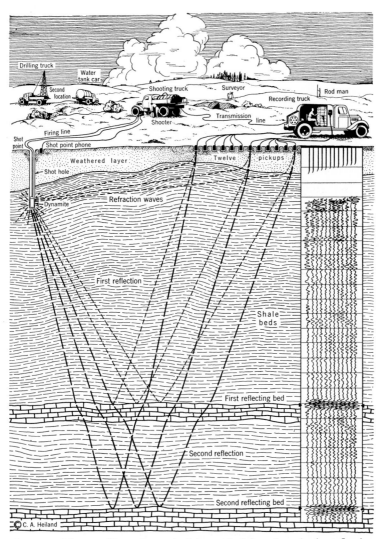

Fig. 185.   Diagram illustrating principles underlying the seismic reflection method.   (After C. A. Heiland, *Geophysical Exploration*, copyrighted by Prentice-Hall, 1940.)

is dissipated and very little is transmitted into the ground.* Consequently, it is usually necessary to place the charges as deep as time and other factors permit.   In reflection shooting it is common practice to place the charges at the ground water table, unless it is extraordinarily deep.   The refraction method usually

Fig. 186.   Portable shallow depth seismograph assembly.   (R. W. Moore, U. S. Public Roads Administration.)

requires heavier charges than the reflection technique, and in long-range refraction work more than a ton of dynamite may be used in a day.

If the charges are to be placed below the surface, some type of drilling machine is usually necessary.   Truck-mounted rotary rigs are the type most frequently employed in seismic work. Drilling is an item of considerable expense, and in rough country may be very costly.   However, in engineering work the charges are often set on or above the surface with entirely satisfactory results.

* There is also a method whereby the charge is placed above ground.

The exact instant of the detonation may be transmitted to the receiver in a variety of ways, among which are: (1) sound waves set up in the air by the explosion, (2) light waves generated by the explosion, or (3) electrical circuits. The latter method requires more or less complicated wiring. Wire transmission is used in short-range refraction and nearly all reflection work, but in instances where the laying of cables is impractical, radio transmission is employed.

In order to transform the ground vibrations into a permanent record for later analysis, some kind of detector and recorder are necessary. The function of a detector is to convert the mechanical vibrations in the ground into electrical vibrations which may be used to drive the recording mechanism. Many types of detectors are possible. In all types of electrical detectors there are two elements: a spring-suspended mass and a transducer. The latter converts the relative motion of the mass and instrument frame into electrical energy. In the induction type of detector, this conversion is accomplished by means of a coil moving in a magnetic field. In the reluctance type, iron armatures surrounded by coils are placed close to the poles of a permanent magnet. In the capacitive type, the mass is mounted near a stationary plate, and the two act together as a variable condenser. Carbon and crystal pick-ups operate on the changing electrical response of these substances to varying pressure.

The small amount of electrical energy supplied by the detector is amplified electronically in a manner similar to that employed in a radio receiver. The energy supplied by the amplifiers is fed to the recording device.

Recorders consist essentially of a recording camera and a bank of galvanometers. In reflection work the reflections must be timed to within $\frac{1}{1000}$ of a second. This is usually accomplished by marking the recorder paper with lines equivalent to $\frac{1}{100}$ of a second and interpolating to within $\frac{1}{10}$ of the interval. The timing apparatus consists of a synchronous motor controlled by a tuning fork or vibrating reeds.

The recording apparatus is usually mounted on a truck, but portable assemblies are also available for use in rough terrains.

Survey Methods and Data Obtained

In both refraction and reflection methods, the locations and elevations of the shot hole and detectors must be known. In refraction work, fan shooting and profile shooting are the techniques usually employed. In the former, the receivers are grouped at equal distances along the circumference of a circle with the shot point at the center. The area is covered by a series of overlapping fans. A high-speed layer is disclosed by a reduction of the normal travel time. In the profile method, three detectors are normally placed in a line at intervals of from 25 to 50 feet. Shots are then fired at constantly increasing intervals from the detector line. As a rule, a shooting distance of three to four times the required depth is necessary.

In the reflection method, profiles are usually taken. In oil prospecting, the preliminary profiles are often arranged in the form of a cross, thus disclosing both the dip and the plunge of concealed structures.

Corrections Applied. In rough terrains, all calculations must be corrected for differences in elevation between the shot point and receiving stations. In precise work, it is also necessary to correct for the increase in velocity in a homogeneous layer caused by increasing pressure with depth. It may also be necessary to correct for a lateral change in velocity such as may be caused by a sandstone grading laterally into a shale. In the reflection method, if the waves do not strike the reflecting surface approximately at right angles it is necessary to apply a so-called "spread correction." In this method, it may also be necessary to correct for the lower velocity of weathered surface materials ("weathered-layer correction").

Plotting of Data. The results of seismic exploration may be plotted as maps or as profiles. Results of refraction work are usually plotted as profiles, but if enough points have been established subsurface contour maps may be constructed. Reflection data may be plotted as profiles or as contour maps. If dip shooting is employed, dips alone may be plotted. In areas where the reflecting beds are persistent laterally and the dips relatively low, it is often possible to correlate the results of reflection work by assuming that reflections of about the same intensity and from about the same depth originated from the same reflecting surface. This hypothetical surface, termed a "phantom horizon," furnishes a basis for the correlation of results with geology and simplifies construction of a subsurface contour map.

Interpretation of Results

Since the depth of exploration can be fairly well controlled and other unknowns either estimated, calculated, or measured, seismic methods ordinarily require a minimum of interpretation. This does not mean that the method is foolproof, however, and a certain amount of interpretation is often necessary. Because interpretation requires a keen sense of geological probabilities, it is advisable for a geophysical party (seismic, gravity, or other) to work in close co-operation with a geologist who knows the area and who has a sound grasp of geophysical principles.

In seismic work much depends on the accurate measurement of wave velocities in natural materials. Because this is difficult to accomplish, it is normal to anticipate minor inaccuracies. Errors arising from inaccurate estimates of velocity will always be systematic, however, and, if the magnitude of the error is known at a few points (determined by core drilling or other direct methods), it is a simple matter to correct the seismic data. For example, if the assumed velocity of the reflected wave is too high, the reflecting horizon will be calculated as deeper than is actually the case. However, when the depth of the reflecting horizon is known at one or more points, the exact velocity can be readily determined. Suitable corrections can then be applied. This characteristic of the seismic method is of great engineering significance, for it makes possible a reasonably accurate determination of underground conditions from a minimum number of drill holes.

Engineering Applications

The chief engineering applications of the refraction method are in the measurement of the depth of overburden at construction sites. The method also offers possibilities of furnishing data bearing on the strength and other physical properties of the underlying rocks. Seismic methods are well adapted to the location of underground water supplies. Water location is a difficult problem, and most geophysical methods offer an indirect, rather than direct, solution. Indirect methods are concerned with the location of aquifers and their structural condition (whether folded, faulted, etc.). Seismographs have long been used in dynamic vibration tests of structures. In view of their broad application to engineering problems, it is unfortunate that seismic methods are usually the most expensive of all geophysical techniques.

SONIC SOUNDING METHODS

In work on water sonic sounding methods are sometimes used. These methods employ a sonic sounding device that differs from the reflection seismograph in that compressional (sound) waves rather than elastic (seismic) waves are employed. A sound impulse is directed toward the bottom and the returning echo is recorded on a chemically sensitized paper. The time elapsing between the outgoing signal and returning echo is obviously a measure of the distance to the reflecting surface.

Sonic sounding has long been used in river, harbor, and marine surveys for determining the depth to bottom. Under favorable conditions, where a hard bottom is overlain by soft mud or silt, two echoes may be received; one from the top of the soft layer and one from the top of the harder surface. Consequently, the method has often been used to measure the rate of silting in rivers and harbors. Sonic sounding has also been used to locate the inlets of leaks in a storage reservoir on the Dix River, Kentucky.

Experience indicates that the method has its difficulties. In some rivers and harbors a suspension of sediment and water called "fluff," fluid enough to permit passage of ships, may return an echo, and much unneeded dredging may be done if fluff is confused with a soft bottom. Furthermore, schools of fish, kelp beds, and temperature discontinuities may also return echoes, but it is usually not difficult to distinguish reflections of this sort from bottom echoes.

COMPARISON OF GEOPHYSICAL METHODS

Occasionally, scientists and engineers lack balance and objectivity. New methods are either ignored or hailed as a universal and infallible solution to a wide range of problems. The history of applied geophysics is replete with examples. A new or improved method is tested with satisfactory results, and in the minds of its partisans soon becomes a sort of supermethod. Before long the technique is applied to situations ill adapted to its use. The results are discouraging, and for a time at least the unfavorable reaction resulting from the experience tends to obscure the real merits of the method. In other words, each method has its characteristic advantages and limitations, and a proper selection depends on a careful balancing of costs, operating conditions, geological conditions, time available, and purpose for which the work is being carried out. Although for any specified location and purpose there is doubtless a best method, no one method is capable of universal application. Fortunately, a large number of instruments and methods is available, and there are probably few areas in which at least one of them will not produce valuable results. The following comments are offered merely as a resumé of the outstanding characteristics of various geophysical methods, and they are not intended as a rigid guide to the solution of any given problem. As a rule, selection of methods must be left to those most familiar with the ground.

Gravity Methods. Gravity measurements are best adapted to land areas of moderate to low relief, but they have also been used in water-covered areas. The most favorable geological conditions are found in areas where materials of markedly different densities lie relatively near the surface and are not too greatly disturbed by folding and faulting. Examples of favorable conditions are: (1) dense igneous or metamorphic rocks overlain by lighter materials, (2) relatively light rocks intruded by denser igneous bodies (plugs, stocks, etc.), and (3) relatively light rocks intruded into denser materials (salt domes). Gravity measurements are relatively cheap, but they may be difficult to interpret geologically.

Magnetic Methods. Magnetic surveys are cheap and well adapted to use in rough and undeveloped country. The development of the airborne magnetometer makes rapid surveys of land and water areas almost equally feasible. The most favorable geological conditions are found in regions of igneous rocks that are strongly magnetic in contact with sediments of less pronounced magnetic properties. Results of such surveys may be difficult to interpret geologically.

Electrical Methods. These methods are characterized by relatively low cost, great flexibility, and a broad scope of application. Resistivity measurements are handicapped by rugged terrains, and they may be difficult to interpret geologically. Geological interpretation usually demands a comprehensive grasp of the physical and geological factors involved. Electrical measurements, probably the most useful single method available, are well adapted to civil engineering purposes.

Seismic Methods. Because of the control of exploration depth inherent in these methods, seismic methods require a minimum of interpretation and yield quantitative results. For these reasons seismic methods are somewhat favored in a broad field of practical work, notwithstanding the comparatively high cost involved. Seismic methods are best suited to land areas. However, work in shallow water has met with considerable success. The most favorable geological conditions are encountered in sedimentary terrains underlain by beds of rather different velocities of wave propagation. If a high-speed bed is on the surface, the refraction method cannot be used, and in some regions reflections cannot be obtained. Loss of the reflected waves is attributable to absorption or scattering by the underlying materials.

Sonic Methods. At present these methods are limited to determination of the depth to bottom of water bodies, and, under favorable circumstances, to the estimation of thickness and character of the bottom deposits. Some research has been conducted with sonic methods in connection with exploration of land areas, but as yet few, if any, practical applications seem possible.

ELECTRICAL MEASUREMENTS IN WELLS

Electrical measurements in wells find their chief application in the oil industry where they have long been used for the following purposes: (1) correlating the strata of adjacent drill holes, (2) determining porosity, (3) determining the inclination of drill holes from the vertical, (4) determining dip of the strata, and (5) estimating rock pressures. These items may also have considerable civil engineering importance, and, as the following discussion will demonstrate, it should also be possible to employ electrical measurements as a means of determining the character of subsurface formations in instances where core losses have been excessive.

Resistivity Measurements. Electrical measurements of the resistivity of strata penetrated by wells are based on the same principles as resistivity methods of surface exploration described in a preceding section. The equipment employed may be designed in a number of ways, the Schlumberger type of device being fairly typical.[9] It consists of three electrodes, separated by a suitable vertical interval, which are lowered into the well. The lowermost electrode is connected with ground, and connected with a battery which causes a current to flow into the walls. Resistivity of the wall rocks is measured by the differences in potential observed between the other two electrodes. If the two upper electrodes are connected to a potentiometer and all three are lowered into the hole at a known rate, a continuous graph of changes in resistivity with depth can be made. Measurements must be made before the casing has been set, and the drilling fluid (mud or water) must be left in the hole.

The boundaries between layers of different resistivities are often marked by sharp discontinuities on the graphical record. It is this characteristic of resistivity measurements that makes them useful in subsurface correlation and structural studies. Very thin interbeds, such as limestone lenses in shale, may be distinguished on electrical logs, and geologic correlation thus resolves itself into matching similar series of peaks and valleys appearing on the graphical record. Under favorable conditions faults can be detected by correlating adjacent logs. A virtue of this method is that faults that do not appear on the surface may nevertheless be located.

Because the resistivity of a rock is a function of the porosity and percentage of dissolved electrolyte in the interstitial water, resistivity measurements alone tell very little concerning the lithology of subsurface materials. However, if the porosity is also measured, it is possible to determine if a bed of high resistivity owes its electrical properties to low porosity, the presence of petroleum or a similar non-conductor, or to low concentration of dissolved electrolytes (such as might be the case in a porous sandstone). Schlumberger [10] states that resistivity and porosity measurements have been successful in detecting coal beds which could not be recovered in drill cores.

Although the electrical logging method does not appear to have been applied to civil engineering work to any considerable extent, it has interesting possibilities. For example, it offers a method for investigating the character of the materials where core losses have been excessive. In this connection, interpretation of the electrical logs can be greatly strengthened by comparison of the electrical properties of the lost intervals with the corresponding properties of those portions of the boring from which the core was obtained.

Procedures for determining the dip of strata, inclination of a well from the vertical, and rock pressure have been described by Schlumberger and Leonardon [9] and need not be repeated here. A discussion of dynamic vibration testing, strain gaging, and other non-geological applications of geophysics may be found in Heiland's treatise on the subject.[3]

REFERENCES

1. American Society for Testing Materials, Symposium of Surface and Subsurface Reconnaissance, *Spec. Tech. Publ.* 122, 1951.
2. Edward B. Burwell, Jr., New Developments in Bore-Hole Photography, *Bull. Geol. Soc. Amer.*, Vol. 61, No. 12, Pt. 2, 1950.
3. C. A. Heiland, *Geophysical Exploration*, Prentice-Hall, New York, 1940.
4. M. J. Hvorslev, *Subsurface Exploration and Sampling of Soils for Civil Engineering Purposes*, Prepared for the Soil Mechanics and Foundations Division of the American Society of Civil Engineers. Distrib-

uted by the Engineering Foundation, New York. Printed by Waterways Experiment Station, Vicksburg, Miss., November, 1949.

5. Karl S. Kurtenacker, Some Practical Applications of Resistivity Measurements to Highway Problems, *Trans. Am. Inst. Mining Met. Engrs.*, Vol. 110, 1934.

6. L. W. LeRoy, *Subsurface Geologic Methods (A Symposium)*, Colorado School of Mines, Golden, 1950.

7. E. D. Lynton, Laboratory Orientation of Well Cores by Means of Their Magnetic Polarity, *Bull. Am. Assoc. Petroleum Geol.*, Vol. 21, 1937.

8. C. and M. Schlumberger and E. G. Leonardon, Electrical Exploration of Water-Covered Areas, *Trans. Am. Inst. Mining Met. Engrs.*, Vol. 110, 1934.

9. C. and M. Schlumberger and E. G. Leonardon, Electrical Coring; a Method of Determining Bottom-Hole Data by Electrical Measurements, *Trans. Am. Inst. Mining Met. Engrs.*, Vol. 110, 1934.

10. C. and M. Schlumberger and E. G. Leonardon, A New Contribution to Subsurface Studies by Means of Electrical Measurements in Drill Holes, *Trans. Am. Inst. Mining Met. Engrs.*, Vol. 110, 1934.

11. C. and M. Schlumberger, Application of Electrical Prospecting to the Study of Dam Sites, *Second Congr. on Large Dams*, Vol. IV, Washington, 1936.

12. E. R. Shepard, The Seismic Method of Exploration Applied to Construction Projects, *Military Engineer*, Vol. 31, No. 179, 1939.

13. B. V. Vacquier et al., Interpretation of Aeromagnetic Maps, *Geol. Soc. Amer. Mem. 47*, 1951.

14. F. A. Vening Meinesz, Maritime Gravity Survey in the Netherlands East Indies; Tentative Interpretation of the Provisional Results, *Proc. Koninkl. Akad. Wetenschap.*, Vol. 33, 1930.

C H A P T E R   19

Dams and Reservoirs

Geology enters into the investigation of dam sites in many ways. It is a major factor in selection of the site and is also important in choosing the type and design of dam to be constructed. It also bears directly on the type and amount of foundation treatment required. In addition, it is the major factor determining leakage from reservoirs and has an important bearing on sedimentation in them. Consequently, the application of geology to investigation of dam and reservoir sites is well established, and few such projects are now undertaken without geological advice. Geological examination of dam sites is seldom so simple as ascertaining whether conditions are "excellent," "good," "poor," or "unacceptable," for some type of dam can probably be designed for almost any type of foundation. Feasibility of the project is almost entirely a matter of economics. It is, therefore, as unwise for a geologist to undertake the study of dam sites without at least a general knowledge of the design characteristics of dams as it is for a civil engineer to design a dam without considering the characteristics of the foundation. For this reason, a brief review of the various types of dams is included in this chapter.

No two construction sites are ever exactly alike, and complete standardization of investigation methods is, therefore, not desirable. Methods of investigation of dam and reservoir sites have grown over a period of many years and are constantly being improved. Geologists and civil engineers should select from the large number of methods and devices available those best adapted to the situation, and should vary procedures or devise new ones, if circumstances so require. In all cases the important thing is to obtain complete and accurate information as cheaply and expeditiously as possible and in the proper order. The last is very

Fig. 187. Remains of the St. Francis Dam, near Saugus, California, October, 1928. Failure is attributable to washing away of the soft conglomerate shown in the lower left. (M. Juul Hvorslev.)

important, for instances are known where failure to obtain sufficient data concerning powerhouse foundations in advance of dam construction has caused considerable trouble and additional expense. It is obviously impossible in a short chapter to give more than a general account of some of the tried and tested methods that are most generally applicable to investigation of dam and reservoir sites. Additional information may be found in the publications cited in the references, especially those by Bryan,[2] Burwell and Moneymaker,[3] Legget,[12] and Mead.[14]

TYPES OF DAMS

Dams may be classified as masonry (concrete or masonry block) and earth structures. In the United States block masonry construction is practically obsolete. Concrete dams may be subdivided as follows: (1) gravity dams, (2) buttress dams (flat deck, round-head buttress, and multiple arch), and (3) arch dams. Occasionally, as at Hoover Dam, a combination arch-gravity design may be used. Earth dams may be subdivided into earth, rock-fill, and a combination earth-rock-fill types.

Solid Gravity Dams. Solid gravity dams derive their stability from weight of the materials composing the structure. Structural strength of concrete is not used except to withstand crushing at the base. As a result, the concrete need not be of the highest grade, but the volume of material required is almost always greater than for other types of design. A typical abutment section of a gravity dam is shown in Fig. 190. Solid gravity dams up to 65 feet in height have been constructed on earth foundations, but solid rock is preferable and is mandatory for higher structures.

Buttress Dams. Buttress dams may be regarded as modified gravity dams with the volume of concrete reduced by taking advantage of the structural strength of reinforced concrete. They may be thought of as gravity dams with the material of certain sections eliminated, the internal stresses resulting being supported by the beam or arch action of reinforced concrete. Stability against overturning and sliding is obtained by designing the upstream face on a slope of about 1 vertical on 1 horizontal, the relatively thin slab constituting the dam proper being supported on the downstream face by massive buttresses. By spreading the buttress footings, lower unit loadings can be obtained than with solid gravity structures. The volume of concrete required is less than for solid gravity designs, but must be of very high grade. The slabs constituting the deck portion must be reinforced. Such dams require more form work and are not always cheaper than solid gravity types. They are best adapted to regions where aggregates are scarce; they may be constructed on foundations that are not suited for solid gravity structures. The three chief types are the flat deck, or Ambursen type, the round-head buttress, and the multiple arch.

Arch Dams. Under a fairly restricted range of conditions it is possible to utilize the structural strength of concrete in combination with the arch principle to lower considerably the volume of material required for a dam. However, as the radius of curvature increases the amount of concrete required per linear foot likewise increases. Furthermore, an arch has a greater length than

Fig. 188. Grand Coulee Dam on the Columbia River, Washington, an example of a large gravity dam. It is 550 feet high and is founded on granite. (U. S. Bureau of Reclamation.)

a straight gravity section. Consequently, there is a limiting radius of curvature where the amount of concrete required is the same as for a gravity dam. This radius is estimated to be about 500 feet. If a central angle of 120 degrees is assumed, this value corresponds to a distance between abutments of approximately 865 feet. Arch dams are thus economic only in narrow gorges. Because the abutments must sustain the stress caused by the water in the reservoir, the additional requirement of strong abutment rock is imposed. Rock foundations are also necessary.

Earth Dams. Dams of this type may be constructed of loose rock, gravel, sand, silt, rock flour, or clay. As in other types of dams, the height is determined by economic factors. In the United States earth dams over 200 feet high are not uncommon.

Fig. 189. Hoover Dam on the Colorado River, Arizona-Nevada. The dam is of the arch-gravity type and is 726 feet high. The foundation is composed of a complex of igneous extrusive rocks. (U. S. Bureau of Reclamation.)

The Cobble Mountain Dam, Massachusetts (255 feet); the Calaveras Dam, California (217 feet); the Fort Peck Dam, Montana (250.5 feet); and Anderson Ranch Dam, Idaho (456 feet) are good examples of high earth dams. The top width may range from about 12 to 40 feet or more. An earth dam will not survive overtopping by water from the reservoir, and sufficient freeboard must be provided as a protection against reservoir waves. The tying in of a concrete spillway with the body of an earth structure involves difficult design problems, and it is desirable to locate it in a side gorge rather than on the dam proper. Because of the large quantities required, construction materials must be obtained from near-by sources. The unit loading pressures being low and the structure flexible, earth dams can be designed for almost any type of foundation. As sites suitable for construction of concrete dams are exhausted, engineers will, perforce, make increased use of this type of structure.

Earth dams are actually porous membranes which serve to deflect seepage from the reservoir to a point at the stream level well within the toe of the dam. If rapid seepage takes place on the downstream face at any point above the toe, safety of the structure is imperiled. To insure that this will not occur, most earth dams are provided with a suitable impervious core and a cut-off wall extending well below the foundation grade.

The principal types of earth dams are: (1) hydraulic fill, where the material is washed from a borrow pit to the site in the form of a suspension in water; (2) roll-fill, in which the material is excavated from a neighboring borrow pit and compacted by rolling; and (3) semihydraulic fill, where the construction methods are a combination of (1) and (2) above. In addition there are rock-fill dams which are constructed of rock spoil or quarried rock. A rock-fill dam usually consists of three elements: (1) a loose rock fill constituting the main mass of the dam; (2) an impervious facing—clay, reinforced concrete, or steel, on the upstream side; and (3) rubble masonry between the first two. Such dams are thought to be more stable than masonry types in the event of earthquakes.

The upstream face of earth dams must be protected against wave and ice action. In small structures light riprap (random placed) or crushed stone may be utilized. In larger structures heavy riprap, pavement, or cement blocks are the materials most frequently employed for this purpose.

Downstream faces are often provided with berms at vertical intervals of 30 to 40 feet. Berms aid in checking erosion by collecting the rain and seepage water into definite channels from which it is directed into catch basins, and eventually into the stream. Berms may also be built into the upstream face. As a protection against erosion by rain, the downstream face may be treated with crushed stone, screened gravel, or simply covered with soil in order to encourage the growth of vegetation.

FORCES ACTING ON MASONRY DAMS

The principal forces acting on masonry dams are: (1) water pressure, (2) earth pressure, (3) ice pressure, (4) weight of the dam, and (5) reaction of the foundation. In certain areas, it may also be necessary to consider earthquake forces.

Water Pressure

Forces originating from water pressure are of three kinds: (1) pressure of the water in the reservoir and the tailwater, (2) pressure of the water discharged over the spillway, and (3) uplift

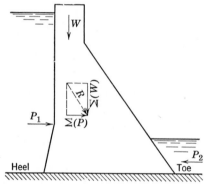

Fig. 190. Abutment section of a gravity dam, showing the principal forces acting on the dam. W is weight of dam; P_1 is pressure of water in the reservoir; P_2 is tailwater pressure; $\Sigma(W)$ is sum of all vertical forces; $\Sigma(P)$ is sum of horizontal forces; R is the resultant of $\Sigma(W)$ and $\Sigma(P)$. (Adapted from Creager, Justin, and Hinds, *Engineering for Dams*, copyrighted by John Wiley & Sons, 1945.)

pressure. The first is essentially hydrostatic and may be computed from hydraulic formulae. It tends to cause the dam to slide on the foundation and to overturn along a line parallel to the axis. In arch dams the tendency to slide is counteracted by the abutments, but in straight gravity or buttress sections it is resisted by friction between the dam and foundation and the shearing strength of the foundation materials. An additional margin of safety is usually provided by leaving the foundation in a rough condition, and in some instances by sloping the foundation grade line upstream, thus introducing a shear component into the foundation rock. Overturning of gravity dams is counteracted by weight of the masonry and the reaction of the foundation. As compared with hydrostatic forces, the dynamic forces resulting from flow of water over the spillway are very small, and for the present purpose they may be neglected.

Uplift Pressure. Uplift pressures result from the permeability of the dam and foundation. Water at the heel of a dam is under hydrostatic pressure proportional to the height of the column impounded. Should it gain access to pores and fissures in the dam or foundation, an equivalent pressure will be exerted. Since under hydrostatic conditions liquids transmit pressure equally in all directions, a resultant upward pressure proportional to the head of water in the reservoir will be transmitted to the dam. At the toe of the structure a similar upward pressure will be exerted by the head of water in the stream channel. It was once thought that, if the openings aggregated say 50 per cent of the area of the foundation surface, an uplift pressure of only one-half the pressure in the interstices would result, but this assumption is now known to be erroneous. The percentage of voids in the surface of the foundation and the contact area between grains in the foundation and dam are very different, and it is along the openings between contacts that uplift pressures operate. For this and other reasons, it is well to assume that uplift pressures act over the entire area of the foundation. Inasmuch as the reservoir level is almost always higher than the elevation of the tailwater, the uplift pressure is greatest near the heel and diminishes toward the toe, the actual uplift pressures depending mainly on the drainage provisions. Uplift pressures tend to lessen the effective weight of the dam and strengthen the tendency toward sliding and overturning. They may be counteracted by increasing the mass of the dam, but it is usually more economical to reduce them by drainage. In modern gravity dams the usual practice is to relieve uplift pressures by means of drain holes located downstream from the grout curtain. With buttress-type dams the problem of uplift pressures in the dam proper may be neglected, but important pressures may exist in the foundation, especially when it is composed of stratified materials of alternately permeable and impermeable character. Arch dams derive their stability from reaction with the foundation and abutments, and uplift pressures are not an important factor in design.

Earth Pressure. If the foundation trench is backfilled or extensive sedimentation occurs in the reservoir, a dam may be subjected to earth pressures. These are usually calculated by the Rankine formula. This and other formulae for estimating earth pressures can be found in all standard books on soil mechanics.

Ice Pressure. If a reservoir is completely frozen over and the temperature continues to drop, the ice cover will contract and tension cracks will begin to form. Water enters the cracks and quickly freezes, with the result that as the temperature rises the expanding ice cover is too large for the reservoir. Theoretically, a sheet of ice 4 feet thick can exert expansion thrust of 500,000 pounds per linear foot. However, this estimate is based on the quick crushing strength of ice, and it is probable that under conditions of slow expansion the value is considerably less.[1] Furthermore, in order for maximum pressure to be exerted, the ice must be of considerable thickness and firmly frozen to the dam and reservoir banks. Allowances for ice pressure vary according to climate; in cold climates they may be as much as 47,000 pounds per linear foot.[6]

Weight of the Dam. Pressures exerted on the base of the dam and its foundation by weight of the dam can be calculated provided the average weight of the construction materials is known. In lack of precise data 150 pounds per cubic foot is the value usually adopted for concrete.

Reaction of the Foundation. The foundation is subject to various types of forces: (1) static weight of the dam, (2) horizontal forces caused by pressure of water in the reservoir, and (3) overturning forces caused by pressure of the reservoir water. For the dam to be in static equilibrium the resultant of all the forces acting on it must fall within the middle third. These forces are counteracted by the foundation, which reacts equally but in an opposite direction. Vertical forces are counteracted by the compressive strength of the foundation materials; horizontal forces are opposed by their shearing strength. Past practice tended toward consideration of sliding friction alone in calculating horizontal resistance, but experience indicates that sliding does not occur until a dam has sheared through a portion of the concrete and/or foundation rock. Present practice is to consider the shearing strength of the foundation in computing horizontal reaction. Both the dam and the foundation are elastic and subject to deformation. In large structures it is, therefore, necessary to take elastic deformation into consideration. The problem of determining the correct coefficient of elasticity of the foundation is often difficult and becomes increasingly so as the variety of rock types in the foundation increases.

Earthquake Forces. Few problems are more thorny than that presented by the possible effects of earthquake forces on dams. Records indicate that well-designed dams have seldom failed, even when subjected to violent shocks. The Crystal Springs Dam is located only a few hundred feet from the fault responsible for the San Francisco earthquake of 1906, but it survived the shock without damage.[10] This structure is built of concrete and is of the arch-gravity type. The Upper Crystal Springs Dam (an earth-fill structure 85 feet in height) was offset laterally by slippage along the fault, but did not fail except in shear.[12] In the Santa Barbara, California, earthquake the Gibraltar Dam (an arch structure) was so badly shaken that the watchman, who happened to be on the top, had difficulty in staying on his feet, but the dam was not damaged. However, there can be no doubt that earthquakes reduce the margin of safety, and conservative judgment demands a careful consideration of the risks involved and possible precautionary measures.

The accelerations produced by earthquakes depend on the intensity of the shock and a complex of local factors. Accelerations approaching that of gravity probably accompany very violent shocks, but their effects are very local. For solid rock at a distance of a few miles from the epicenter accelerations probably seldom exceed 0.1 that of gravity. On loose materials the value may be somewhat higher. In the United States it is common practice to adopt a safety factor of about 0.1 that of gravity for horizontal forces. Vertical accelerations are neglected. If the dam is close to an active fault, or founded on loose materials, higher safety factors are probably necessary. Dams should not be built in such locations if the consequences of failure are likely to be serious. Faulted sites should also be avoided, but, unless there is good evidence that the fault is active, such sites may be made acceptable by suitable treatment and structural design.

The Pine Canyon Dam near Pasadena, California, is situated over a fault thought to be inactive, but as an additional factor of safety a slip type of joint has been built into the masonry overlying the fault zone.[17] Should displacement of the fault occur, the flexibility provided by the joint should prevent, or minimize, cracking of the concrete.

The Rodriquez Dam in Mexico is situated over a fault which passes through the stream bed.[8] As an additional margin of safety a special type of arch was built into the lower portion of the downstream face. The purpose of the arch is to equalize the portion of the load carried by the weakened rock in the stream bed and to transfer the remainder to the sound rock of the valley walls. As favorable sites are exhausted, conditions of the type described above will demand increased attention.

SEEPAGE PROBLEMS

Even concrete dams are permeable, but seepage through the dam can usually be ignored. Seepage through the foundation is normally a far more serious matter, and its control requires very close co-ordination of geologic and engineering knowledge. No naturally occurring material is impermeable under the hydraulic pressures existing beneath dams, and seepage is an ever-present problem. Seepage is generally undesirable because it permits loss of water from the reservoir, and, if the velocity is sufficiently high, it may lead to piping (removal of loose particles from definite channel ways) and may result in failure of the structure. It is not possible completely to eliminate seepage through the foundation, and the so-called preventive measures are merely ways of checking the velocity so that seepage presents no serious economic loss or safety hazard.

Preventive Measures. Seepage through foundations may be checked by reducing the permeability of the foundation materials, by lengthening the path of percolation, or both. The usual method of reducing permeability is by pressure grouting, which consists of forcing liquid cementing material into the voids. Grouting not only serves to reduce permeability but also increases the bearing strength of the foundation.

In earth dams on soil foundations constructed for purposes in which loss of water is relatively unimportant, it is often best not to attempt complete control of seepage, but to remove the water in such a manner that piping cannot occur. This may be done by installation of drainage, or relief, wells near the downstream face. Carefully designed sand screens placed in the wells permit free flow of water without removal of soil particles.

Fissured or weak foundations are often sealed and strengthened by a procedure known as consolidation, or blanket, grouting. The base area of the dam is covered with a series of shallow drill holes. Neat cement, or other suitable bonding material, is pumped into the grout holes under pressure in the hope that all cavities will be sealed off. Care must be taken to employ pressures sufficient to produce adequate penetration of the grout without uplifting the foundation or widening the fissures. Grouting should never be approached without a thorough knowledge of the location and condition of joints, fissures, and other zones of weakness or permeability.

Cavities in limestone are frequently coated with a film of residual clay, and if a tight bond between the grouting material and wall rock is to be secured, the clay film must be removed. At Norris Dam this was accomplished by flushing the cavities with water pumped into selected grout holes under high pressure.

Porous sandstones are not amenable to treatment with cement grout, the cement merely piles up around the periphery of the hole and seals off circulation to more distant areas. In such materials some success has been achieved with chemical grout consisting of two separate solutions. The first merely penetrates the material, whereas the second reacts with the first to form an insoluble material.[13] Large cavities are not susceptible to pressure grouting but have been successfully treated by filling with clay,[11] concrete, or asphalt.[5]

The most effective method of lengthening the path of percolation is by providing a suitable impervious cutoff on the upstream side of the dam. This may consist of a concrete wall, sheet piling, or a grout curtain. Grout curtains are placed along a line of drill holes spaced at suitable distances apart (usually about 4 to 8 feet) near the heel of the dam. For best results, the curtain should extend to an impervious bed. In other words, depth of the grout curtain should be determined by local geological conditions, and not arbitrarily. Experience indicates that, for a cutoff extending 90 per cent of the distance to an impervious layer, seepage is still 35 per cent of what it would be without a cutoff.[8]

Other methods of checking seepage are the placement of impervious blankets on either the upstream or the downstream portion of the dam. The first will tend to reduce uplift pressure, whereas the second will tend to increase it. These methods, especially the latter, are used mainly with earth dams.

Seepage through Earth Dams. The saturation line (upper surface of seepage) in an earth dam is not a straight line, as was

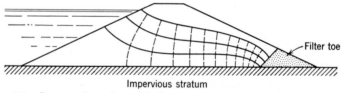

Impervious stratum

Fig. 191. Seepage through an idealized homogeneous earth dam with filter toe, full reservoir. Flow lines indicating direction of seepage shown by solid lines; dotted lines are lines of equipotential. Flow lines and lines of equipotential form what is known as a flow net. (After Terzaghi and Peck, *Soil Mechanics in Engineering Practice*, John Wiley & Sons, New York, 1948.)

formerly thought to be the case, but a parabola.[4] This configuration complicates the problem of preventing moisture from reaching the toe. Because it is highly desirable to keep the saturation line well within the dam in order to prevent excessive loss of water and possible piping, most earth dams are provided with rather elaborate devices such as impervious cutoffs, blankets, and filters.

The action of impervious cutoffs has already been mentioned. In earth dams (especially those founded on soil) a cutoff extending from the level of the impounded water to a fairly impervious layer in the foundation is often provided. This usually consists of an impervious core in the dam and a cutoff wall in the foundation soils.

The downstream configuration of the percolating moisture stream may be altered by offering it an easy way to leave the dam through a highly permeable material built into the body of the structure. Moisture reaching the surface of such a material may be compared with a very fine dew. It simply falls down, and the tangent to the streamline reaching such a surface is vertical. A filter reaching one third or more of the width of the dam draws the streamlines away from the downstream face and has the same effect as flattening of the latter. One drawback to the use of pervious filters is the possibility of clogging by silting. In this event, the streamlines will return to the downstream face, and stability of the structure will be impaired. A downstream rock-fill toe may also be used to divert the moisture streamlines.

EROSION BELOW SPILLWAYS

When the spillway is in use the velocity of the descending water is often great enough to cause marked erosion of the stream bed, and more or less elaborate stilling basins, baffles, and similar devices are often necessary to prevent serious damage. The water falling over a spillway is very effective in searching out and scouring weak zones such as joints, faults, and bedding planes. Gently dipping laminated shales may "unravel" upstream with the result that the foundation may be undermined. If this possibility exists, it should be carefully investigated.

GEOLOGICAL EXAMINATION OF DAM SITES

Usually, the first studies are made by civil engineers, and at least rough topographic maps are available. Suitable sites for the dam are then selected primarily on the basis of topography, and it is the geologist's function to investigate them from the geological point of view. This involves careful geological mapping of all the sites under consideration, and investigation of natural construction materials available near the sites (gravel, sand, earth, potential quarries, etc.). Faults and joints should be carefully mapped, and attention should be given to evidences of former floods (driftwood, gravel deposits at high levels, etc.). If an arch dam is under consideration, careful attention should be given to the orientation of joints and weak zones in the abutments. Where such zones coincide with lines of potential shear resulting from the thrust of the dam, stability of the structure may be imperiled, especially where the topography is unfavorable.

If one or more of the sites is overlain with alluvium, an estimate of the thickness should be made. As Bryan [2] has pointed out, if the stream is aggrading, the depth of alluvial fill may be very great. If the stream is degrading its channel, the alluvium is removed by floods and redeposited at lower stages. This process is known as scour and fill. The depth of scour and fill is a function of the volume and velocity of floods and type and quantity of alluvial material available. It was possible to predict the depth of alluvium in the San Juan River Canyon with surprising accuracy by proceeding on the assumption that depth to bedrock did not exceed the range of scour and fill. However, it is seldom possible from geological observations alone to predict the depth of alluvium with any accuracy. Where earth dams are under consideration, careful attention should be given to the composition of the alluvium, its stratification, and its permeability.

The purpose of the geological reconnaissance is to classify the sites in order of relative suitability when all topographic, geologic, and engineering factors have been taken into account. A full report should be submitted with an estimate of the number and locations of core borings and/or test pits necessary for the next stage of investigations.

Predesign Borings. Core borings should always be started at the preferred site. During predesign stages sufficient core borings should be completed to determine the depth of overburden, depth of weathered rock to be excavated, character and structure of the bedrock, and permeability of the foundation materials. At most sites cores $2\frac{1}{8}$ inches in diameter will be sufficient at this stage. Drilling inspection should follow the general procedure set forth in Chapter 18. If at an early stage unfavorable subsurface conditions are disclosed, the possibility of shifting the explorations to an alternate site should be seriously considered.

Permeability of rock foundation materials may be determined by pressure testing. A pressure testing device consists of a length of perforated pipe equipped with a rubber packer at each end. The pipe is secured in the hole by pumping water into the packers. Water is then forced into the interval between the packers by means of a pump. The pressure used is ordinarily about 50 pounds per square inch. A hole may be tested in 5- or 10-foot intervals, or the entire length may be tested by removing the plug from the bottom of the pipe. The amount of water pumped into the foundation is measured by reading a water meter placed in the system. Usually, the intake will be greatest at first and will decrease to a constant, or semiconstant, value within a few minutes. Readings of the water meter should be taken every 30 seconds, and a careful and detailed record of the tests should be submitted. It is also advisable to grout the holes that have been pressure-tested in order to give the foundation engineer preliminary information concerning practical aspects of the grouting problem. Permeability of soil foundations may be determined by laboratory tests of samples or by field tests.

Selected core samples should be collected for laboratory tests. Opinions differ as to the type of specimens to be selected. Many engineers believe that neither the best nor the poorest examples should be chosen, but it seems advisable to include a fair sampling of the more doubtful materials. The best, and obviously suitable, materials ordinarily need not be tested. The tests performed include shearing strength, compressive strength (confined and unconfined), bond strength with concrete, resistance to alternate wetting and drying, alternate freezing and thawing, and in some instances the magnesium sulphate test. The last is thought to be somewhat more rigorous than the freezing and thawing test. All specimens to be tested in the laboratory should be wrapped in wax paper and coated with paraffin immediately after being taken from the core barrel. Otherwise, deterioration is likely to result.

The predesign studies should furnish the engineers with sufficient data to enable them to decide on the type of dam best adapted to the site, but not necessarily enough to permit decisions concerning details of the structure. They should also permit the working out of tentative programs for foundation treatment, including such matters as the type and extent of grouting required, and, if diversion tunnels are to be used, some estimate of tunneling conditions should also be made. An estimate of the amount of excavation necessary to reach sound rock should also be made at this stage.

Design Borings. Design borings are essentially a continuation and elaboration of earlier investigations. A considerably greater number of small-diameter core holes are normally put down; they will cover at least the entire base area of the dam rather than

Fig. 192.   Minor fault encountered during excavation for the Pine Flat Dam, California.   (Corps of Engineers, U. S. Army.)

merely the center line, as is usually the case during earlier studies. The powerhouse foundations and conditions in the vicinity of the intake towers should also be carefully investigated. A still larger portion of the holes is normally pressure-tested and grouted. If grouting studies are made, it is well to use colored grout for light-colored rocks and to investigate the thoroughness of the results by drilling a number of large-diameter holes (36 inches or more), thus enabling a close visual inspection of foundation conditions and the effectiveness of the grouting procedure contemplated. The design studies should be complete enough to permit detailed design of the structure and appurtenances and a careful estimate of the foundation treatment required.

Studies during Construction. When the foundation is excavated, the conditions revealed should be carefully mapped by a geologist on a scale sufficient to show all details. The information gained in this manner is very important in determining details of foundation treatment and in checking design assumptions. For example, the designer might wish to lower his estimate of the foundation bearing capacity if the weak zones are closely spaced and relatively wide, whereas, if the fractures are very narrow and far apart, he would be justified in considering the strength of the foundation as a whole as about equal to the unconfined crushing strength of the average sample tested in the laboratory. The designer will, of course, have been furnished with the geologist's estimate of these factors long before, but only rarely will the excavated site conform exactly with the assumptions drawn from earlier investigations. The cores from all grout holes drilled through the dam from the grouting gallery should be inspected by a geologist, and the data secured should be used to check the adequacy of the proposed foundation treatment. The geologist should be present during grouting operations as a consultant to the foundation engineer.

GEOLOGY OF DAM SITES

Each site is a special case and must be investigated accordingly. However, with certain simplifying assumptions it is possible to classify dam sites according to the type of rock represented. The best treatments of the geology of dam sites are those of Mead,[14] and Burwell and Moneymaker.[3] The following discussion owes much to these publications.

Sites on Igneous Rocks

In general, igneous and pyroclastic rocks are as strong as, or stronger than, concrete. Many tuffs and agglomerates are weak, however, and weathering and alteration may reduce greatly the bearing capacity of normally very strong rocks. For this reason, it is wise to examine very carefully even presumably strong rocks when important structures are contemplated. Furthermore, all rocks are elastic, and, if their elastic properties differ considerably from those of concrete, this factor must be taken into account.   In high dams this involves laboratory testing of the modulus of elasticity and measurement of the resident strain in the rock while actually in place.   This can ordinarily be done in tunnels and galleries appurtenant to the dam.

Leakage in rocks of this class takes place mainly along joints, shear zones, faults, and other fissures. Igneous rocks are not soluble, and solution cavities do not occur.   Extrusive rocks, such as basalt, may possess a high proportion of gas cavities.   These cavities are not normally interconnected, and the permeability is usually low as compared with the porosity.   Ordinarily, the seepage channels in igneous rocks can be treated by cement grout.   As a rule, igneous rocks are among the most satisfactory materials for foundations of dams.

Metamorphic Rocks

Rocks of this class resemble igneous rocks so far as their foundation characteristics are concerned.   Many schists are soft and unsuited to foundations of high concrete dams.   The orientation of schistosity is important.   Steeply dipping foliation paralleling the axis of the proposed structure is the most favorable alignment; an orientation at right angles to the axis is the least favorable.   Marble is soluble and may possess large solution cavities similar to those encountered in limestone.   As a rule, foundations on metamorphic rocks are susceptible to treatment with cement grout.

Sandstone

Sandstones range all the way from quartzites, which have considerably higher crushing strength than concrete, to soft and friable materials.   In general, the strength of sandstones depends on the type and amount of cementing material present.   Assuming equal consolidation, silica cements produce stronger materials than carbonates and similar substances.   The crushing strength and elastic properties should always be determined in the laboratory.   It should be noted, however, that even poorly cemented sandstones are not subject to plastic deformation.

As in other types of rocks, the resistance to sliding is essentially a function of the shearing strength.   Pure sandstones usually have high shearing strengths, but if interbedded with clay or shales the latter may form zones of potential sliding, especially if at shallow depth below the base of the structure.

Sandstones may permit seepage not only along joints and other fissures but through the intergranular pores as well. Many sandstones possess a very high intergranular porosity, which is difficult to treat with cement grout. As previously mentioned, chemical grout, used for this purpose in Europe, may offer a solution to the problem.

Limestone

Limestones range all the way from very firm, strong rocks to soft and weak coquinas. The majority of these rocks are as strong as, or stronger than, concrete. In thin-bedded limestones with shale or clay partings there is a certain possibility of sliding within and along the softer layers. The bearing strength may also be reduced by such partings.

The chief problem presented by limestones is caused by their solubility, which may lead to the formation of enormous underground cavities. Exploration of such cavities by drilling is very costly and time consuming. As mentioned in Chapter 18, it may be that geophysical methods will eventually be found useful, but careful geologic studies are always necessary. Frequently, solution is confined to a relatively thin zone in an otherwise homogeneous formation, and when this can be demonstrated its engineering importance needs no elaboration. In other instances, the cavities can be definitely related to the present or past configuration of the water table. Furthermore, it is common for solution cavities to be localized along fissures formed by folding or faulting. Frequently, the open-tension fractures on the crests of anticlines are widened by circulating ground water, whereas the tightly compressed joints along the axes of synclines are not affected. When this relationship can be established, it is obvious that synclines offer better sites than anticlines. Many of the dams built by the Tennessee Valley Authority rest on cavernous limestones, and the reports of this organization contain a great amount of valuable information concerning the construction of dams and reservoirs in limestone regions. The procedure followed at Norris Dam is particularly worthy of study.[9]

Shale

As mentioned in Chapter 3, shales are of two types: the so-called cementation variety and the compaction type. However, there is no sharp demarcation between the two, and hybrids are fairly common. Compaction shales slake and disintegrate when subjected to alternate wetting and drying. The rapidity and completeness of disaggregation are a function of the thoroughness of the previous drying. Specimens taken from below the water table may also crack and fall apart when placed in water. Mead [14] is of the opinion that this reaction may be due, at least in part, to elastic expansion of the material.

Cementation Shales. Thoroughly cemented shales do not disaggregate when subjected to wetting and drying. Intermediate types may crack and fall apart, especially along the bedding planes, but they do not disintegrate completely as compaction shales do.

Compaction Shales. Compaction shales contain a high proportion of water, and when subjected to pressure they undergo plastic deformation. Cementation shales do not flow plastically when subjected to similar pressures. Compaction shales may be compressed by the load of overlying strata. If this load is removed by erosion or excavation, the rock will expand, and open joints and fissures may be formed. These tend to weaken the rock still further, and also to furnish channels along which water may penetrate and disaggregate the constituent particles. For these reasons it is important to investigate the elastic properties of shale foundations for masonry dams. The relatively high elasticity of shale must be taken into account in the design of the structure in order that deflection of the foundation under various conditions of loading brought about by variations in the reservoir levels can be allowed for. Otherwise, the differential movement between the dam and the foundation may disrupt the bond between the concrete and foundation rock, open grouted seams, and impair the effectiveness of the grout curtain. At Conchas Dam, New Mexico, it is estimated that about 4 inches of settlement of the compaction shale forming a part of the foundation will take place during a period of years. A special joint has been provided to permit differential movement between the section founded on shale and the section founded on less compressible materials.[7]

In general, the bearing strength of compaction shales increases as the water content decreases, and there are good reasons for believing that most of the settlement of shale foundations under load is caused by the squeezing out of water from between the constituent particles. Cementation shales generally have higher bearing strengths than the compaction variety, and are less subject to failure. Their strength is comparable to concrete, but their elasticity is considerably greater.

Strength of Shales. Shales are subject to slow plastic deformation under loads considerably below their shearing strength, provided the pressure is applied for a sufficient length of time. For this reason, laboratory tests of the compressibility and shearing strengths may not furnish a reliable basis for design. Conservative judgment should consider the possibility of settlement of the foundation under loads somewhat less than the compressive and shear strengths indicated by laboratory tests. The shear strength of shales is relatively low, and it is necessary to give careful attention to the possibility of sliding. In case of doubt, it is well to slope the foundation grade line upstream in order to throw potential shear lines into the foundation.

Foundation Preparation. The preparation of foundations on compaction shales presents special problems. Exposure of such materials to other than a water-saturated atmosphere involves loss of moisture from the shale with consequent formation of shrinkage cracks. If concrete is applied to such a partially desiccated surface, the surface layer absorbs water from the concrete with the resulting formation of a mud or soft clay seam between the concrete and the foundation rock. It is obvious that every possible measure should be taken to prevent, or minimize, this condition. Various procedures are possible. One is to specify that the foundation be excavated to within a suitable distance from grade and that final excavation of the remainder should be carried out immediately before the pouring of concrete. As an additional precaution, the surface should be kept wet by means of a water spray. Another method is to cover the surface with a layer of asphalt or other waterproofing material immediately after excavation is completed. Because compaction shales tend to expand upon removal of superincumbent load with the resulting formation of tension cracks, it is well to permit only a minimum time interval between excavation and pouring of concrete, regardless of the type of foundation treatment adopted.

Shales are usually fairly watertight, and this is especially true for compaction types in which the plasticity under load tends to prevent the formation of open joints and fissures. Grouting pressures should be kept to a safe minimum in order to avoid splitting open of the bedding planes, widening of joints, and uplift of the foundation. In no case should the grouting pressure exceed the load imposed by the overlying column of rock.

Foundations of Earth Dams

Earth dams on rock present no problems different from those already discussed for concrete structures. Many earth dams are constructed on alluvial deposits, and it is necessary to consider the problems presented by this type of foundation.

Alluvial deposits may consist of materials ranging in size from large boulders to clay. Foundations for earth dams on materials consisting of boulders, gravel, sand, or mixtures of these materials involve only seepage problems, because the compressive and shearing strengths of these materials are superior to those of the dam. Such materials are permeable, and extensive field and laboratory tests are necessary to estimate the seepage problem and to devise means of seepage control.

Foundations on flood plain deposits consisting of very fine sand, silt, and clay present special problems, the chief of which is settlement under pressure. These materials are generally highly porous, and the pores are ordinarily filled with water. Consolidation of these materials involves escape of the water contained in the voids. With slow application of load and the opportunity for escape of the interstitial water, consolidation of the porous mass occurs with consequent increase in the bearing strength. If the load is applied too rapidly to permit the escape of water, or if the escape of water is prevented, the unconsolidated material tends to flow from under the loaded area. Consequently, it is important to make sure that the water contained in sands and silts can escape from the mass. When impervious interbeds are present, provision should be made for upward drainage of the water liberated by consolidation of the foundation materials. In extreme cases, the removal of the impervious beds from the entire area of the dam may be necessary.

Clay foundations are especially troublesome. This material is plastic and remains so even after consolidation. When a certain critical height is reached, clay tends to fail under loads imposed by dams with side slopes suitable for other types of foundations. The initial movement may be very slow and gradual, but after a certain point is reached a sudden acceleration in rate of failure may take place. Many clays have an initial capacity to resist failure, but, when loaded beyond this limit so that movement once takes place, the resistance to continued failure is considerably less than before the initial movement occurred. Experience indicates that earth dams exceeding a certain height on clay foundations often require such flat side slopes that their construction is uneconomical.

RESERVOIRS

The purpose of a reservoir is to store water, and in order to fulfill this function it is obvious that seepage from the reservoir must be less than flow into the reservoir area. Although it might be assumed that so elementary a requirement would always be assured prior to construction, it would not be difficult to cite instances where a dam has remained high and dry, simply because seepage losses exceeded the flow into the reservoir. As Bryan [2] has pointed out, seepage from reservoirs is a problem in ground water hydrology, and is, therefore, susceptible to geologic study and analysis. The usual procedure is to determine the position and movements of the ground water under normal conditions in the reservoir area, and from the information thus acquired to estimate the effects of the new conditions that will be brought about by filling the reservoir. Consequently, the position of the water table under normal conditions is of fundamental importance.

High Water Table. Figure 193a shows a cross section of a reservoir underlain by relatively impervious materials. Consequently, the water table is fairly close to the surface, and the ground water divides follow the surface topography. If the reservoir is filled to a level below that of the top of the ground water divides, there will be no seepage loss; but, on the contrary, a certain amount of ground water will flow into the reservoir from storage in the banks. Similarly, the unconsolidated soil

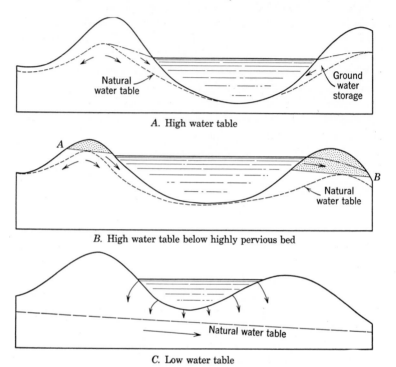

A. High water table

B. High water table below highly pervious bed

C. Low water table

Fig. 193. Hypothetical sections illustrating effect of position of the water table on seepage from reservoirs. Direction of subsurface water movement indicated by arrows. (Adapted from Kirk Bryan.[2])

and weathered materials mantling the banks will afford a certain amount of bank storage. The quantity of ground water storage will depend on the volume and permeability of the materials in question.

If the reservoir level is raised above the level of the ground water divides, a certain amount of seepage loss may result. This will depend on the permeability of the materials and the width of the divides; it is likely to be greatest where the divides are narrow or broken by closely spaced joints and/or faults. It should be noted that low saddles and narrow divides may be genetically related to the structural features mentioned above, and for this and other reasons they should be carefully examined as possible zones of excessive seepage. Figure 193*b* illustrates a region of a high water table overlain by a highly pervious bed. If the reservoir level is brought above the line *A–B*, seepage losses will be excessive.

Low Water Table. In permeable materials the water table is usually deep. Bryan [2] lists the following rock types in which this condition may be expected: (1) soluble rocks (limestone, gypsum, salt), (2) scoriaceous basalt and flow breccia, (3) faulted and shattered rocks of almost any type, and (4) coarse boulder beds. As shown in Fig. 193*c*, if a reservoir is constructed in an area where the water table is deep, there will be a downward movement of water from the reservoir to the zone of saturation. Water reaching the zone of saturation will then move laterally in the direction of the regional gradient, and there will be a net loss from the reservoir. Losses from the reservoir will be proportional to the hydraulic gradient and permeability of the materials involved. In absence of precise data it is necessary to exercise judgment as to whether the losses will be excessive. Experience with reservoirs built under comparable natural conditions is perhaps the best guide, and an extensive examination of all pertinent literature should be made. The reports of the Tennessee Valley Authority furnish a wealth of information concerning seepage and methods of seepage prevention in limestone reservoirs.

Where permeable sands and gravels fill channels extending well below the reservoir level, extremely difficult seepage problems may be encountered. This situation is well illustrated by the Le Sautet hydroelectric project in the French Alps. Here the reservoir is impounded by an arch dam over 500 feet in height, and the full reservoir is thus very deep. An old glaciated valley filled with very permeable sands and gravels outcrops in the reservoir and extends well below the level of the full reservoir. Consequently, seepage through this channel is very considerable, especially at high reservoir levels. It has been found that seepage cannot be checked by coating the reservoir outcrop of these materials with clay, for when the pool is drawn down the reverse circulation of water contained in the sands and gravels washes out the impervious cover.   The only really effective way of controlling this type of seepage appears to be by constructing a suitable cutoff wall, which is unfortunately very expensive.

Perched Water Tables.   The complications that may ensue from perched water tables are well illustrated by the Carlsbad irrigation project,[15] where investigations disclosed that the Pecos River runs on top of dry rock in a natural flume formed by deposition of calcium carbonate from the river water.   The actual water table was found to be about 60 feet below the river level.   Since a perched water table of this kind usually has only a limited lateral extent, if a reservoir is constructed part of the water will eventually spill over the sides of the perched zone and percolate down to the main water table.   In other words, the presence of a perennial stream does not invariably indicate that the water table is high.

Control of Seepage from Reservoirs.   If the zones of seepage are relatively restricted, as is usually the case with faults and shear zones, grouting is both feasible and effective.   The openings in limestone reservoirs may be so widely dispersed that it is not feasible to locate and treat all of them before the filling of the reservoir.   In many instances it is probably preferable to let the reservoir fill, and if important leakage is noted the reservoir may be drawn down and the trouble spots plugged.   Objections to this procedure are that it adds an element of uncertainty to the costs, and residents of the area may become unduly alarmed. Furthermore, the localization of seepage zones may prove difficult even with a full reservoir.   Leakage through saddles and narrow divides can usually be checked by grouting, pavements, or cutoff walls.

Instances are on record where natural silting has greatly lessened the volume of seepage,[19] and attempts have been made to cut reservoir losses by dumping fine-grained materials into tributary streams in order to accelerate silting.   Willow mats, and even bales of hay, have been used in attempts to check seepage. It should be noted that in soluble rocks, such as gypsum and rock salt, seepage may actually accelerate with time.   In nearly all other types of materials, however, the usual tendency is toward a lessening of seepage through the years.

Sedimentation in Reservoirs. Reservoirs are subject to sedimentation which in time will completely destroy their usefulness. The greatest volume of sedimentation occurs at the points where the main stream and tributaries enter the reservoir. Here deltas are formed which gradually extend themselves downstream into the deeper parts of the reservoir. If much fine material is carried by the streams, a fairly uniform blanket of silt and mud may be deposited over the entire reservoir area. Sedimentation cannot be controlled, but it can be minimized by proper selection of the reservoir site. The factors involved in the problem are well summarized in reference 18.

Oil and Mineral Deposits in Reservoir Areas. If mines or oil fields occupy a reservoir site the problem of estimating their value is an important one. Normally, their value will be large enough to make construction of the reservoir uneconomic. However, after a period of depletion, the reservoir project may become economically feasible. It is a common practice for property owners to make extravagant claims concerning oil and mineral prospects on their lands. Although in most instances the possibilities are exceedingly remote, these claims must be taken seriously, and careful examination of all claims should be made. With regard to mineral prospects careful attention must be given to the possibility of salting, i.e., the ingenious placing of high-grade ore from another locality in an otherwise barren prospect, fraudulent assays, etc. For these and other reasons, it is advisable to have mineral claims examined by specialists who are experienced with the methods and procedures of land evaluation.

REFERENCES

1. Ernest Brown and George C. Clarke, Ice Thrust in Connection with Hydroelectric Plant Design, *Eng. J.*, January, 1932.
2. Kirk Bryan, Geology of Reservoir and Dam Sites, *U. S. Geol. Survey, Water Supply Paper 597*, 1928.
3. Edward B. Burwell, Jr., and Berlen C. Moneymaker, Geology in Dam Construction, in *Application of Geology to Engineering Practice* (Berkey Volume), Geol. Soc. Amer., 1950.

4. Arthur Casagrande, Seepage through Dams, *J. New Engl. Water Works Assoc.*, Vol. 51, 1937.

5. G. W. Christians, Asphalt Grouting under Hales Bar Dam, *Eng. News-Record*, Vol. 96, 1926.

6. William P. Creager, Joel D. Justin, and Julian Hinds, *Engineering for Dams*, John Wiley & Sons, New York, 1945.

7. Irving B. Crosby, Engineering Geology Problems at Conchas Dam, New Mexico, *Proc. Am. Soc. Civil Engrs.*, Vol. 65, 1939.

8. *Eng. News-Record*, Unique Cutoff Construction and Arched Foundation Features of Rodriquez Dam, Vol. 105, 1930.

9. *Eng. News-Record*, Thousands of Holes Grouted under Norris Dam, Vol. 115, 1935.

10. Alfred D. Flinn, Arch Dam Investigation, Part IX, Earthquakes, Ice Pressure, and Deterioration of Concrete, *Proc. Am. Soc. Civil Engrs.*, Vol. 54, 1928.

11. F. H. Kellogg, Clay Grouting at Madden Reservoir, *Eng. News-Record*, Vol. 109, 1932.

12. Robert F. Legget, *Geology and Engineering*, McGraw-Hill Book Co., New York, 1939.

13. J. D. Lewin, Grouting with Chemicals, *Eng. News-Record*, August 17, 1939.

14. W. J. Mead, Engineering Geology of Dam Sites, *Second Congr. on Large Dams*, Vol. 4, Washington, 1936.

15. O. E. Meinzer et al., Geology of No. 3 Reservoir Site of Carlsbad Irrigation Project, New Mexico, with Respect to Water-tightness, *U. S. Geol. Survey, Water Supply Paper* 580, 1926.

16. Thomas A. Middlebrooks, Earth Dams, in *Applied Sedimentation*, John Wiley & Sons, New York, 1950.

17. S. B. Morris and C. E. Pearce, Concrete Gravity Dam for Faulted Mountain Area, *Eng. News-Record*, Vol. 113, 1934.

18. *Ohio Water Resources Board*, Sedimentation of Reservoirs in Ohio, *Bull.* 17, Columbus, April, 1948.

19. W. L. Strange, Reservoirs with High Earthen Dams in Western India, *Proc. Inst. Civil Engrs. (London)*, Vol. 132, 1898.

20. Karl Terzaghi, Effect of Minor Geologic Details on Safety of Dams, *Am. Inst. Mining Met. Engrs., Tech. Publ.* 215, 1929.

C H A P T E R   2 0

Tunnels

The construction of tunnels requires the application of geology to a degree rivaled by few other types of civil engineering work. A knowledge of rock and soil conditions is basic not only to selection of the route (provided more than one is possible) but also to choosing the proper construction methods and supports as well. It is also a major factor entering into the estimates of cost which determine whether the project is economically feasible.

Although many tunnels have been completed despite great unforeseen difficulties, the cost was almost invariably higher than it would have been had accurate and complete information been available. For this reason, modern practice is to allocate more funds to preliminary work and less to contingencies which may result from poor or incomplete geologic knowledge. Prior to construction of the tunnels along the Pennsylvania Turnpike, for example, a year was spent in making very complete surveys of the proposed tunnel routes. Exploratory core boring, areal geologic mapping, and other preliminary studies were carried out in such detail that the position of every change of rock was predicted to within a few feet. Faulted zones were discovered, and a zone of water-bearing sand and clay was located months in advance of construction. Acid water conditions were spotted, and suitable provisions were made for protection of the tunnel linings. In one instance the geologic studies disclosed that the type and structural characteristics of the rock lent themselves to the construction of a deep cut instead of a tunnel. Although the initial cost of an open cut was greater, the savings resulting from elimination of tunnel maintenance overbalanced the difference in original cost. The correctness of this decision has been demonstrated since the road was opened.

Although most of the work involved in tunnel investigations is clear-cut structural, stratigraphic, petrographic, and ground water geology, the geologist should know something about tunnel terminology, problems of tunnel support, and related topics in order to present his conclusions in language understandable to engineers.

Tunneling methods and associated engineering problems have been described by Proctor and White,[9,10] and the influence of geologic conditions on tunnel operations has been analyzed by Terzaghi.[14,15] Preparation of this chapter is greatly indebted to these publications.

ENGINEERING CONSIDERATIONS

In tunnel terminology earth pressure is defined as the thrust exerted by cohesionless or plastic materials against the tunnel supports. If the walls are not supported, the material moves into the tunnel and eventually fills the entire bore. Material exerting this type of pressure is known as "earth," and the methods employed in construction are known as earth-tunneling methods. Rock load is the pressure exerted on the roof of the tunnel by the material which tends to fall from the roof. If no support is furnished, the material falls piece by piece, and in time the tunnel roof assumes the shape of an irregular vault. When this stage is reached, the roof is supported by the arching effect of the materials involved. Tunnels in such materials are known as rock tunnels, and are constructed by rock-tunneling methods. Rock loads depend chiefly on the spacing and orientation of joints, whereas earth pressures depend on the physical characteristics of the material surrounding the tunnel.

Tunnel Supports

Very few tunnels require no supports. Even those in chemically intact and non-fractured rock often need support at some

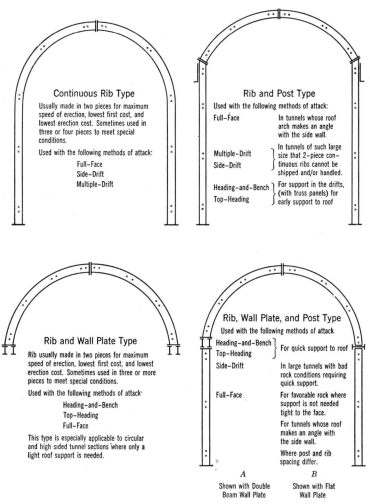

Continuous Rib Type

Usually made in two pieces for maximum speed of erection, lowest first cost, and lowest erection cost. Sometimes used in three or four pieces to meet special conditions.

Used with the following methods of attack:

    Full–Face
    Side–Drift
    Multiple–Drift

Rib and Post Type

Used with the following methods of attack:

Full–Face	In tunnels whose roof arch makes an angle with the side wall.
Multiple–Drift Side–Drift	In tunnels of such large size that 2–piece continuous ribs cannot be shipped and/or handled.
Heading–and–Bench Top–Heading	For support in the drifts, (with truss panels) for early support to roof

Rib and Wall Plate Type

Rib usually made in two pieces for maximum speed of erection, lowest first cost, and lowest erection cost. Sometimes used in three or more pieces to meet special conditions.

Used with the following methods of attack:

    Heading–and–Bench
    Top–Heading
    Full–Face

This type is especially applicable to circular and high sided tunnel sections where only a light roof support is needed.

Rib, Wall Plate, and Post Type

Used with the following methods of attack

Heading–and–Bench Top–Heading	For quick support to roof
Side–Drift	In large tunnels with bad rock conditions requiring quick support.
Full–Face	For favorable rock where support is not needed tight to the face.
	For tunnels whose roof makes an angle with the side wall.
	Where post and rib spacing differ.

A
Shown with Double
Beam Wall Plate

B
Shown with Flat
Wall Plate

Fig. 194. Various types of steel tunnel supports. (After Proctor and White, *Rock Tunneling with Steel Supports,* copyrighted by Commercial Shearing and Stamping Co., 1946.)

place or time during construction. A few decades ago wood was the material most commonly used to support tunnel walls, but the present trend is toward the adoption of steel supports. If the tunnel is lined, the steel ribs are usually allowed to remain and are embedded in the concrete. Invert struts are beams placed between posts at the bottom of the tunnel to resist mild side pressures. If these pressures are severe, full-circle ribs are used. Wall plates serve as sills for the arch ribs. They transmit the load from the ribs onto the rock. In order to prevent material from falling out between the ribs, wood or steel planks known as lagging are placed across the ribs. Shifting and side buckling of ribs and posts are prevented by lateral braces and tie rods. Crown bars are beams placed above the arch ribs parallel to the center line. After blasting, they are slipped forward to support the freshly exposed roof. Truss pannels are temporary supports for the arch ribs while the bottom bench is being taken out. They are replaced by posts after the bench is removed.

Tunneling Methods

In full-face procedure the tunnel cross section is blasted out to full size, each shooting round. In this and other methods the sequence of operations is: (1) drilling, (2) blasting, (3) ventilating, (4) mucking, and (5) erecting supports. Formerly, large-diameter bores were constructed by the heading-and-bench method, but development of a drilling platform, or jumbo, which permits the employment of an adequate number of drillers at two or more levels of the face, has made it possible to use full-face methods in large-diameter tunnels. This method may be used in materials in which the bridge action period (time elapsing between excavation of the tunnel and sloughing of the roof and walls) is long enough to permit ventilating and mucking.

Heading-and-Bench Method. In the heading-and-bench method a top heading is carried forward ahead of the underlying bench for a distance of about 1.5 times the length of one drilling and shooting round. The heading extends for the full width of the tunnel. Since the invention of the jumbo, this method is now used only in situations where the bridge action period of the rock is short and support is required immediately after blasting. Under such conditions it has the following advantages: (1)

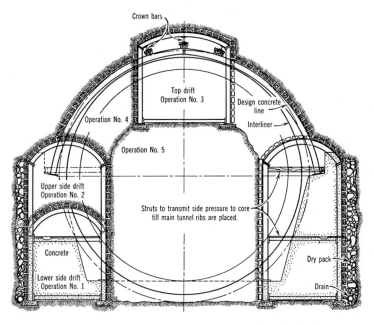

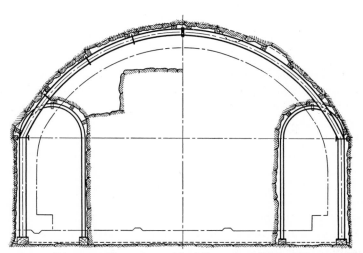

Fig. 195. Diagrams illustrating: side-drift and multiple-drift tunneling methods. (After Proctor and White, *Rock Tunneling with Steel Supports,* copyrighted by Commercial Shearing and Stamping Co., 1946.)

The bench serves as a convenient work platform. Posts can be placed very quickly under rock that threatens to fall, and crown bars can be slid forward immediately before the placing of steel ribs. (2) Mucking can proceed while roof steel is placed in the heading. Both the bench and the heading are shot during each round.

Top-Heading Method. In a variation of this procedure known as the top-heading method the top heading is driven completely through the length of the tunnel during the first phase of operations, whereas the bench is removed during the second phase. This method is useful in situations where bad roof conditions are known to exist for practically the entire length of the tunnel. The chief advantage is that difficulties are overcome in a smaller portion of the bore, and subsequent operations can be carried out at a considerably faster rate. Holes can be drilled vertically for excavation of the bench with resulting powder economy. However, the roof loads are permitted to build up for a longer time, and when the bench is removed more load is transferred to the posts than in the heading-and-bench method.

Side-Drift Method. In large-diameter bores in bad rock the side-drift method is frequently employed. A drift is driven ahead on each side of the tunnel, and posts and wall plates are erected. If the rock conditions permit, the operation may be full face for a short distance back of the drifts, but where conditions are very bad breakups are made to the crown, leaving a central core. Temporary posts may be placed between the core and the roof to support caving blocks, or the crown bars may be slid forward. Roof ribs can then be placed on the wall plates, securely blocked, and the temporary supports can be removed. After this the core may be removed or left in place until after concreting, during which time the core furnishes support to the forms. Under certain conditions it is advantageous to drive the drifts through the entire length of the bore and to use the space opened for construction of the permanent concrete lining below the spring line (line at which the arching of the roof begins). The main tunnel ribs are then installed as an arch on the concrete abutments (see Fig. 195).

Multiple-Drift Method. The multiple-drift method is employed in penetrating bad rock or earth. It is a combination of the side- and top-drift methods. A drift is run at subgrade along each side of the tunnel, and a concrete wall is placed in each drift. If the height of the side walls is too great to permit this to be done in a single operation, a second side drift is run above the first, and the concrete side walls are carried up to the spring line. A top center drift is then driven through with the roof support sufficiently far above the projected position of the main tunnel ribs to provide space for the crown bars over the ribs. A short section of the roof is blocked on the crown bars, and the side posts are removed. The top drift is then widened and connected with the roofs of the side drifts. The main arch ribs are placed on the concrete side walls, lagged, and packed. The crown bars supporting the roof members of the top center drift are securely blocked to the ribs, and the next advance is made.

Other Methods. There are many other methods of attack. For example, it may be desirable to run a bottom drift ahead of the top in order to provide drainage and to explore the rock. A center drift may also be driven to explore the rock, and later it may be widened to the desired diameter. Powder savings may result, for radial shot holes are more efficient than holes normal to the working face.

In tunneling through water-logged clays, silts, sands, and gravels use is often made of the shield method in which the water is held back by air pressure and the face is supported by a heavy steel shield which is forced into the materials to be excavated. The walls are supported by heavy steel which also functions as a part of the permanent lining. In Europe, chemical grouting methods have been successfully employed in tunneling through unconsolidated materials. Water-logged sands and gravels may also be handled by draining them in advance of tunneling operations.

The type and design of supports required in the various tunneling procedures are described in the treatises by Proctor and White [9,10] listed in the references.

At times it may be advantageous to construct a pioneer or pilot tunnel ahead of, or adjacent to, the main drift. Pilot drifts may be very useful in exploring rock conditions and transporting men and equipment to and from the working face, but, owing to their smaller diameter, they may fail to furnish information of value relative to the type of supports required by the main tunnel.

Pressure Tunnels

Tunnels that are subject to both internal and external pressures are known as pressure tunnels. Where the rock cover is light, external pressures are disregarded, and the lining is designed as a reinforced concrete conduit. Where it is appreciable, earth and rock pressures are taken into account in design of the lining. Because rock pressures vary with the type and physical condition of the material, very close attention must be given to geology in planning the linings of pressure tunnels. It should be noted that even when the lining is grouted to the wall rock it cannot be assumed that danger of rupture is eliminated, for if the wall rock is compressible the internal pressure may cause fracturing of the lining.

Overbreak

The inside diameter of the lined tunnel plus twice the thickness of the lining furnishes the minimum diameter of the bore, and the contractor is usually required to remove all earth and rock up to this line, which is known as the neat line. It is common practice to pay for excavation and lining up to a line called the pay line, which is somewhat beyond the neat line. All excavation beyond the pay line is done at the contractor's expense, and the material removed is known as overbreak. In view of the fact that contractors insure themselves against loss by adding their estimate of overbreak quantities to their unit prices, it is to the interest of both the owner and the contractor to arrive at intelligent specifications concerning the pay line. Because overbreak is largely a function of the type and physical condition of the material, it is apparent that estimation of overbreak requires an understanding of the behavior of various types of rock encountered in tunnels.

GEOLOGICAL CONSIDERATIONS

In most long tunnels both earth and rock conditions will be encountered. Earth conditions are met not only in unconsolidated deposits such as sands and gravels but also in soft shales, and even in hard rocks like granite adjacent to fault and shear zones, especially in instances where the fractures have served to localize circulating waters which have altered the rock chemi-

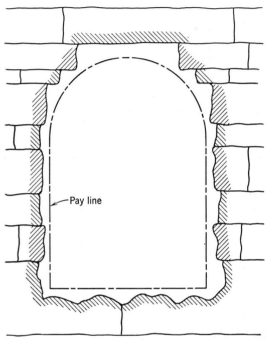

Fig. 196. Diagram illustrating overbreak. (After Terzaghi in *Rock Tunneling with Steel Supports*, copyrighted by Commercial Shearing and Stamping Co., 1946.)

cally. In short, an estimate of rock defects is as important as an accurate appraisal of the rocks themselves. Since the best geologist in the world cannot hope to obtain conclusive and absolutely accurate information regarding many features of importance in advance of construction, Terzaghi[9] recommends that the report should set forth the most favorable and the least favorable conditions anticipated.

Stresses in Intact Rocks

Theoretically, the stresses in the wall rocks decrease rapidly with increasing distance from the walls of the tunnel. At the walls the radial stress acting at right angles to the wall is zero, and the circumferential stress acting in a circumferential direction is approximately equal to twice that which acted at the same point prior to excavation of the tunnel. As the distance from the walls increases the radial stress increases and the circumferential stress decreases. At a distance equal to about the diameter of the tunnel the stress conditions in the rock are practically unaltered. The walls will not fail until the circumferential stress equals the unconfined compressive strength of the rock. In rocks that are not affected by tectonic forces the circumferential stress does not exceed approximately twice the overburden pressure; hence failure by crushing can occur only at considerable depth. In porous sandstone the critical depth is about 5000 feet; it is 35,000 feet for granite and other very strong rocks.[9]

Popping. Popping is the term applied to the sudden detachment of thin slabs from a tunnel or quarry face. Popping is encountered only in hard and brittle rocks. It invariably happens that the detached slabs do not fit the surface from which they separated, indicating that the rock to which the slab belonged is in a state of intense elastic deformation. Popping may be caused by tectonic stresses or by release of pressure near the surface by erosion. Its common occurrence in deep mines suggests that static pressure may also be a cause. In the Simplon Tunnel, where popping was very troublesome, the overburden reached a maximum of 7000 feet. Spalling may take place with no great violence, but in many instances considerable force is involved. Popping is more common from the walls than from the roofs of tunnels. It may be prevented by reasonably strong support and careful backpacking.

Tunnels in Stratified and Metamorphic Rocks

The planes of weakness in stratified rocks consist of bedding planes and joints. Of the latter, the transverse sets, normal to the bedding, are especially troublesome. In schists and gneisses the planes of foliation or schistosity and planes of lineation have the same inherent weakness as the bedding planes of sedimentary rocks. In rocks in which the principal planes of weakness lie horizontal, or nearly so, a bridging effect is present. When the transverse joints are spaced farther apart than the width of the tunnel, the blocks act as solid beams subject only to a bending moment. If the bending moment is less than the flexural strength of the rock the roof is stable; if the bending moment is greater, the roof requires support. Fig. 196 illustrates the advantages derived from an arched roof, where the sides of the arch form corbels that reduce very considerably the free span of the roof slabs.

Overbreak. The chief causes of overbreak are: (1) joint spacing, (2) shattering effect of explosives, (3) the distance between the working face and roof support, and (4) length of time elapsing between removal of natural support and the placement of artificial supports. The closer the joint spacing, the more pronounced is the overbreak. Experience shows, however, that breakage above the top of the pay line will seldom exceed a height of one-half the width of the tunnel. This value determines the maximum load that the supports will have to bear. In highly folded and fractured shales, however, as much as 22 feet of overbreak has occurred. The shattering effect of explosives opens incipient fractures, widens joints and stratification planes, and thus tends to increase the amount of overbreak.

Arch Action. In tunnels where roof support is required, the shorter the linear distance from the supports to the working face, the less rock is likely to fall from the roof after a blasting round has been fired. If tunnel supports are placed and backpacked immediately after shooting, frictional forces acting along the fractures between the roof and the natural vault above the roof transfer a portion of the weight against the rock located beyond the sides of the vault. The vault is the natural shape of the void above the roof that would form with no roof support. Hence the load over an adequately backpacked roof support is likely to be much less than the maximum of one-half the tunnel width. Arch action takes place not only normal to the center line but also in a direction parallel to the last rib support and the tunnel face; hence the unsupported section left at the end of a blasting round constitutes a half dome. A half dome can carry a somewhat heavier load than an arch of the same span, and for this reason the load on the tunnel supports in the immediate vicinity of the face is somewhat less than at some distance back. However, if the half dome is left unsupported for any appreciable time after a blast, the bridging action steadily deteriorates until general collapse occurs. The bridge action period lasts from

Fig. 197. Diagram illustrating formation of a natural arch in moderately jointed rocks. *B* indicates the design width of the tunnel. (After Terzaghi in *Rock Tunneling with Steel Supports,* copyrighted by Commercial Shearing and Stamping Co., 1946.)

the time a blast is fired until the first blocks of rock begin to fall from the roof of their own accord.

When supports are properly erected, blocked, and wedged during the period of bridging action, the rock load on the supports continues to increase for the following reasons: (1) as the tunnel is advanced half-dome action is succeeded by arch action; (2) backpacking and wedging do not cause a halt to all rock adjustments above the roof. As a result of blasting, joints above the roof are opened, some more, some less. The resulting shattering is accompanied by downward movement of the entire roof. Proper supports stop downward movement of the lower portions of the roof rock, but progressive readjustments continue to take place above. As the joints in the lower portion of the shattered mass tend to close, those higher up tend to open. Consequently, the total load on the supports continues to increase until all re-adjustment ceases.

Closer spacing of the roof supports, by reducing the distance from the face to the first support rib, increases the bridge action period; hence, with increasing frequency of joints, or in the case of progressively deteriorating rock, closer spacing of the support ribs is necessary.

In vertically dipping strata where the strike is normal to the tunnel center line, the load depends chiefly on the transverse joints. However, because joint surfaces are very rarely smooth, the friction may be greater than the weight of the blocks, and the roof support required is, therefore, comparatively slight. The more massively bedded the strata, the less support is required. Where massively bedded vertical strata were encountered in the Pennsylvania Turnpike Tunnels, little timbering was necessary. Less satisfactory conditions are encountered where the strata strike parallel to the center line of the tunnel, despite the fact that under these conditions a more satisfactory bridge action is afforded between the working face and the nearest roof supports. Under these conditions the roof supports may be expected to carry the weight of the volume of rock loosened and shattered by the explosives.

Under certain conditions, usually resulting from weathering or ground water circulation, the joint surface may be lubricated by clay films. In this event, the joint blocks are not supported by friction, and the weight on the supports may be very great. In Chapter 6 it was mentioned that this condition was encountered in one of the tunnels of the Los Angeles-Colorado River Aqueduct. When a foreman spotted the yellow color of clay in the muck coming out of the tunnel, he started for the face, knowing there was trouble ahead.

In tunnels where there are inclined strata overbreak tends to produce an irregularly peaked roof, and blocks tend to slide into the tunnel along downsloping bedding planes (see Fig. 198). Eccentric loads are imposed on the supports, which must be properly backpacked and designed to carry such loads.

Fig. 198. Diagram illustrating overbreak in inclined strata. (After Ter-zaghi in *Rock Tunneling with Steel Supports*, copyrighted by Commercial Shearing and Stamping Co., 1946.)

Tunnels in Massive Rocks

In moderately jointed massive rocks, such as granite or dia-base, the overbreak and loads are generally similar to those in stratified rocks. A vault-shaped roof develops much the same as in stratified rocks, but usually the blocks located between joints (unless the joints are slickensided or weathered) are so intimately interlocked that they do not fall from place. Usually, under these conditions the roof is the only place where blocks are likely to become detached.

Tunnels in Crushed Rock

When thoroughly competent rock, such as quartzite or granite, has been subjected to intense deformation (shearing in a fault zone, for example) the fracturing may be so extreme that the rock loses the capacity to form an unsupported vault. In the Kittatinny Tunnel of the Pennsylvania Turnpike the rock locally was reduced to loose sand and powder. However, according to Terzaghi,[9] even in sand and totally crushed but chemically intact rock the load on the roof supports will not exceed a small fraction of the weight of the rock above the roof. Furthermore, if the depth of the overburden is greater than approximately 1.5

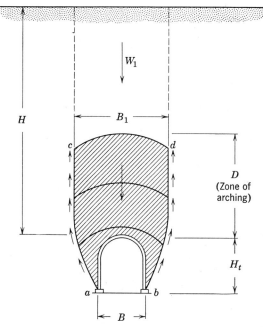

Fig. 199. Diagram illustrating ground-arch action. (After Terzaghi in *Rock Tunneling with Steel Supports,* copyrighted by Commercial Shearing and Stamping Co., 1946.)

times the combined width and height of the tunnel, the rock load is practically independent of depth. The cause of this condition is known as ground-arch action. Ground-arch action indicates the capacity of the material above a tunnel roof to transfer most of the total weight of the overburden onto the rock on both sides of the tunnel. The body of rock which transfers the load is known as the ground arch. The mechanics of ground-arch action are illustrated in Fig. 199. Model tests indicate that the ground arch (indicated by the shaded area) is largely supported by friction along the faces *a–c* and *b–d*. The pressure on the roof of the tunnel, H_p, has been found to conform closely to the equation below, where C is a constant the value of which depends on the compactness of the material and the distance which the crown of the ground arch subsided before the roof supports were installed. Other factors being equal, the value of H_p is proportional to the width of the ground arch.

$$H_p = C \times B_1$$

When the tunnel is below the water table, and the crushed rock or sand is water-saturated, the load exerted by entrance of water into the material approximately doubles the height of the layer exerting the load on the roof. The arching action is not destroyed, however. When ground water seeps into a tunnel through the floor, the bearing capacity is greatly reduced from the capacity in the dry state. In such tunnels the support of the footings of the roof supports is difficult. The working face may require breasting, and rock-tunneling methods may have to be supplanted by methods used in water-bearing sands.

Rock Loads on Tunnel Supports in Blocky and Seamy Rock

In tunnel language, the terms blocky and seamy imply that the blocks of rock separated by joints are not interlocked or firmly held together. The joints may be open or filled with decomposed material. With random orientation of the joints, the roof load may be associated with a horizontal pressure on the sides of the tunnel support. In these materials, the load H_p on the roof supports in tunnels of considerable depth is independent of depth and increases in direct proportion to the sum of the width B and the height H_t of the tunnel. Therefore, if H_{p10} is the load on the roof of the tunnel with a width and height of 10 feet, the corresponding load on the roof with any width of B feet and any height of H_t feet through the same rock is:

$$H_p = H_{p10} \left(\frac{B + H_t}{20} \right)$$

Values of H_{p10} are listed by Terzaghi.[9]

Clays and Decomposed Rock in Tunnels

As used in tunnel work, the term clay applies not only to the sedimentary deposit but to decomposed rock of any kind in which significant clay development has taken place. Under certain conditions a rock may be completely converted into clay, but more commonly alteration is limited to certain mineral constituents only. Transformation may occur throughout the entire mass or only along fissures. However, the altered rock has entirely different physical properties than the parent material, and it commonly reacts to stresses much like clay. As in other types of materials, clay is characterized by ground arching and bridge action. However, development of the ground arch in clay and decomposed rock is accompanied and followed by phenomena that are absent in other types of materials. In clay and decomposed rock the bridge action period is commonly longer than in crushed rocks, and breasting is rarely required. However, the load on the supports is likely to increase for a period of weeks and even months. Tunneling conditions in clay and decomposed rock have been studied by Terzaghi,[9] who has reached the following conclusions.

The resemblance between tunneling conditions in clay and in decomposed rock is caused by the low permeability, high compressibility, and other peculiar properties of clay minerals. Although particles of these minerals may have the same size and shape, their physical properties are very different. Two sands of the same particle size are very much alike in physical properties, but two clay samples of equal grain size may have little in common. The montmorillonites (especially bentonite) are the worst troublemakers because of their tendency to absorb water with an increase of up to 10 times the original volume. Hydration pressures of montmorillonite may be as high as 10 tons per square foot. In tunnel work the important physical properties of clay are swelling associated with removal of pressure, the relationship between pressure and shearing resistance, the rate at which the material reacts to changes in stress conditions, and volume changes resulting from hydration of montmorillonite minerals.

Swelling Due to Removal of Load. If a clay is subjected to pressure, water is squeezed from the voids and a decrease in volume takes place. Terzaghi [9] has shown that, if the pressure on a consolidated clay is decreased and a source of water provided, the material will increase in volume. For any specified material, the increase in volume resulting from the removal of load increases with the intensity of the load under which the material was consolidated. Hence, clays that have been consolidated under a great thickness of overlying strata (which may have been subsequently removed completely or partially by erosion) will expand more than those consolidated under more moderate loads. Water is drawn from the surrounding material, which is under greater pressure, and fills the voids of the lower-pressure areas. It is commonly believed that the water responsible for the softening of clays in tunnels is drawn from the moist tunnel atmosphere, but this opinion is erroneous. The moisture is drawn from materials located at some distance from the tunnel walls. This conclusion has been proved by experiments in which samples taken from the heading of swelling ground were exposed to the atmosphere of the tunnel for several days, during which time they dried out rather than softened. In the same tunnel the water content of the clay 15 feet from the walls was found to be well below that for the same rock before tunneling operations.[9,15] Although any clay will swell upon removal of load, materials containing a high percentage of montmorillonite are the most susceptible.

Relationship between Pressure and Shearing Resistance. In sands and other granular materials the effect of pressure is to increase the shearing resistance. In clays, however, the effect of pressure on shearing strength is less marked. The decrease of pressure on a clay is associated with a small increase in the water content. Because of the low permeability, even a sudden decrease of pressure produces only a slight increase in moisture content, and only a slight decrease in shearing resistance. The resistance will show very little change for hours or even days. However, after a certain point is reached the shearing resistance decreases rapidly until it reaches a new value, representing ultimate cohesion. Because of the effect of time, the load on tunnel supports always increases considerably but at a decreasing rate. Experience indicates that the load-increase period ranges from several weeks to many months.

Slaking of Tunnel Walls and Bridge Action Period. Many stiff clays may contain a network of closely spaced hair cracks. If an unsupported column of this material is loaded it disintegrates into small fragments. In tunnel terminology this process, called slaking, is often wrongly attributed to the effects of exposure to the air. As soon as the clay above the roof begins to slake it loses what little strength it had and masses begin to drop from the roof. The time elapsing until the roof disintegration starts is the bridge-action period and ranges from several hours to several days. Roof supports must be installed before the bridge-action period expires. In soft clays, the term bridge-action period has no meaning. Such materials do not disintegrate, they merely squeeze.

Squeezing and Swelling Ground in Tunnels. Squeezing ground is material that contains a large amount of clay. This may be present originally as clay shale, or it may be altered materials. The rock may be mechanically intact, jointed, or crushed. The clay fraction may be dominated by the less offensive members of the kaolin group, or it may be dominated by the montmorillonites. Hence the properties of squeezing ground may vary through much the same range as clay.

If the squeeze of a material is mainly due to swelling, the material is called swelling ground. The bridge action period of swelling ground depends on the same factors as for ordinary squeezing materials. Failure of tunnel supports is associated with an almost instantaneous relaxation of pressure. If a new support is placed, the pressure once more increases, but the ultimate intensity is less than in the first instance. It is a common occurrence that ribs capable of withstanding the pressure of swelling rock gradually penetrate the material with which they are in contact. At the same time the bottom of the tunnel heaves and displaces tracks and conduits, and it is advisable to use full-circle ribs.

In squeezing ground, crushing combined with chemical alteration may produce a material comparable with river silt. The pressure phenomena in some decomposed rocks, both igneous and metamorphic, are similar to those found in stiff swelling clay. When squeezing ground is so soft that it does not require blasting, the same methods must be used as are commonly employed in tunneling through clay of similar consistency. When a tunnel is being made through decomposed rock, the construction difficulties generally increase with the swelling capacity of the material.

In moderately jointed and seamy rock, careful backpacking reduces the load-increase period and the ultimate pressure on the tunnel supports. Even in squeezing ground with a low swelling capacity, the presence of open spaces between the supports and wall rock is likely to do more harm than good. Providing an opportunity for the material to squeeze into the tunnel should be adopted only as a last resort, i.e., if the tendency to swell is very marked. Therefore, before a material is considered as swelling ground its swelling properties should be carefully investigated. Terzaghi [9] recommends that only when samples of freshly exposed rock taken from the face or cores show an increase of over 2 per cent in volume during immersion in water should the material be considered to have swelling properties. Unless the material is known to possess swelling properties, the tunnel supports should be tightly wedged.

When anhydrite is encountered in tunnel workings it should be treated with great respect, because hydration of this material to gypsum involves a 30 per cent volume increase. In nature, anhydrite occurs only between impervious strata; hence, when encountered, it should be sealed off as quickly as possible. Bentonite should also be protected from wetting to the greatest possible degree.

Summary of Definitions of Rock Conditions

Intact rock contains neither joints nor hair cracks. Consequently, when fractured it breaks across sound rock. Owing to the effect of blasting, the roof and walls may spall for hours or days after a round has been fired. Popping conditions may be encountered, involving the spontaneous and often violent detachment of rock fragments from the roof and walls.

Stratified rocks consist of individual strata that possess the tendency to separate along the bedding planes. In metamorphic rocks the planes of lineation or schistosity may have the same characteristics as stratification. The strata may or may not be weakened by transverse joints more or less normal to the bedding or schistosity. Spalling conditions are common.

Moderately jointed rock possesses joints and hair cracks, but the blocks between joints are so interlocked or secondarily cemented that vertical tunnel walls do not require support. Both spalling and popping may be encountered.

Blocky and seamy rock consists of chemically unaltered rock fragments, either entirely separate from each other or only imperfectly interlocked. In this type of material vertical tunnel walls may require support.

Crushed, but chemically intact, rock has the characteristics of crushed stone. When the majority of the constituent particles are as fine as small sand grains and no secondary cementation has occurred, crushed rock below the water table has the characteristics of water-bearing sand.

Squeezing ground advances slowly into the tunnel with no perceptible increase in volume. The primary requisites for squeezing are high proportions of microscopic and submicroscopic particles of micaceous and clayey minerals with low swelling capacity. All parts of the tunnel require support, and full-circle ribs may be necessary.

Swelling ground advances into a tunnel principally because of volume expansion. The capacity to swell is chiefly limited to those materials which contain clay minerals of the montmorillonite group that have a high swelling capacity. All parts of the tunnel require support.

If a rock formation consists of a sequence of horizontal layers of sandstone or limestone and of immature shale, the excavation of the tunnel is commonly associated with a gradual compression of the rock on both sides of the tunnel, involving a downward movement of the roof. Furthermore, the relatively low resistance against slippage at the boundaries between the so-called shale and rock is likely to reduce very considerably the bridging capacity of the rock located above the roof. Hence, in such rock formations, the roof pressure may be as heavy as in a very blocky and seamy rock.

Tunnels in Soft Ground

Terzaghi [15] distinguishes between the following types of soft ground: firm, raveling, running, flowing, squeezing, and swelling. The last two have already been discussed, and it is only necessary to review Terzaghi's definitions of the remaining categories.

Firm Ground. Materials in which a tunnel can be advanced without any roof support, and the permanent lining can be constructed before the walls begin to move, are known as firm ground. Loess above the water table and calcareous clays of low plasticity, known as marls, are good examples of firm ground.

Raveling Ground. The materials known as raveling ground flake or spall some time after they are exposed. This process may begin in a few minutes, in which case the materials are known as fast-raveling. Fast-raveling materials may be encountered in residual soils and in sand with a clay binder below the water table. Above the water table, they may be slow-raveling or firm.

Running Ground. The materials known as running ground flow on slopes greater than 34° to the horizontal, and flow continues until the slope becomes about equal to the limiting value of 34°. Such conditions are generally found in clean loose gravel and clean coarse-to-medium sand above the water table.

Flowing Ground. The materials known as flowing ground move like viscous liquids. They may invade the tunnel not only from the roof and sides but through the floor as well. A rush of flowing ground into a tunnel is known as a "blow." Flowing conditions prevail in any material with an effective grain size (10 per cent of the particles are finer and 90 per cent are coarser than the effective grain size) in excess of about 0.005 mm when encountered below the water table. Above the water table, the same ground acts as raveling or running material.

Tunneling Methods in Soft Ground. The methods used in tunneling through soft ground have been well described by Terzaghi,[15] and no extended discussion is needed here, except to point out the importance of the position of the water table. Water saturation may change a material from firm to flowing ground, and if a continuous layer of water-saturated sand is encountered a dangerous blow may result. Therefore, in exploration for tunnels in soft ground it is absolutely necessary to have undisturbed samples of the materials and accurate determinations of the position of the water table. In clean sands and

Fig. 200. Flowing ground in the Kittatinny Tunnel, Pennsylvania Turnpike, caused by grinding of quartzite by earth movements along a fault. This seam of water-bearing sand was located in advance and deliberately broken into and allowed to flow. (Pennsylvania Turnpike Commission.)

gravels the water table may be determined by long-term observations of the water level in bore-holes, but in silty sands and silts it is necessary to install in the bore-holes piezometer tubes with diameters less than ½ inch When the materials consist of fine silt or clay, no accurate information on the water table is necessary, for their performance depends only on their water content and consistency.

TUNNEL HAZARDS

Terzaghi [9] uses the term "tunnel hazard" to indicate an un-anticipated source of expense and delay. Hazards are usually caused by the departure in behavior of the rock in question from the statistical average of similar materials. Such departures may be the cause of either higher or lower costs than those antici-pated.

Limestone and Sandstone. Above the water table limestone and sandstone rocks present few difficulties. Below the water table limestone may contain large quantities of water in solution cavities, which when opened give rise to heavy concentrated flows. Crushed zones in limestone are commonly narrow. Where the rock is not crushed, the rock load is always moderate. CO_2 and H_2S are uncommon, except in regions of volcanic activity.

Shale. The character of shale may range all the way from sound rock to swelling clay. However, in any given material tunneling conditions are likely to be rather uniform. The rock load is either consistently high or low. Usually the influx of water is small, but large flows may be encountered where a porous formation rests on shale. Shales may be associated with coal and anhydrite. In coal-bearing shales the explosive gas methane (CH_4) may be encountered. Unless protected from water, anhydrite swells. Water in contact with anhydrite contains calcium sulphate which is injurious to concrete. It may also contain H_2S which is lethal even in moderate concentrations. H_2S may be especially abundant if, as frequently happens, the rock contains appreciable quantities of pyrite and marcasite. Oxidation of these minerals and associated organic matter gives rise to H_2S, sulphuric acid, and considerable heat. Tunnels in pyritiferous shales are often abnormally hot, and the water may be acidic. Acidic water is deleterious to ordinary concrete. In general, shale should be considered hazardous, especially when the overburden is considerable.

Schist. Unaltered schists are characterized by negligible to moderate rock loads. In general, the hazards are rather slight. In chemically altered schists, however, squeezing and swelling conditions may be encountered. These conditions are associated with very high rock loads. Water discharge may be very high, especially in altered materials.

Intrusive Rocks. In chemically intact intrusive igneous rocks the hazards are very slight. However, where such rocks are

Fig. 201. Roof failure in West Allegheny Tunnel, Pennsylvania Turnpike, caused by settlement of temporary wall plate sills on a soft clay. (Pennsylvania Turnpike Commission.)

altered chemically, squeezing and swelling conditions may be encountered as in the Moffat Tunnel.[8] In many instances the existence of abnormal conditions cannot be detected from the surface. Consequently, these rocks should be considered treacherous.

Extrusive Rocks. Extrusive rocks, such as rhyolite, are commonly associated with tuffs and breccias. These materials may be in an advance stage of decomposition, or even in an unconsolidated state. Montmorillonites are commonly associated with altered tuffs. Large quantities of water may be present in porous extrusives such as basalt, and in young extrusives harmful gases such as CO_2 and CO may be encountered. In general, extrusive rocks should be considered extremely hazardous. The experience with the Mono Craters Tunnel, California, may be regarded as typical.

Influence of Geologic Structure. In tunneling through synclines (either parallel or normal to the axis) water troubles are likely to be encountered wherever porous beds are found, and artesian conditions may also be encountered. Furthermore, the joint blocks tend to form inverted keystones and thus require immediate and heavy support. In anticlines the blocks tend to form normal keystones and thus are more nearly self supporting. Anticlines are less likely to cause water trouble, especially where overlain by an impervious bed. If the rock is intersected by numerous open fissures, attention should be given to surface water conditions in the vicinity of the tunnel line. For example, in the Moffat Tunnel a lake 1400 feet above the tunnel drained into the workings through open joints and fissures.

Faults are a common source of difficulty in tunneling. Fault hazards are proportional to the amount of mechanical and chemical alteration attending the movement rather than to the amount of movement that has taken place. Although it is usually possible to detect the presence of faults in advance of tunneling operations, it is much more difficult to predict the width of the zone affected and the hazards involved. If possible, tunnels should always be driven at right angles to a fault; parallel alignments should be avoided. Among the tunnels that have encountered faults are the Rondout Tunnel, New York City-Catskill Aqueduct;[1] the Simplon Tunnel in the Alps;[4] the Moffat Tunnel, Colorado;[6,8] and the San Jacinto Tunnel of the Los Angeles-Colorado River Aqueduct. The last is of exceptional interest because of the water troubles associated with the fault zone. Large open fissures were encountered which gave rise to veritable underground rivers, and so much water was pumped from the tunnel that the water table was considerably lowered. After extensive litigation substantial damages were recovered by neighboring ranch owners. Water troubles are more common in faulted brittle rocks than in softer materials.

Methods of Coping with Tunnel Hazards. The methods of coping with tunnel hazards are generally improvised on the spot;

they range all the way from the simple expedient of allowing squeezing material to move into the tunnel, where it is removed at periodic intervals until the movements subside, to the traveling cantilever device used in the heavy ground of the Moffat Tunnel. However, it is easier and cheaper to solve the problems if they have been anticipated to some extent and a suitable assortment of equipment is on hand, such as large-capacity pumps for handling water, heavy steel supports, and core drilling rigs. Under very bad conditions it may be necessary to core-drill ahead of the face or drive a small pilot bore to explore rock conditions.

The most effective way to meet tunnel hazards is to diagnose the cause of the trouble. If this is done successfully, the correct treatment will suggest itself and futile expedients will be avoided. For example, it is useless to spray squeezing shale with cement (gunnite), since, as has been previously indicated, the instability of such materials is not caused by exposure to the air. However, swelling ground containing anhydrite or bentonite can, and should be, treated with gunnite. Gunnite may be effective in checking water flows from minute fractures under rather low hydraulic head, but it is futile if the fissures are wide and the water is under high hydraulic head. The most effective way to treat important water-bearing fissures is to grout them off in advance of tunneling. This can be done by drilling ahead horizontally from the face, a technique that was successfully employed in the San Jacinto Tunnel. As the above discussion indicates, tunnel hazards can be met most effectively by close co-operation between the geologist and engineers. The geologist must not only explore the tunnel route in advance of construction but also map the geology as exposed during the work. As Terzaghi [9] has pointed out, in this way his capacity to make successful predictions constantly improves, and one or two accurate predictions may more than cover the expense involved in retaining a geologist for all phases of the work.

GEOLOGICAL EXAMINATION OF TUNNEL ROUTES

As in other phases of engineering work, the basis of geological examination of rock tunnel routes is an accurate and detailed geologic map. In no other phase of engineering work is a thorough surface geologic study more vital, for deep borings are usually too expensive to permit the use of an adequate number to secure complete subsurface information. However, if properly selected and spotted at vital points, deep core borings may justify their cost. In important projects the application of one or more of the geophysical methods described in Chapter 18 may be advisable. Usually, there is little choice in alignment, but so far as conditions permit the geologist should indicate the most favorable alignment practicable.

During construction the geologist should map the geology exposed in the tunnel walls on a scale sufficient to show all details. This information may not only be valuable in overcoming hazards, but it may also be useful in planning the lining and in maintenance work.

Soft-Ground Tunnels. In constructing soft-ground tunnels adequate soil maps are seldom available, and detailed subsurface exploration is imperative. As a rule, tunnels in soft ground are not so deep below the surface as rock tunnels, and a larger number of borings are, therefore, feasible. The chief functions of the geologist are to aid in locating borings and in interpreting the results. In many instances the methods of aerial photographic interpretation of soils described in Chapter 16 can be used to great advantage in locating borings; in the interpretation of boring data the geologist can bring to bear his knowledge of the processes by which the materials were formed. As Terzaghi[15] has demonstrated, the geologist engaged in this type of work should be conversant with the principles of soil mechanics.

REFERENCES

1. C. P. Berkey and J. F. Sanborn, Engineering Geology of the Catskill Water Supply, *Trans. Am. Soc. Civil Engrs.*, Vol. 86, 1923.
2. J. V. Davies, The Astoria Tunnel under the East River for Gas Distribution in New York City, *Trans. Am. Soc. Civil Engrs.*, Vol. 80, 1916.
3. Frank F. Fahlquist and C. P. Berkey, Geology of the Quabbin Aqueduct and Reservoir in Central Massachusetts, *Pub. Doc. 147, Common. Mass.*, November 30, 1935.
4. F. Fox, The Simplon Tunnel, *Proc. Inst. Civil Engrs. (London)*, Vol. 168, 1907.

5. G. Haskins, The Construction, Testing and Strengthening of a Pressure Tunnel for Water Supply of Sydney, N.S.W., *Proc. Inst. Civil Engrs. (London)*, Vol. 234, 1932.

6. R. H. Keays, Construction Methods on the Moffat Tunnel, *Trans. Am. Soc. Civil Engrs.*, Vol. 92, 1928.

7. Robert F. Legget, *Geology and Engineering*, McGraw-Hill Book Co., New York, 1939.

8. T. S. Lovering, Geology of the Moffat Tunnel, Colorado, *Trans. Am. Inst. Mining Met. Engrs.*, Vol. 76, 1928.

9. R. V. Proctor and T. L. White, *Rock Tunneling with Steel Supports*, The Commercial Shearing and Stamping Co., Youngstown, Ohio, 1946. (Contains a discussion of geologic and soil conditions by Karl Terzaghi, also published separately as item 14 cited below.)

10. R. V. Proctor and T. L. White, *Earth Tunneling with Steel Supports*, The Commercial Shearing and Stamping Co., Youngstown, Ohio. (Section I consists of *Principles of Earth Tunneling* by Karl Terzaghi.) In preparation.

11. James Sanborn, Engineering Geology in the Design and Construction of Tunnels, in *Application of Geology to Engineering Practice* (Berkey Volume), Geol. Soc. Amer., 1950.

12. Karl Terzaghi, Shield Tunnels of the Chicago Subway, *Jour. Boston Soc. Civil Engrs.*, Vol. 29, 1942.

13. Karl Terzaghi, Liner-plate Tunnels on the Chicago (Ill.) Subway, *Trans. Am. Soc. Civil Engrs.*, Vol. 108, 1943.

14. Karl Terzaghi, Rock Defects and Loads on Tunnel Supports, *Harvard Univ. Graduate School of Engineering, Bull.* 418, 1946.

15. Karl Terzaghi, Geologic Aspects of Soft-Ground Tunneling, in *Applied Sedimentation*, John Wiley & Sons, New York, 1950.

16. Lazarus White, *The Catskill Aqueduct*, John Wiley & Sons, New York, 1913.

CHAPTER 21

Highways and Airfields

From an engineering point of view, highways and airfields have much in common, and the applications of geology to these two types of construction are also essentially alike. Therefore, highways and airfields are discussed together. Geologists have been employed in highway work for many years, and since the early 1940's there has been a growing appreciation of the value of geology in airfield work. The methods of aerial photographic interpretation of soils described in Chapter 16 are finding increasing acceptance in airfield site selection and highway alignment; and soil mechanics, some of the basic principles of which are discussed in Chapter 17, is now generally regarded as indispensable in highway and airfield construction. It follows that the geologist employed in this type of work must know something about soil mechanics and the methods used in the design and construction of pavements. The engineer engaged in this type of construction may also benefit from a knowledge of geology. Chapters 6 through 12 contain geological information bearing on highway and airfield construction, and the following chapter discusses concrete aggregates. The chief purpose of this chapter is to correlate engineering principles applying to highways and airfields with some of the more important geological considerations bearing on these types of construction.

Geology is obviously useful in prospecting for construction materials, and it is also important in selecting airfield sites and highway alignments. It also has numerous applications to drainage problems, determination of stable slopes for cuts and fills, selection and processing of fill materials, location of borrow pits, classification of materials to be excavated, slide studies, and subsurface exploration. Most of these subjects are not susceptible to generalization, and in a short chapter it is not possible to give a really adequate idea of all the problems and complexities that may arise in practice. This chapter will serve its purpose if it gives the engineer a better appreciation of the uses of geology in airfield and highway work and introduces the geologist to some of the basic engineering procedures used in highway and airfield construction.

PAVEMENT DESIGN

Proper design of pavement is, of course, basic in highway and airfield construction, but until the 1930's pavement design was generally by rule of thumb. Heavier highway traffic and the rapidly increasing weights of aircraft have imposed serious design problems, and standardized methods of pavement design have been developed to meet these problems. Research pointed toward improvement of existing design methods, or the development of better methods, is being carried out by various organizations, and considerable modifications of present practices are to be expected.

Design of Rigid Pavements

The design of rigid pavements is generally based on the value of the subgrade modulus K and the thickness and flexural strength of the concrete used for the pavement. The subgrade modulus may be defined as the ratio of unit load to subgrade deflection. For example, if a load of 10 pounds per square inch causes a deflection of 0.05 inch, the value K is 10/0.05, or 200 pounds per inch. In the procedure used by the Corps of Engineers, U. S. Army, the subgrade modulus is determined by field loading tests in which a circular plate, 30 inches in diameter, is first seated by a load of 1 pound per square inch applied in 30 seconds, immediately released, and the zero readings of the dial gauges recorded. A load of 7070 pounds is then applied in 10 seconds and maintained until complete deformation of the subgrade has occurred. The deflection is measured and recorded, and the uncorrected value of K is found from the formula, $K_u = 10/d$, where d is the total movement which took place during application of the load. Consolidation tests are made on representative samples of the material in both an undisturbed and a remolded condition. One set of samples is tested at the natural moisture content and the other at the saturation point. The results are plotted and used in the determination of the corrected value of K from the equation, $K = d/d_s \times K_u$, where d is the deformation of the specimen at natural moisture content and d_s the deformation of the saturated specimen, both under a load of 10 pounds per square inch. The thickness of pavement required for a given value of K, loading condition, and concrete flexural strength may either be calculated or found from curves constructed for this purpose.

The above procedure, or the various modifications introduced by other organizations, give generally satisfactory results, but it should be added that spacing and design of joints are perhaps the most difficult part of rigid pavement design. Poorly designed joints may crack and spall; and, unless adequately sealed, any type of joint will permit water to enter the subgrade, where it causes loss of bearing strength and may also lead to pumping of fine soil particles through joints and cracks as the pavement is deflected by the passage of heavy loads. With continued pumping, cavities are formed under the pavement and cause it eventually to fail. In brief, the design of rigid pavements largely resolves itself into problems associated with joints and joint sealing materials.

Design of Flexible Pavements

Flexible pavements differ from rigid types in that the load is not distributed over the subgrade by the beam action of the pavement itself. Consequently, to prevent failure of the subgrade, a greater or smaller thickness of crushed rock or gravel base course is usually required in order to spread the load. The design of flexible pavements is not as standardized as that of rigid types. Present designs used in the United States and Canada may be considered as falling into the following rather broad groups, based on the chief factor, or factors, employed: (1) soil profile; (2) mechanical analyses and Atterberg limits; (3) triaxial test; (4) plate bearing test; (5) California Bearing Ratio test; and (6) other more or less specialized tests such as the stabilometer test used in California. This is not the place to discuss the relative merits of the methods, but it might be observed that the method used by the highway department of a given state gen-

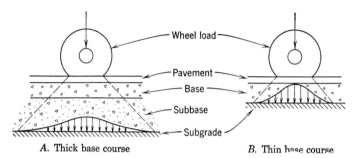

Fig. 202. Diagram illustrating spreading of the load transmitted to the subgrade by a granular base course. (After Corps of Engineers, U. S. Army.)

erally incorporates a vast amount of experience with the soils, climate, etc., of the state and, therefore, furnishes a logical reference point for highway and airfield design in that region. Local conditions may, of course, require special measures, but regional practices should not be disregarded without reason. Neither should practices suitable for a given region be extended into another, unless it can be shown that all pertinent factors are essentially comparable.

Methods Based on the Soil Profile. The Michigan State Highway Department [10] has developed a standard pavement section for use throughout the state, and design is largely directed toward improving the subgrade in order to permit use of the standard section. Pavement performance has been correlated with pedology (see Chapter 6), and the design recommendations are based on the pedological classification of the soil. The uses of geology in this method should be obvious.

Methods Based on Mechanical Analyses and Atterberg Limits. The Civil Aeronautics Administration has developed a method based on mechanical analyses, the liquid limit, and plasticity index of the subgrade soils, with allowances for drainage and frost conditions. Soils are classified into 13 groups, one of which is considered to be unsuitable, which when correlated with drainage and frost conditions are subdivided into various subgrade classes, R1a, F1a, etc. Design charts have been constructed correlating the subgrade soil classes with thicknesses of base courses and pavements required for various types of airfield facilities under various conditions of loading.[8]

The so-called group index method, developed by the Highway Research Board, is used in many states. The group index is calculated from the grain size, liquid limit, and plasticity index of the subgrade soil and ranges in numerical value from a fraction of 1 to 20. Fractional indices indicate "good" subgrade soils; a value of 20 indicates "very poor" materials. Thickness of pavement has been correlated with group indices on the basis of service records.

Methods Based on Triaxial Tests. The Kansas Highway Commission has developed a procedure for flexible pavement design based on triaxial compression tests of subgrade soils. The samples are tested in a moisture-saturated condition, but for regions of low rainfall allowances are made by introducing "saturation coefficients." The pavement and base course thicknesses required for a given volume of traffic with wheel loads not exceeding the legal limit for the state are calculated from formulae. The State of Texas also employs the triaxial test in pavement design.

Methods Based on Plate Bearing Tests. One of the better-known applications of plate bearing tests to flexible pavement design is that developed by the Canadian Department of Transport. It is based on investigations of Canadian airfields, during which repetitive bearing tests were made using plates 12, 18, 24, 30, 36, and 42 inches in diameter. The loads were applied to give deflections of approximately 0.05, 0.2, and 0.5 inches. For flexible pavements, the critical deflection was considered to be 0.5 inch, and for rigid types 0.05 inch. Deflections were plotted against repetition of loading, and it was decided that for capacity operation a deflection of 0.5 inch for 10 repetitions of applied load should be the criterion for flexible pavement design. An equation was developed relating pavement and base course thickness to wheel loads and subgrade support. A set of design curves was then constructed showing thickness of pavement and base course required for a given wheel load with a specified contact area. The combined thickness of pavement and base course may be somewhat reduced by taking into consideration the comparatively greater strength of the asphaltic concrete.[8]

California Bearing Ratio Method. The California Bearing Ratio method, usually shortened to CBR, is used to design flex-

Fig. 203. Truck-mounted CBR apparatus for taking in-place measurements. Piston is operated by the hand crank; load readings are registered by the dial inside the proving ring; penetration is registered by the dial resting on the beam to the right of the surcharge weights. (Corps of Engineers, U. S. Army.)

ible pavements constructed by the Corps of Engineers, U. S. Army; in more or less modified forms it is also employed by the highway departments of several states. In the procedure developed by the Corps of Engineers, representative samples of subgrade soils are compacted in molds 6 inches in diameter to a density corresponding to that which should be obtained by good construction methods. The upper surface of the specimen is loaded with a weight corresponding to that of the base course and pavement and immersed in water for 4 days. The specimen is then removed from the water and allowed to drain for a period of 15 minutes. The surcharge weight is then replaced, and the penetration resistance of the material is determined by forcing a circular piston slightly under 2 inches in diameter into the specimen at a rate of 0.05 inch per minute. Load readings are recorded at deformations of 0.025, 0.050, 0.075, 0.1, 0.2, 0.3, 0.4, and 0.5 inch. The moisture content of the material is then determined and the CBR value found as described in Chapter 17. The CBR value selected is usually that at 0.1-inch penetration, but if that at 0.2 inch is greater it is the value selected. The thickness of base course and pavement required for a specified loading condition is found from curves especially constructed for that purpose.[8]

The CBR value of any given soil depends on the type of compaction employed, the density of the compacted material, and the water content at which the specimen was molded. Consequently, in any given construction problem families of CBR curves showing the effects of these variables must be run. The best construction procedures and the proper thickness of base course and pavement can then be determined. The effect of increased moisture content is to lower the CBR, and this effect is usually the most pronounced in fine-grained cohesive soils where a change of a few per cent in moisture content may cause a serious loss in CBR value. It is largely for this reason that the 4-day soaking period is required for the test specimens. Clean well-graded gravels and crushed rock have the highest CBR values, often reaching 60–80 per cent or more. Poorly graded gravels and gravel and sand mixtures with little or no fines have values ranging from 25 to 60 per cent; silty and clayey gravels have values ranging from 20 to 80 per cent. Well-graded sands or gravelly sands with little or no fines have values of 20 to 40 per cent; silty and clayey sands range from 10 to 40 per cent; silts and clays generally range from about 5 to 15 per cent; some highly plastic clays may have values as low as 3 per cent.[5] Some glacial clays and silts may have, when fairly dry, CBR values as high as 50 per cent.

The CBR method of design has been criticized as leading to excessive base course and pavement thicknesses, but a comparison of airfields constructed by the plate bearing and CBR methods indicates that the charge has not been sufficiently established.

The CBR method is admittedly conservative, but the rapidly increasing use and heavier loading of highway and airfield pavements apparently justifies conservatism. The most conservative feature is undoubtedly the 4-day soaking period leading to water saturation of the test specimens. In other words, the CBR method of design generally assumes eventual saturation of the subgrade, although in instances where saturation is improbable reduction in base course thickness is permitted. The reasons for moisture accumulation under pavements are not fully understood, but it cannot be assumed that the base course below a watertight pavement with good surface drainage will not become saturated, even in an arid climate.[14] Studies by the Waterways Experiment Station [12] indicate that the degree of saturation of airfield subgrades and base courses with plastic fines tends to vary with rainfall, the lowest values occurring in low rainfall zones. The degree of saturation of clayey gravels, silty gravels, clayey sand, silty sand, and fat and lean clays was found to be within the range of values obtained by the 4-day soaking period in high and medium rainfall zones, but somewhat below it in the low rainfall zone. Well-graded sands and gravels without fines showed moisture contents in the high and medium rainfall zones well below those obtained by the 4-day soaking period. Where absorption of water is accompanied by densification under traffic, CBR values well below those produced by the 4-day soaking period may occur. Additional tests are needed before it can be stated conclusively where conditions depart significantly from those indicated by the 4-day soaking period.

The CBR method is frankly empirical; that is, it is based on extensive laboratory tests and investigations of pavements that have failed and those that have not failed, supplemented by data obtained from specially constructed and controlled test sections. These data have been combined in the design curves. As is true for all empirical methods, if it is to be used at all the procedure selected (Corps of Engineers, U. S. Army, the Colorado method, Wyoming method, or other) must be rigidly followed, for the slightest deviation at any point may completely invalidate the entire procedure. Soils containing large fragments of rock give erratic CBR values, and it is necessary to repeat the tests until a representative value can be estimated, but the testing equipment, procedures, and design curves should never be varied arbitrarily in order to suit what is thought to be a special condition. It should always be remembered that the CBR procedures and design curves incorporate a wider range of experience than a single individual could compress into several lifetimes. Perhaps some day a procedure based on analysis of stresses and strains in soils will be perfected, but until then the only modifications in CBR procedures and design curves permissible are those resulting from continued research.

Other Methods. Space does not permit discussion of the various modifications of the above procedures introduced by various organizations or the more or less special methods employed in some regions. The California Division of Highways uses the stabilometer method of design. A good discussion of the method may be found in reference 8 cited at the end of this chapter.

The geologist familiar with the requirements for highway and airfield subgrades can be of considerable assistance in laying out the alignment of highways and selecting airfield sites. He should recognize that uniform soils of only mediocre bearing capacity generally permit more economical construction than rapid alternations of very good and very poor subgrade materials, and he should strive to find the most uniform, as well as the most favorable, subgrade conditions available for the project.

DRAINAGE PROBLEMS

In the field of drainage problems the geologist can often greatly aid the civil engineer, especially during early planning stages before a highway route or airfield site has been selected. It is important to discriminate carefully between surface water, perched water, ground water, and capillary water, for different treatments are usually required in each eventuality. Surface water is generally comparatively easily disposed of; perched water may also be fairly easily handled, but ground and capillary water may present very difficult problems, and it may be more economical to design on the basis of subgrade saturation than to attempt to drain the subgrade.

Drainage systems for highways are of three principal types: (1) longitudinal ditches and gutters for protecting and removing water from the pavement surface; (2) culverts and bridges for conveying water through an embankment; and (3) various types of pipe and stone-filled trenches (French drains) for keeping the roadbed free of water.

Underdrains are placed at a level below the road surface for either of two purposes: (*a*) lowering the ground water level in order to improve stability of the roadbed; (*b*) the interception and removal of seepage from side-hill cuts and springs. When subdrains are required they are usually located on both sides of the road. At times, however, they are carried from the cut side to the downhill side only.

In soils that would otherwise retain moisture, lateral subdrains may also be constructed for leading the water to the sides of the road. The lateral drains may discharge into longitudinal subdrains, but where the lateral drains are shallow they may enter the longitudinal gutters or ditches. The lateral drains may be laid out at right angles to the center line or in a herringbone pattern at an angle of about 60° to the center line.

Subgrade drainage cannot always be solved by typical or systematic layouts. A common-sense plan for intercepting the ground water and removing it by the easiest and shortest path should be the guiding principle. If there are springs or unusually wet places in the subgrade, drains should be placed so as to carry off the water in the most direct manner. Many specifications stipulate a longitudinal drain under the middle of the road to which lateral drains are fed from the shoulders. The chief disadvantage of this type of drain is that if repairs are required the roadbed must be torn up. In situations where the ground water table rises steeply on either one or both sides of a highway or flight strip, the subgrade cannot be kept dry by shallow under drains. The only really effective solution generally consists of lowering the water table over a considerable area, and this may not be economical. Special difficulties may also arise when the ground water table is exposed in deep cuts some distance below the crests of hills. However, the road is usually on the down grade at such points, and the ground water can be led off downslope. Attention to proper subdrainage and suitable surface drains may often result in substantial savings in maintenance costs.

For drains to be effective, the subsurface water must have access to them. Hence, subdrainage efficiency depends in a large measure on the permeability of the soil. In pervious materials the water flows to the drains by gravity, and usually it is not difficult to keep the roadbed dry. However, many types of soils tend to retain water, and in others water may rise considerable distances by capillary action. The height to which water will rise by capillarity depends mainly on the grain size of the soil, and in fine-grained clays it may be as much as 40 feet. In practice, however, about 15 feet is generally taken as the limit of effective capillary action. The only really effective way to deal with capillary water is to lower the water table, and this is often too costly to be adopted.

Generally it is better to remove organic deposits and backfill the space than to attempt to improve their bearing capacities by drainage. Peat should always be removed, and the space should be backfilled with free-draining materials. This is especially true in airfields and modern superhighways.

Deep cuts that are subject to rapid erosion should be protected by diversion ditches constructed along the top of the cut. Generally, diversion ditches should be lined, and this is especially true where the soils are granular. Otherwise, the water will seep through the bottom of the ditch and impair the stability of the cut slope. Diversion ditches are also effective in improving the stability of natural slopes that might otherwise be subject to sliding.

CUTS, FILLS, AND GRADING OPERATIONS

The purpose of grading is to prepare the subgrade for the surfacing materials and provide smooth, well-drained, gently sloping surfaces. Grading involves the removal of obstructions, clearing and grubbing, trimming of high spots and the filling of low areas, excavation for culverts and other structures, the shaping of surfaces of cuts and embankments, and the preparation of shoulders and drainage ditches. Satisfactory grading is essential to good construction. Experience indicates that cracks in pavements are often caused by settlement of fills, poor compacting, poor drainage, or other unstable conditions in the subgrade.

Excavation

Excavation is usually classed as common, rock, structure, and borrow. Common excavation includes the removal of all decomposed rock, soil or earth, and other materials not classed as rock, structure, or borrow. Rock excavation includes the removal of boulders over ½ cubic yard in volume and all materials that cannot be removed advantageously except by blasting. Controversies often arise concerning "rock" that was blasted from a cut but required treatment as earth when placed in fills. For example, many clay shales require drilling and blasting for excavation, but when fragmented they absorb moisture and disintegrate to such an extent that within a few hours or days they must be rolled into a fill in layers not over 8 inches thick, precisely as in the case of soil or earth. In such instances, the contractor is entitled to payment for "rock" excavation and for "earth" placement in a fill. If clay shales are placed in a fill in 14-inch layers, as is done with rock, subsequent disintegration will result in serious settlement.

Structure excavation includes all excavation for foundation of bridges, culverts, and other structures.

Borrow excavation consists of material obtained from pits located beyond the limits of the right of way. Such excavation is necessary when the volume of material from cuts does not balance that required for fills. The geologist experienced in construction work can be of considerable assistance in estimating and classifying excavation quantities as well as in aiding in the location of borrow pits.

Design of Cuts

Stable excavation slopes in soil are generally determined by methods described in all standard textbooks on soil mechanics. Where rock is involved, however, much depends on the physical characteristics of the material and structural details. Furthermore, the problem is often complicated by faulting, or the occurrence of exceptionally soft strata above or below harder layers. Where a considerable thickness of soft rocks requiring flat excavation slopes occurs near the bottom of a deep cut, any hard rocks overlying these materials must be cut to the same slope as the soft materials, or the slope will eventually fail through shearing and deformation of the soft materials. Design of deep cuts in rock is not susceptible of standardization, and for best results the design should be based on a thorough understanding of the local geology.

Fig. 204. "Little Panama" a deep cut on the Pennsylvania Turnpike. This cut is 163 feet deep at the deepest point and over 1,130,000 cubic yards of rock were removed. The rock is interbedded thin layers of sandstone and shale. (Pennsylvania Turnpike Commission.)

Many deep rock cuts were made during construction of the Pennsylvania Turnpike. The steepest slopes were ½ to 1; that is, for every foot vertically the slope was laid back ½ foot horizontally. The average slopes were 1 vertical on 1 horizontal. The deepest cut, popularly known as "Little Panama," is through Clear Ridge about ½ mile east of the town of Everett. This cut is 2600 feet long, 380 feet wide at the top, 88 feet wide at the base, and 163 feet deep at the deepest point. More than 1,130,000 cubic yards of material were removed. The original plans called for a tunnel, but core borings and a careful study of the rock revealed the feasibility of a cut. In a cut of this depth, if the slopes were not broken by benches, or berms, traffic would be exposed to the hazard of falling rock. Therefore, two berms, the lower 23 feet wide and 30 feet above the road grade, and a higher one 85 feet above grade were provided for protective and drainage purposes. Access was provided for the removal of accumulated debris.

The rock through which this cut was made consists of sandstone and shale interbedded in layers 1 to 18 inches in thickness. The bedding trends at right angles to the center line of the highway and dips 53° to the southeast. Considering the character of the rocks and their structural attitude, it is thought that the ultimate in safety consistent with economy of excavation was achieved. The slope from the base to the first berm is $\frac{1}{2}$ on 1, that between the first and second berms is $\frac{3}{4}$ on 1, and from there to the top it is 1 on 1.

In cuts through relatively flat-lying strata consisting of thick-bedded hard rock and clay or soft shales, undermining of the hard rock by disintegration of the clays or shales should be anticipated. Disintegration will be accelerated where the ground water table has been intersected. Benching may help in maintaining stability, especially if seepage from overlying beds is caught and led away by drains. However, the benches may also become undermined. The only sound solution is to flatten the slopes and plan for constant maintenance.

Materials like loess, which has a strong tendency to stand on nearly vertical slopes, are generally cut practically vertical. However, experience indicates that water penetrating along cracks and fissures tends to soften the material and eventually causes blocks to slough away from the cut banks. Sloughing is especially frequent after prolonged and heavy rains. Consequently, steep cuts in loess should be made wide enough so that failure of the banks does not block traffic. In the case of very deep cuts in loess, the slopes should still be cut nearly vertical, but in order to prevent shear failure one or more berms should be introduced. Adequate drainage of the berms is very important, as is good surface drainage of the area along the top of the cut.

Slopes above side-hill cuts should always be examined for indications of ancient slides, excessive surface creep, sheetwash, and similar phenomena. Highway alignments should be kept as far from unstable hillsides as possible, and when this cannot be done every effort should be made to divert surface and subsurface water from the slopes above a cut.

Fills

Fills may be constructed either with earth or with rock. Rock may be placed in a fill in layers about 14 inches thick; earth fills are generally rolled in layers not over 8 inches in thickness. As previously mentioned, sometimes material may be classed as "rock" for excavation purposes and as "earth" when placed in fills.

Fig. 205. A high fill on the Pennsylvania Turnpike. (Pennsylvania Turnpike Commission.)

Consideration must be given to the change from the original volume in the cut to the final volume in the fill after shrinkage and settlement have taken place. When the material is broken into pieces of various sizes during excavation, it nearly always increases in volume when placed in a fill. As a result of compaction during placement of a fill, or natural causes operating over a longer period of time, most soils decrease in volume after being placed in a fill. This phenomenon is known as shrinkage. It varies with different types of soil, but in highway construction about 15 per cent is generally allowed.

Settlement. This is the total decrease in height of an embankment from the time it was first placed to the time it becomes stabilized. It is caused by a gradual consolidation of the fill material, consolidation of the foundation materials, or both. Settlement varies with the type of soil and the methods used in placing and compacting the fill. It may continue over a period of years, although the greater part occurs during the first year or so.

The geologist has little to do with problems involving the settlement of the fill materials themselves, but can be of considerable assistance in problems involving settlement of foundations. Peat and certain other types of highly organic soils may permit excessive settlements of fills of only moderate heights, and if possible these materials should be removed from under fills. Where this is not possible, special methods of construction, such as the placing of blankets to spread the load and the gradual construction of the fill over a period of years, must be adopted. Explosives have proved very useful in aiding the settlement of fills on soft and unstable soils. In marshy or meadow areas the vegetative mat is broken up. The fill is then placed, and explosive charges are set beneath it. The amount of explosive, number of holes and their depth depends on the depth of the soft materials and the height of the fill. The explosion pushes the soft foundation material to the sides, and the fill then settles into place. The explosion performs two functions: (1) it produces a cavity for the reception of the fill, and (2) it stirs up and semi-liquefies the material left in the cavity and thus aids rapid settlement of the fill. In a variation of this method, especially adapted to clayey soils, ditches are shot on either side of the fill, and the fill gradually pushes aside the soft foundation material left between the ditches. These methods require the assistance of explosive experts.

Shoulder Material. Shoulder materials available vary with the type of rock, soils, or industrial wastes found in the locality. Sodded shoulders are often satisfactory, but for high-speed roads more stable shoulders are desirable. Crushed rock chinked with quarry spoil and fines, or primed with asphalt, is often used. Siliceous shales are satisfactory, but clay shales are not. Clay shales disintegrate rapidly and absorb moisture, resulting in a clay-like product. Culm, material from burned-over coal mine dumps, may be used. Slag is acceptable. In glaciated regions or in areas where fluvial, lacustrine, or marine gravels are available, suitable shoulders may be constructed from this material. Chert makes excellent material, and in eastern Missouri plentiful supplies of chert, locally called "chat," are widely used for high-way shoulders.

CONSIDERATIONS APPLYING PARTICULARLY TO AIRFIELDS

The factors influencing airfield site selection and construction are generally similar to those pertaining to highways. However, the stresses imposed by heavy aircraft are much more severe than those caused by highway traffic, and good subgrade conditions are thus even more important for airfields than for highways. Extensive tests and observations indicate that the stresses caused by impact of aircraft landings are not as severe as those caused by taxiing. Therefore, the ends of runways, taxiways, and parking aprons are more likely to fail than the central portions of runways. This consideration is sometimes overlooked in airfield site selection and design.

Site Selection

No two airfield sites, or any two construction sites, for that matter, are ever exactly alike. Therefore, in each case the topographic features, soils, depth to ground water, surface drainage, vegetative cover, accessibility of construction materials, and other local conditions must be taken into consideration. The direction of the prevailing wind and other meteorological factors may often be decisive. Freedom of the flight strip from obstructions, natural or artificial, that would inhibit the safe approach or departure of aircraft is essential.

Topography. The topography of the site affects the costs involved in grading, drainage, and surfacing. Sufficient slope to provide good surface drainage is desirable, but this feature must be weighed against grading costs. Ideally, an airfield should have an even, well-drained surface sufficiently firm to permit the construction of the pavement at reasonable cost. As is explained in the following paragraph, soil conditions are not invariably subordinate to topographic factors.

Soils. As mentioned earlier, soils differ greatly in their bearing capacities, the best material being well-graded gravels and coarse-grained materials in general. These materials not only have good bearing capacities but also are generally free-draining. Therefore, they are to be preferred for airfield subgrades. In the past, it was a rather general practice to base site selection largely on topography, but the very heavy planes now in use impose such severe stresses, which in the case of certain types of aircraft extend to depths of several feet, that soil conditions often rank equally with or of even more importance than topography. A layer of soft material hardly more than 1 inch thick below a flexible pavement may lead to failure of the subgrade. Consequently, soil conditions should be explored to a depth of at least 5 feet, or more, depending on the uses to which the field will be placed. The position of the water table should also be determined.

Surface and Subsurface Water. Good surface drainage and a low water table are important, for a high percentage of pavement failures can be traced to poor drainage or a high water table. Sites subject to flooding should be avoided, for floods not only make the field inoperative for a more or less extended period but also tend to cause saturation of the subgrade with attendant loss of bearing capacity. This effect is the more serious for, once a subgrade is saturated, there is usually no feasible way to restore the bearing strength of the flight strip, except the costly method of adding to the base course and/or pavement thickness. Sites near dikes and levees are also to be avoided, if possible, for they are subject to flooding if the protecting structures fail, and at times of high water considerable artesian pressures may be transmitted to pervious strata in the subgrade. In such situations, relief wells may be necessary.

Pockets of perched water are sometimes encountered. When caused by a thin impervious bed overlying free-draining materials, perched water can be disposed of by the simple and cheap expedient of perforating the confining bed. Ground water drainage generally poses more serious problems, and in impervious soils it may not be feasible. Seasonal fluctuations of the water table should be determined in advance of construction, when possible.

INFLUENCE OF CLIMATE AND GEOGRAPHY

The influence of climate and associated factors is essentially comparable for both airfields and highways. Therefore, a discussion of one will serve reasonably well for the other. Construction of highways and airfields in arctic, tropical, and other more or less isolated and poorly developed regions is mainly of fairly recent date and much remains to be learned. However, the following generalizations appear to be warranted.

Arctic Airfields

The adverse climatic conditions of the Arctic are the basic factor affecting airfield construction. Not only do they make soil conditions unique, but the effect of low temperatures on personnel and machinery must also be considered. High wind velocities are common and constitute an important factor in the design and construction of airfields and appurtenant structures. High wind resistances are necessary for the structures and buildings, and special care must be taken to prevent stripping of soil from runways and shoulders. Wind erosion is particularly likely to occur during excavation of cuts and the placement of fills. If the vegetative mat is removed from large areas, the soil may dry out very quickly and be removed by the wind. Such conditions are especially common in the Aleutian Islands, where it has been found necessary to protect fills with sandbags and sod.[3] The presence of permafrost, discussed in Chapter 11, is perhaps the most important single factor in Arctic airfield design and construction.

Tropical Airfields

The United States forces learned much concerning airfield construction in the tropics during World War II. In many instances there was no choice concerning site location, availability of suitable construction materials, or even favorable orientation of runways with respect to the prevailing wind. The rush of airfield construction forced many varied improvisations on the engineers; many were successful, and some were not. One rehabilitated and enlarged field on Guadalcanal Island (Henderson Field) was successful in large part because of the favorable location and the presence of granular soils and construction materials. Another airfield on the same island built on fine silts was a failure because of lack of appreciation of drainage problems involved.

In tropical regions where the rainfall often exceeds 150 inches per year, adequate drainage of airfields is very important. Large ditches made with back-hoes, draglines, carryalls, and similar equipment are the fastest and cheapest to construct, but, when feasible, underdrains, pipe drains, and other more elaborate methods are used. Where the rainfall is exceptionally heavy, open ditches must have large cross-sectional areas. During World War II, culverts under roads, taxiways, and runways were often made by welding together heavy fuel drums.

Construction on Coral Islands. During the rush of wartime construction it was not possible to base selection of construction materials on experience or tests. One used the materials available. Because of their availability, abundance, hardness, and favorable drainage characteristics, the abundant coral deposits found on many Pacific islands became one of the most commonly used construction materials. According to Stearns,[13] coral suitable for construction purposes ranges from pockets that look and feel like flour to hard underwater reefs that must be blasted before handling. The principal criterion for suitable coral materials is the presence of cementing agents; for example, 5 to 10 per cent of clay or humus usually makes the materials usable when wet. Cementing materials may be recognized by the fine white deposit left on the hands when a sample is kneaded by the fingers. Coral from which cementing substances have been removed, such as coral sand produced by wave action, or badly weathered coral, will not set up. Material blasted from a reef often appears to be deficient in cementing materials because of the washing incidental to handling, but when crushed it generally produces sufficient cementing substance to form a set. Ancient coral reefs found above sea level are usually overlain by a mat of earth and vegetation which may serve to delay weathering to such an extent that the materials are not usable.

The most satisfactory coral feels slightly sticky, like mortar, when properly mixed with water. When obtained from below water level, it has a slight hydrogen sulphide odor. Coral with desirable setting qualities is said to be "live" and is apparently activated by the crushing and grinding incident to breaking up the mass and working into base courses or pavement surfaces. Under favorable moisture conditions the setting produces a compact slab that turns water very satisfactorily.

Stearns groups coral deposits into three main types: (1) reef, (2) beach, and (3) lagoon. They may exist well above sea level as compact limestones that have been raised to their present position, or as loose or often well-cemented deposits near or below sea level. The occurrence and age of coral, whether found on dry land or below the sea, a few years or thousands of years old, do not determine its suitability for construction purposes.

For concrete aggregates heavy nodular coral with a crystalline calcite center and a specific gravity of about 2.7, preferably broken by wave action to form fragments less than 6 inches in diameter, is the most suitable. Screening and crushing are usually necessary, however. Such materials are found along the windward side of atolls and in adjacent portions of lagoons.

Calcareous beach and dune sands can be used in making cement, but such materials have proved to be too porous and uniform in grain size to set up in runways and roads.

Coral deposits from fringing reefs and lagoons are commonly obtained by power shovels and draglines. These materials are a mixture of coral, sand, shells, algae, organic debris, and limy mud in various proportions. Generally placed in bank-run condition and saturated with sea water, they are one of the best natural mixes for quick-bonding surfaces of runways and roads.

The factors causing coral materials to "set" are still not completely understood, but this property seems to be caused chiefly by the presence of fine limy material of silt size. The occurrence of decomposing marine organic matter also seems to assist cementation by the release of organic acids which react with the lime. Rainfall containing ocean salts and carbon dioxide from the air and admixtures of foreign matter may also play an important role. Alternate wetting and drying of loose coral blocks will cement them into firm aggregates.

Many of the coral-built runways were constantly sprinkled with sea water to lay the dust. Layers of coral placed on runways and having large open interstices, or only a small per cent of fines, cement slowly if at all. However, in such situations traffic may increase the proportion of fines to such a degree that the surface eventually becomes satisfactory. The rate of bonding may be increased by mud tracked onto the runways. Volcanic muds have, in some cases, been used for filling the interstices. These muds are often bentonitic and thus absorb and retain large quantities of water.

With proper treatment almost any type of coral material can be used effectively, but occasionally some factors are beyond control. Failures may be caused by lack of rainfall and sprinkling, lack of impurities, or lack of proper mixture of coarse and fine fractions. Coral from submerged sources containing fair proportions of limy muds is the most economically handled and sets up rapidly. Best results are obtained with salt-water sprinkling.

Maintenance of well-constructed coral runways requires less manpower and equipment than strips constructed of other local materials or various types of steel mats. Fresh raw material with the proper moisture content is all that is needed for repair. Maintenance is most effective when carried out after a rain while the surface is still wet. Low spots are filled with bonding materials after which they are rolled. When the rainfall is adequate to lay the dust a busy coral strip may go as long as 3 months with no maintenance. During prolonged dry weather coral runways and roads must be sprinkled to hold down dust and prevent raveling.

Construction on Volcanic Islands. Many oceanic islands are entirely volcanic, and various types of volcanic materials must, therefore, be used in construction. Experience indicates that various types of volcanic tuffs, cinders, and bombs make suitable base course materials. Dense lavas, such as basalt, make excellent sources of crushed rock for base courses and concrete and bituminous aggregates. Scoriaceous and highly vesicular types of lavas are less desirable. They lower the flexural strength of concrete, and when they are used for bituminous aggregates proper proportioning of the mix is difficult to impossible. The cavities of scoriaceous lavas are often filled with zeolites and other reactive minerals that cause deterioration of concrete.

The Pacific island of Iwo Jima contains volcanic ash and cinders that proved exceptionally satisfactory for airfield construction.[4] Much of the island is covered with deposits of loose black cinders referred to as "black sand" and a consolidated buff-colored ash sometimes called "sand rock." The "black sand" proved easy to work and made stable and inexpensive fills. The porous cinder grains are generally angular in shape and are thought to have fairly high frictional resistances in any plane of potential shear. Embankments constructed from this material are sufficiently pervious to drain well. The buff-colored ash proved to be well suited for "clay" surfacing of airstrips. A layer at least 1 foot thick was placed on all cinder fills, rolled with a sheepsfoot roller, bladed, and smooth-rolled to grade. This material also proved suitable as a base course under bituminous pavements. The buff ash has little stability, however, when the optimum moisture content is exceeded.

REFERENCES

1. *Amer. Soc. Civil Engineers*, Development of CBR Flexible Pavement Design Method for Airfields, *Trans.*, Vol. 115, pp. 453–589, 1950.
2. E. F. Bean, Engineering Geology of Highway Location, Construction, and Materials, in *Application of Geology to Engineering Practice* (Berkey Volume), Geol. Soc. Amer., 1950.
3. *Civil Eng.*, Peculiarities of Aleutian Military Design, Vol. 15, p. 403, 1945.
4. *Civil Eng.*, Seabees Encounter Unusual Soils in Iwo Jima, Vol. 15, pp. 453–454, 1945.
5. Corps of Engineers, U. S. Army, The Unified Soil Classification System, Appendix B, Characteristics of Soil Groups Pertaining to Roads and Airfields, *Waterways Experiment Station, Tech. Memo. 3-357*, Vicksburg, Miss., March, 1953.
6. H. K. Glidden, H. F. Law, and J. E. Cowles, *Airports; Design, Construction and Management*, McGraw-Hill Book Co., New York, 1946.
7. Seward E. Horner and John D. McNeal, Applications of Geology to Highway Engineering, *Colo. School Mines Quart.*, Vol. 45, No. 1B, 1950.
8. Robert Horonjeff and John Hugh Jones, The Design of Flexible and Rigid Pavements, *Univ. Calif. Syllabus Series*, No. 319, 1951.
9. Robert F. Legget, *Geology and Engineering*, McGraw-Hill Book Co., New York, 1939.
10. Michigan State Highway Department, *Field Manual of Soil Engineering*, revised February, 1946.
11. Public Roads Administration, Federal Works Agency, *Principles of Highway Construction as Applied to Airports, Flight Strips and Other Landing Areas for Aircraft*, June, 1943.

12. J. F. Redus, Jr., and C. R. Foster, Moisture Conditions under Flexible Airfield Pavements, *Highway Research Board, Special Report* No. 2, pp. 126–146, Washington, 1952.

13. H. T. Stearns, Characteristics of Coral Deposits, *Eng. News-Record,* Vol. 133, No. 2, 1944.

14. Hans F. Winterkorn, Climate and Highways, *Trans. Am. Geophys. Union,* 1944.

C H A P T E R   2 2

Concrete Aggregates

Aggregate is the general term applied to materials both fine and coarse which when bonded together by cement form concrete. It may consist of natural sand and gravel, crushed rock, blast furnace slag, cinders, or other inert substances possessing hard durable surfaces free of adherent coatings. The success of concrete structures depends among other things on a thorough understanding of the properties of the substances used as aggregates. Concrete research has become a fairly exact science, and the study of aggregates is an important subdivision. Although concrete aggregates are not the only type of natural construction material concerning which the geologist can aid the engineer, such materials as building stone,[5] clay, and cement,[10] have been adequately discussed elsewhere and need not be treated here.

The specifications for fine and coarse aggregates vary with the uses to which they are to be put. For example, resistance to abrasion is an important factor in highway construction, whereas in buildings and dams compressive and shear strengths are important. The crushing strength of concrete is largely determined by the strength of the aggregate and the texture of the finished product. Thus in drafting the specifications for mass concrete for large gravity dams it is customary to specify very coarse aggregate (4 inches in diameter and above) consisting of materials of high crushing strength so placed that the void space between the particles is a minimum. The voids between the coarse aggregate particles should be completely filled by fine aggregate, and the remaining voids filled by cement. In other words, the texture of high-grade concrete should resemble that which nature imparts to a massive (unstratified) conglomerate. During placement, care must be taken to insure that no stratification of fine and coarse materials occurs. This is usually accomplished by holding the water content to a minimum and stirring the

Fig. 206. Large aggregate processing plant, Columbia Basin Project, Washington. (U. S. Bureau of Reclamation.)

newly placed materials with mechanical vibrators. Because cement is the weakest material involved and, therefore, the limiting factor in determining strength of the finished product, any film of cement which may rise to the surface (laitance) is removed by an air jet prior to placement of the overlying lift Flat, pencil- and disk-shaped particles tend to prevent close packing, and hence are undesirable in high-grade concrete. References to specifications for concrete aggregates may be found in the publications of the American Society for Testing Materials, the American Association of State Highway Officials, the Bureau of Reclamation, and the Corps of Engineers, U. S. Army.[12]

As in other branches of engineering, the selection of concrete aggregates is governed by economic factors. Transportation costs are a very important item, and in large projects involving millions of cubic yards of aggregates the availability of suitable materials within an economical hauling distance may be the controlling factor determining feasibility of the project. Frequently, the engineer must use the only aggregates available although they are lacking in many desirable qualities. This is particularly true in isolated areas and in undeveloped countries where transportation facilities are primitive or unavailable.

Much of the data contained in the following pages has been adapted from the *Concrete Manual* issued by the U. S. Bureau of Reclamation.[11]

GEOLOGICAL CONSIDERATIONS

The geologic and physiographic history of a region is often intimately reflected in the character of aggregate deposits. The geologic history is indicated by: (1) kind of rock, (2) physical characteristics of the rock, and (3) secondary structures within the rock such as joints and cleavage. The physiographic history determines the geologic processes that have been and are now acting on the aggregates and are responsible for their alteration, transportation, and deposition. These processes affect the size, shape, location, uniformity, rounding, and grading of the material, and thus a general knowledge of geological principles leads to a more intelligent search for aggregates. Such knowledge also contributes to more effective sampling procedures.

Occurrence and Types of Aggregate Sources

Aggregates occur in natural deposits of sand and gravel or may be manufactured from quarries developed in bedrock. Where natural sands and gravels are present in sufficient quantity they are often the most economical source of supply. Such deposits are found chiefly in river, beach, glacial, delta, and alluvial fan areas of deposition. Though wind-blown sands are ordinarily too fine, they may be successfully blended to satisfy specific requirements.

River Deposits. River sands and gravels are frequently the most abundant and most suitable materials, provided they are not too remote from their source of supply. In that event the coarser sizes may be limited in volume and layer thickness, or they may be entirely lacking. In general, the individual particles are well rounded, well sorted, and of relatively uniform hardness. Abrasion during stream transportation greatly reduces or eliminates the less durable materials. Terrace deposits are often good sources of sand and gravel. On the Susquehanna River at Middletown, Pennsylvania, terraces formed during the Ice Age were exploited for use in the Pennsylvania Turnpike, whereas the river channel was found to contain relatively little usable sand and gravel.

Alluvial Fans. Alluvial fans, formed where streams flowing from mountainous areas debouch onto flatter land, are often excellent sources of aggregate. In Trinidad, B.W.I., for example, such sources have been exploited. However, because of the limited distance of transportation natural processing of the material is frequently not satisfactory, and extensive plant processing is necessary.

Glacial Deposits. Glacial deposits are restricted to regions of mountain glaciation or areas formerly covered by continental ice sheets. They supply aggregates of heterogeneous characteristics. Because they may have been at least in part ice-transported, the size range is great, the degree of sorting limited, and particle angularity marked. The softer materials are often retained. Consequently, plant processing is more difficult than with water-transported materials. Fluvial-glacial deposits, in which ice-carried debris has been worked over, sorted, and abraded by stream action, may be as suitable as normal river sands and gravels.

Talus Debris. Talus debris accumulates at the foot of cliffs and results from mechanical weathering. Such materials usually consist of coarse angular fragments. Virtually no natural processing has taken place, but if the blocks were derived from initially suitable parent materials they may be prepared for use as aggregate.

Eolian (Wind-Blown) Materials. Usually eolian materials consist of fine sands that are frosted and well rounded. They are normally too fine for use except for blending purposes. The individual grains are usually quartz or other resistant minerals that have survived the intense attrition to which eolian materials are subjected.

Crushed Rock. Fine and coarse aggregates may be produced by quarrying and processing suitable bedrock in case natural sands and gravels are not available. Such materials are subject to the same specifications as naturally occurring deposits. A frequent defect of crushed rock is the presence of incipient fractures. In addition, there may be an undesirably high proportion of flats and fines in the crushed material. Crushed-rock aggregates are usually more brash than natural sands and gravels, and therefore are more difficult to place in a condition of close packing.

Physical Characteristics of Various Rock Types

The data contained in Chapters 2 and 3 are very useful in classifying and identifying materials that may be considered in a search for aggregates. Based on their origin, rocks are divided into the following principal groups which are further subdivided according to mineral constituents and physical characteristics.

Igneous Rocks. In general, igneous rocks make good aggregates. However, volcanic tuffs and many lavas are porous and consequently have low strength, light weight, and high absorption. Such materials are not satisfactory for aggregates.

Sedimentary Rocks. Sedimentary rocks cover a wide range of physical characteristics which range from soft to hard, heavy to light, and from dense to porous. Their adaptability as aggregates varies accordingly. Conglomerates tend to break down into progressively smaller-sized particles when processed, and hence they are rarely satisfactory. When hard and dense, sandstones often make good aggregates Friable sandstone is normally too porous and weakly cemented to be satisfactory. Argillaceous sandstones are usually unsatisfactory because of their softness and high absorption. Hard, dense limestones make good aggregates. Shaly limestones with a large clay content, on the other hand, are undesirable because of their softness and high absorption. Both limestones and sandstones may grade into limy and arenaceous shales. Shale rarely makes good aggregate. Usually it is too soft, light, weak, and absorptive. Furthermore, shales are characteristically thinly laminated and tend to produce flat to pencil-like fragments when processed into fine and coarse aggregate. Chert and flint are used as aggregates, but service records of many cherts are so unsatisfactory that these materials must be regarded with suspicion until their individual service records and petrographic characteristics prove their use to be justified.

Metamorphic Rocks. Metamorphic rocks vary greatly in their physical characteristics. Marbles and quartzites are usually dense and massive and possess satisfactory toughness and strength. Gneisses may be tough and durable, but because of lineation and foliation they often exhibit the unsatisfactory features of schists. Schists as a rule are strongly foliated and tend to break into thin irregular sheets and small fragments. The mica content is high, and they are usually rather soft. Even though extremely variable in physical characteristics, they cannot be entirely ruled out as aggregate materials. Slates are usually thinly laminated and possess good cleavage. They rarely make suitable aggregates. However, they may be used as shoulder material.

Alteration. Fresh rocks are frequently altered when exposed to secondary processes such as weathering and the action of circulating ground waters. Furthermore, igneous rocks may be altered by action of hot solutions derived from the cooling magma or lava. These agencies may subtract elements from the mineral constituents and change them completely from their original state. Secondary agencies may also carry material in solution and deposit it as a coating on the rock or in pores and fractures. Similarly, cementing materials may be deposited or removed throughout the rock. The secondarily deposited materials may be deleterious, or objectionable, because they make processing more difficult. For example, the basalt of Sosa Hill, Canal Zone, contains up to 20 per cent of clay minerals, presumably montmorillonite of secondary origin. Obviously, it is unsuited for use in high-grade concrete. The aggregate used in Parker Dam consisted of altered andesite. This structure is of the arch type, and after completion it was found to be moving up rather than downstream. This behavior has been found to be caused by reaction of the aggregate and the cement with resulting increase in volume. Secondary changes may, on the other hand, have the opposite effect and actually improve the material.

CHEMICAL STABILITY AND CONTAMINATION OF AGGREGATES

The U. S. Bureau of Reclamation and the Corps of Engineers, U. S. Army, continue to conduct research relating to deleterious aggregates. In large part this section has been adapted from the *Concrete Manual* published by the Bureau of Reclamation.

Fig. 207. Pattern cracking caused by cement-aggregate reaction, Highway overpass, Laramie, Wyoming. (U. S. Bureau of Reclamation.)

Some objectionable materials may not manifest their undesirable nature for years, whereas others may expose themselves within a very brief period of time. Aggregates that react with high-alkali cements may prove quite satisfactory in concrete made with low-alkali cements. Whereas some chemical changes brought about by reaction of cement and aggregate may actually be beneficial, others are definitely deleterious. Chemical changes that may take place in concrete include: (1) reactions between the aggregate and the cement, (2) removal of soluble constituents, (3) oxidation, and (4) complicated processes that interfere with the normal hydration of cement. Volume changes of clays by absorption and hydration are included under item 1. The obvious results of deleterious materials are manifested by expansion, cracking, and general deterioration of the concrete. The outstanding offender with regard to reactions of this type is opal (hydrous, amorphous silica). Very small amounts of opaline silica in an aggregate are sufficient to cause rapid deterioration and expansion of the concrete. It may be present as a coating on the aggregate particles or as an actual rock constituent.

Other minerals and rocks that have bad service records with respect to reaction with cement alkalies are: volcanic rocks of medium to high silica content; silicate glasses, artificial or natural, with the exception of the basic types such as basaltic glass; chalcedonic rocks, including most chert and flint; some phyllites; and tridymite. For some unknown reason, aggregates containing very high proportions of chert may not be deleterious. The Bureau of Reclamation specifies that aggregates petrographically similar to known reactive types, or which because of unsatisfactory service records or laboratory tests are suspected of reactive tendencies, should be used only with low-alkali cements. In some cases deleterious reactions may be eliminated by limiting the alkalies (Na_2O plus K_2O) to 0.5 to 0.6 per cent of the cement.

Pyrite and marcasite are iron sulphides that oxidize rapidly, and thereby cause reduction in strength and unsightly staining of concrete. The acid products resulting from this reaction are injurious to concrete and many steels.

Coal in any form is undesirable. It lowers the resistance of the concrete to freezing and thawing and lends an objectionable appearance. Nearly all organic materials contain acids which inhibit the hydration of concrete. Usually, aggregate specifications prescribe standard tests for dark-colored minerals suspected of possessing organic coatings and/or constituents. Because not all organic materials are objectionable, a specific determination is desirable before an aggregate is rejected.

Sulphates, chlorides, carbonates, phosphates, and other chemical salts may be present in aggregates in variable combinations. Their chemical activity may seriously retard or modify the setting up of concrete. In combination with objectionable silt or clay fractions they are doubly injurious. Such salts may also cause unsightly efflorescence. Silts and clays in a "mix" tend to increase the water requirements, thereby reducing durability and adding to the unsoundness of the finished product. Some fine-grained siliceous materials, on the other hand, such as ground pumice or diatomaceous earth are occasionally used to increase the workability and strength through their pozzuolanic action.

Contaminating substances may be present in both fine and coarse aggregates. Especially common in arid regions are chemically precipitated coatings on sand grains or in rock fractures or pore spaces. Such coatings may be so thick as to constitute a cement between the grains. They may also occur as layers in sand and gravel deposits.

Washing generally removes mica, readily soluble salts, most organic matter, silt and clay coatings, and coal. If the fragments are too large, coal may be retained in the washed product. Certain types of coatings and encrustations require special abrasive processing for removal. Others cannot be removed except at prohibitive cost.

PROSPECTING FOR AGGREGATES

The purpose of the search for aggregates is to find materials in quantity and quality suitable for concrete. The cost of processing the materials and transporting them to the site should be as low as possible. Prospecting for aggregates is treated in nearly all concrete manuals, engineering handbooks, and other publications. The *Concrete Manual* published by the U. S. Bureau of Reclamation contains an especially competent treatment of the subject.

Ideal conditions are seldom found in natural sand and gravel deposits. Suitable materials may be interbedded with undesirable substances, particle shapes may be flat or pencil-like, and the deposit may be contaminated by clay, silt, or organic material. The ground water table may be so high as to impede development of the site, or the overburden may be excessive. Whereas plant processing of the material may eliminate many undesirable features, it is usually necessary to map and explore thoroughly a deposit prior to its selection. Accessibility, proximity to the construction site, and workability of the deposit must also be carefully evaluated. Quantity of aggregate available, the area involved, depth of the deposit, and its grading must likewise be determined. Whereas the initial examination may leave much to be desired, the approximate area of the deposit can be determined, and some idea of the type of deposition, particle type, origin, size, and durability, can be ascertained. If the preliminary examination and sampling are satisfactory, plans must be made for an accurate survey. Generally, a grid system is laid out, and the deposit is mapped on this grid. Carefully planned test pits, trenches, and test holes are then opened. Spacing of the pits, trenches, and holes varies in accordance with characteristics of the deposit and judgment of the engineer in charge. Nearly all Federal agencies have detailed specifications relating to procedure and techniques. Independent investigators will do well to follow the time-tested methods that have proved satisfactory with these organizations. Standard methods of sampling and testing may be found in references 1 and 2.

TESTING OF AGGREGATES

The chief tests for determining the quality of aggregates are grading, unit weight, specific gravity, absorption, silt content, organic content, and soundness. In addition, some laboratories perform petrographic examinations and chemical tests for potential reactivity.[7]

Physical and chemical tests do not always yield results compatible with service records. Hence, freezing and thawing tests are desirable. Rhoades [8] recommends petrographic examination, when possible, of all aggregates. This is chiefly a visual appraisal, and it aids interpretation of the physical and chemical tests to which the material may have been subjected. In addition, it provides information relating to such properties as pore size, interconnection of pores, and bonding characteristics which cannot be ascertained by any of the standard tests. Chemically reactive elements can be identified, and new aggregates may be compared with those for which service data are available. Such an examination may indicate the necessity of supplementary tests.

Grading tests are performed with sieves made of wire cloth. Mesh spacing conforms to rigid standards which provide uniformity in particle sizes passing through the screen. Sets of sieves permit ready analyses of grain size. Field tests are conducted by shaking the sieves manually, but laboratories are usually provided with a Ro-Tap (mechanical sieve shaker). Specifications

Fig. 208. A portion of the Concrete Research Laboratory of the Waterways Experiment Station. Crushing and screening plant near left wall. Technician at right is lowering samples in drum onto a variable-speed rolling device used in a special test of the abrasion resistance of concrete. (Corps of Engineers, U. S. Army.)

generally require that a fine aggregate shall be well graded from coarse to fine material, and that when tested with standard sieves it should conform to the requirements as given in Table XIII.

TABLE XIII. GRADATION REQUIREMENTS FOR FINE AGGREGATES IN PER CENTS

Passing a ⅜-inch sieve	100
Passing a No. 4 sieve	95–100
Passing a No. 16 sieve	45–80
Passing a No. 50 sieve	10–30
Passing a No. 100 sieve	2–10

Should the fine aggregate fail to meet the minimum requirement for material passing the No. 50 or the No. 100 sieve, it may be used provided that a satisfactory inorganic fine material is added to correct the grading deficiency. It is further required that fine aggregates from different sources of supply should not be mixed or placed in the same storage pile, or used in the same class of construction or mix, without the expressed permission of the engineer.

The American Association of State Highway Officials states that the amount of deleterious substances shall not exceed those listed in Table XIV.

TABLE XIV. PERMISSIBLE PERCENTAGES OF DELETERIOUS SUBSTANCES IN CONCRETE AGGREGATES

	Recommended Permissible Limits, % by Weight	Maximum Permissible Limits, % by Weight
Clay lumps, not more than	0.5	1.0
Coal and lignite	0.25	1.0
Material passing No. 200 sieve		
(*a*) In concrete subject to surface abrasion, not more than	2.0	4.0
(*b*) All other classes of concrete, not more than	3.0	5.0
Other deleterious substances (e.g. alkali, mica, coated grains, soft and flaky particles)	...	As specified

The soundness test determines the resistance of aggregates to saturated solutions of sodium sulphate or magnesium sulphate. It affords data that aid in judging the soundness of aggregates with respect to weathering action, and it is especially helpful when service records of the aggregate in question are not available. When an aggregate fails to meet the specifications established for the soundness test, at the option of the engineer it may be subjected to alternate freezing and thawing tests, after which it may be accepted, provided the weighted per cent of loss at the end of a specified number of cycles does not exceed a previously determined per cent by weight. The soundness test may be eliminated when the aggregate is to be used in structures, or those portions thereof that will not be exposed to weathering.

Organic impurities are determined by a chemical test in which to a quartered 1-pound sample is added 2.5 ml of a 2 per cent solution of tannic acid in 10 per cent alcohol mixed with 97.5 ml of a 3 per cent solution of sodium hydroxide. The color of the clear liquid above the sample is compared with the color of a reference color solution (see American Association of State Highway Officials Designation T 21-42). In place of the standard color solution the color of the clear liquid may be compared with the colors given in Figures 1 to 5 of the color plate accompanying the American Society for Testing Materials Method C 40-33. The value of this test lies in that it may serve as a warning that additional tests are necessary.

Coal, lignite, and other substances of low specific gravity are determined by means of flotation. In this test materials having specific gravities greater than coal and other vegetable matter settle out in the heavy solution (see American Association of State Highway Officials Designation T 113-45).

The absorption test is devised for the determination of bulk and apparent specific gravities and absorption of fine aggregates after 24 hours' saturation in water at room temperature. Bulk specific gravity is the value generally desired in relationship to Portland cement.

Apparent specific gravity is "the ratio of the weight in air of a given volume of the impermeable portion of a permeable material (that is, the solid matter including its impermeable pores or voids) at a stated temperature to the weight in air of an equal volume of distilled water at a stated temperature ⋯." [2]

Bulk specific gravity is "the ratio of the weight in air of a given volume of a permeable material (including both permeable and impermeable voids normal to the material) at a stated temperature to the weight in air of an equal volume of distilled water at a stated temperature ⋯." [2]

In addition to those cited above, relatively simple tests for the determination of the per cent of material passing a No. 200 sieve (American Society for Testing Materials Designation C 117-37), the per cent of light-weight material in aggregates (American Association of State Highway Officials Designation T 10-35), and the per cent of soft pebbles in gravel (American Association of State Highway Officials Tentative Specification T 8-24) have been devised.

REFERENCES

1. American Association of State Highway Officials, *Standard Specifications for Highway Materials and Methods of Sampling and Testing,* 1947.
2. American Society for Testing Materials, *A.S.T.M. Standards Including Tentatives,* 1946.
3. R. F. Blanks and H. S. Meissner, Deterioration of Concrete Dams due to Alkali-Aggregate Reaction, *Trans. Am. Soc. Civil Engrs.,* Vol. 111, 1946.
4. R. F. Blanks, New Concepts Applied to Concrete Aggregates, *Proc. Am. Soc. Civil Engrs.,* 1947.
5. E. O. Bowles, *The Stone Industries,* McGraw-Hill Book Co., New York, 1934.
6. D. McConnell, R. C. Mielenz, et al., Cement-Aggregate Reaction in Concrete, *J. Am. Concrete Inst.,* Oct., 1947.
7. R. C. Mielenz and K. T. Greene, Chemical Test for Reactivity of Aggregates with Cement Alkalies; Chemical Processes in Cement-Aggregate Reaction, *J. Am. Concrete Inst.,* Nov., 1947.
8. Roger Rhoades and R. C. Mielenz, Petrography of Concrete Aggregates, *Proc. Am. Concrete Inst.,* 1946.
9. Roger Rhoades and R. C. Mielenz, Petrographic and Mineralogic Characteristics of Aggregates, *U. S. Dept. Int., Bur. Reclamation, Research & Geol. Div., Petro. Lab. Rept.* No. Pet-92, 1948.
10. H. Ries and Thomas L. Watson, *Engineering Geology,* John Wiley & Sons, New York, 1936.
11. U. S. Dept. Interior, Bureau of Reclamation, *Concrete Manual,* 1949.
12. Waterways Experiment Station, Corps of Engineers, *Handbook for Concrete and Cement,* Vicksburg, Miss., 1949.

Author Index

Place Name Index

Subject Index